THE CAMBRIDGE COMPANION TO
THE "ORIGIN OF SPECIES"

The *Origin of Species* by Charles Darwin is universally recognized as one of the most important science books ever written. Published in 1859, it was where Darwin argued for both the fact of evolution and the mechanism of natural selection. The *Origin of Species* is also a work of great cultural and religious significance, in that Darwin maintained that all organisms, including humans, are part of a natural process of growth from simple forms. This Companion commemorates the 150th anniversary of the publication of the *Origin of Species* and examines its main arguments. Drawing on the expertise of leading authorities in the field, it also provides the contexts – religious, social, political, literary, and philosophical – in which the *Origin* was composed. Written in a clear and friendly yet authoritative manner, this volume will be essential reading for both scholars and students. More broadly, it will appeal to general readers who want to learn more about one of the most important and controversial books of modern times.

Michael Ruse is the Lucyle T. Werkmeister Professor of Philosophy and director of the Program in History and Philosophy of Science at Florida State University. The author or editor of more than thirty books, including *Can a Darwinian Be a Christian?* and *Darwinism and Its Discontents*, he is a Fellow of the Royal Society of Canada and the recipient of several honorary degrees.

Robert J. Richards is Morris Fishbein Professor of the History of Science and director of the Fishbein Center for the History of Science and Medicine at the University of Chicago. He has held major fellowships for work in the history and philosophy of biology and is the author of many books, including *Darwin and the Emergence of Evolutionary Theories of Mind and Behavior* and *The Tragic Sense of Life: Ernst Haeckel and the Struggle over Evolutionary Thought*.

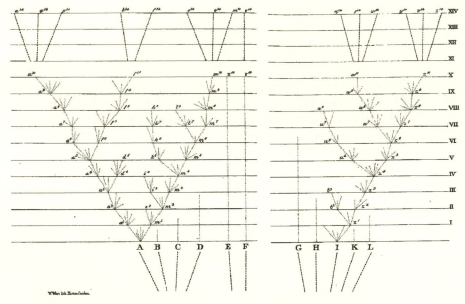

Darwin's diagram of species (marked A to L) supposedly descending from one genus (not seen). The intervals (marked by roman numerals) represent one thousand or, perhaps, ten thousand generations. Varieties are represented by lowercase letters. At level fourteen, we may suppose the original varieties have become species. "Thus, as I believe, species are multiplied and genera are formed"(*Origin*, 120).

The Cambridge Companion to

THE "ORIGIN OF SPECIES"

Edited by

Michael Ruse
Florida State University

Robert J. Richards
University of Chicago

CAMBRIDGE
UNIVERSITY PRESS

CAMBRIDGE UNIVERSITY PRESS
Cambridge, NewYork, Melbourne, Madrid, Cape Town, Singapore, São Paulo, Delhi

Cambridge University Press
32 Avenue of the Americas, New York, NY 10013-2473, USA

www.cambridge.org
Information on this title: www.cambridge.org/9780521691291

© Cambridge University Press 2009

First published 2009

Printed in the United States of America

A catalog record for this publication is available from the British Library.

Library of Congress Cataloging in Publication Data

Ruse, Michael.
The Cambridge companion to the Origin of species / Michael Ruse,
Robert J. Richards.
  p. cm.
Includes bibliographical references and index.
ISBN 978-0-521-87079-5 (hardback) – ISBN 978-0-521-69129-1 (pbk.)
  1. Darwin, Charles, 1809–1882. On the origin of species   2. Evolution (Biology)
3. Natural selection.   I. Richards, Robert J.   II. Title.
QH365.08R865 2009
576.8′2 – dc22   2008000484

ISBN   978-0-521-87079-5 hardback
ISBN   978-0-521-69129-1 paperback

CONTENTS

CONTRIBUTORS

Naomi Back received her Ph.D. from the University of Paris 1 (Panthéon-Sorbonne) in 2005. She is currently employed as Assistant Professor in the Social Sciences Collegiate Division at the University of Chicago.

Gillian Beer is Edward VII Professor Emeritus at the University of Cambridge. She is a Fellow of the British Academy and of the Royal Society of Literature and a Foreign Honorary Member of the American Academy of Arts and Sciences. Among her books are *Darwin's Plots* (second edition 2000), *Open Fields: Science in Cultural Encounter* (1996), and *Virginia Woolf: The Common Ground* (1996). She has recently been writing about rhyming and about the *Alice* books, as well as preparing new work on Darwin and consciousness. She is the president of the British Comparative Literature Association and has twice been a judge for the Booker Prize.

Peter J. Bowler is Professor of the History of Science at Queen's University, Belfast, Northern Ireland. He is a Fellow of the British Academy and a former president of the British Society for the History of Science. He has published widely on the history of evolutionary theory and is now completing a book on popular science in early twentieth-century Britain.

John Hedley Brooke held the Andreas Idreos Professorship of Science and Religion and directorship of the Ian Ramsey Centre at Oxford University from 1999 to 2006. He is an Emeritus Fellow of Harris Manchester College, Oxford, and Honorary Professor of the History of Science at Lancaster University and in 2007 was Distinguished Fellow at the Institute of Advanced Study, University of Durham.

His books include *Science and Religion: Some Historical Perspectives* (Cambridge University Press, 1991), *Thinking About Matter: Studies in the History of Chemical Philosophy* (1995), and (with Geoffrey Cantor) *Reconstructing Nature: The Engagement of Science and Religion* (Clark, 1998). He is currently president of the UK Forum for Science and Religion. His most recent publications include *Heterodoxy in Early Modern Science and Religion,* coedited with Ian Maclean (2005) and *Religious Values and the Rise of Science in Europe,* coedited with Ekmeleddin Ihsanoglu (2005).

David J. Depew is Professor of Communication Studies and Rhetoric of Inquiry at the University of Iowa. He works in the history, philosophy, and rhetoric of biology, both ancient and modern. He is the coauthor, with Marjorie Grene, of *Philosophy of Biology: An Episodic History* (Cambridge University Press, 2004) and, with Bruce H. Weber, of *Darwinism Evolving: Systems Dynamics and the Genealogy of Natural Selection* (1994). With Weber, he has coedited a number of collections, most recently *Evolution and Learning: The Baldwin Effect Reconsidered* (2003). He has written articles on the bearing of Aristotle's biological treatises on his social and political theory and on how Aristotle and Darwin can most accurately be compared. With John P. Jackson, he is currently working on a book about how Darwinians have intervened in American social and political controversies.

Sandra Herbert is Professor of History at the University of Maryland–Baltimore County and a Fellow of the American Association for the Advancement of Science and in 2006–07 was Distinguished Visiting Scholar at Christ's College, Cambridge. Her book *Charles Darwin: Geologist* (2005) has received awards from the American Historical Association, the Geological Society of America, the History of Science Society, and the North American Conference on British Studies.

Chris Kohler became an antiquarian bookseller in 1961 when he was eighteen, and *Michèle Kohler* joined his firm after their marriage in 1973. They put together collections of books that they sell to university and national libraries throughout the world. They spent twenty years building the most comprehensive collection of books and autograph letters by and about Charles Darwin that has ever been assembled, which they sold to the Natural History Museum in London.

David Kohn is General Editor of the Darwin Digital Library of Evolution at the American Museum of Natural History and Robert Fisher Oxnam Professor of the History of Science Emeritus at Drew University. He has edited Charles Darwin's Transmutation Notebooks and is currently writing an intellectual history of Darwin's botany.

Mark A. Largent is an historian of biology. He is Assistant Professor of Science Policy and directs the Science, Technology, Environment and Public Policy Specialization at James Madison College at Michigan State University. He earned his Ph.D. from the University of Minnesota's Program in History of Science and Technology and has taught American history and history of science courses at Oregon State University and the University of Puget Sound. His research and teaching focus on the role of biologists in public affairs and in the history of nineteenth- and early twentieth-century biology. He has published on the history of the evolution/creation debates, evolutionary theory, and the American eugenics movement; his most recent book is *Breeding Contempt: The History of Coerced Sterilization in the United States* (2007).

Tim Lewens works in the Department of History and Philosophy of Science at the University of Cambridge. He is the author of *Darwin* (2007) and *Organisms and Artifacts* (2004).

A. J. Lustig is Assistant Professor of History at the University of Texas at Austin. She is the coeditor, with Robert J. Richards and Michael Ruse, of *Darwinian Heresies* (2004) and is currently writing a book on explanations of altruism in biology and society since Darwin.

David Norman is working on describing the early Jurassic dinosaurs *Heterodontosaurus* and *Scelidosaurus*. He is currently directing a major new exhibition entitled Charles Darwin the Geologist at the Sedgwick Museum of Earth Sciences, University of Cambridge. He is also developing exhibitions on Charles Darwin's life as a student at Christ's College, Cambridge, and is a member of the committee organizing the Darwin 2009 Festival at the University of Cambridge. Further, he is coordinating research on Darwin's historical work as a geologist and new geological research on material collected from the Galápagos Islands in 2007. He has written several prize-winning books on dinosaurs, their evolution and biology, as well as numerous scientific papers on related topics, including a few on the history of science with respect to fossil reptiles. He was the Asher Tunis Distinguished Research Fellow (Paleobiology) at the Smithsonian Institution, Washington, D.C., for 2000–02 and a Visiting Scholar at St. John's College, Oxford, in 2006. He is the director of the Sedgwick Museum of Earth Sciences, University of Cambridge; an Odell Fellow in Natural Sciences at Christ's College, Cambridge; and a University Reader in Vertebrate Palaeobiology at the University of Cambridge.

Lynn K. Nyhart teaches history of science at the University of Wisconsin–Madison. She is the author of *Biology Takes Form: Animal Morphology and the German Universities, 1800–1900* (1995), and of a forthcoming book, *Modern Nature*, on natural history reform movements in nineteenth-century Germany. Her specialty is the history of evolutionary thought in its various guises.

Robert Olby is a retired professor of the history of science at Pittsburgh University. He is the author of many books on the history of genetics, including the classic *The Path to the Double Helix: The Discovery of DNA* (1994).

Richard A. Richards is Assistant Professor of Philosophy at the University of Alabama and Associate Professor of Philosophy at Yeshiva University in New York City. He is the author of articles on Darwin and on taxonomy. For the first decade of his adult life, he was a professional ballet dancer.

Phillip R. Sloan is Professor in the Program of Liberal Studies and in the Graduate Program in History and Philosophy of Science at the University of Notre Dame. His research specializes in the history and philosophy of life science from the early modern period to contemporary molecular biology. His writings include studies of Buffon, Darwin, and Richard Owen and on the history of classification in biology. He is currently working on a book on the conception of life in contemporary biophysics.

Vassiliki Betty Smocovitis is Professor of the History of Science in the Departments of Zoology and History at the University of

Florida and is affiliate in the Botany department. She is the author of *Unifying Biology: The Evolutionary Synthesis and Evolutionary Biology* (1996). Her research interests include the history of plant evolutionary biology; she is currently completing a biography of G. Ledyard Stebbins.

NOTE ON CITATIONS

Some works are cited often, and for convenience their titles have been abbreviated and used in the text.

Barrett, P. H., P. J. Gautrey, S. Herbert, D. Kohn, and S. Smith, editors. 1987. *Charles Darwin's Notebooks, 1836–1844*. Ithaca, NY: Cornell University Press. (*Notebooks*)

Darwin, C. 1859. *On the Origin of Species by Means of Natural Selection, or the Preservation of Favoured Races in the Struggle for Life*. London: John Murray. (*Origin*)

Darwin, C. 1909. *The Foundations of the Origin of Species: Two Essays Written in 1842 and 1844*. Edited by F. Darwin. Cambridge: Cambridge University Press. (*Foundations*)

Darwin, C. 1958. *The Autobiography of Charles Darwin, 1809–1882*. Edited by N. Barlow. London: Collins. (*Autobiography*)

Darwin, C. 1959. *The Origin of Species by Charles Darwin: A Variorum Text*. Edited by M. Peckham. Philadelphia: University of Pennsylvania Press. (*Variorum*)

Darwin, C. 1975. *Charles Darwin's Natural Selection, Being the Second Part of His Big Species Book Written from 1856 to 1858*. Edited by R. C. Stauffer. Cambridge: Cambridge University Press. (*Species Book*)

Darwin, C. 1985–. *The Correspondence of Charles Darwin*. Cambridge: Cambridge University Press. (*Correspondence*)

Darwin, F. 1887. *The Life and Letters of Charles Darwin, Including an Autobiographical Chapter*. London: John Murray. (*Letters*)

Darwin, F., and A. C. Seward, editors. 1903. *More Letters of Charles Darwin*. London: John Murray. (*More Letters*)

FOREWORD

One hundred and fifty years past its publication, I believe we can safely say that the *Origin of Species* is the most important book of science ever written. Indeed, given its importance to all of humanity and the rest of life, it is the most important book in any category. No work of science has ever been so fully vindicated by subsequent investigation, or has so profoundly altered humanity's view of itself and how the living world works. The theory of natural selection continues to gain relevance to the things that matter most to humanity – from our own origins and behavior to every detail in the living environment on which our lives depend. Little wonder that the adjective "Darwinian," sometimes lowercased to "darwinian" as a tribute to its fixity, far outranks "Copernican," "Newtonian," and "Mendelian" in the frequency of usage.

The *Origin* won the day quickly for such a revolutionary proposal, so much so that Darwin could confidently publish *The Descent of Man* only twelve years later. It succeeded not just for the mass of evidence adduced to support evolution but because of the clarity and authority of its text. The quality of the mind that erected it did not come from the blue. For nearly three decades, extending from the departure of *HMS Beagle* from Plymouth on December 31, 1831, to the day in 1859 the *Origin* was sent to press, Darwin remained almost continuously absorbed in scientific natural history. He inhabited this subject, and he lived it. And fortunately, the middle of the nineteenth century was a time that so little was known about nature in the rest of the world, so few unifying concepts existed to guide the collection of data, that every fact, every specimen was valued. Darwin's mind was an open vessel. By absorbing with little discrimination those domains of natural history most relevant to geology and evolutionary biology, he became enormously learned.

In preparing my recent anthology entitled *From So Simple a Beginning* (2006), I read for the first time in chronological order all four of Darwin's greatest books, *Voyage of the Beagle* (1845), the *Origin of Species* (1859), *The Descent of Man* (1871), and *The Expression of the Emotions in Man and Animals* (1872). These are, I can assure you, the four to read, and straight through if you can. It is impossible to imagine a higher quality of original intellectual exposition. Taken in sequence, the four books reveal the development of a mind priming itself to address the greatest of subjects during the most opportune of times to do so. Darwin must have been continuously exhilarated by what he had come upon. Galileo had his telescope. Leeuwenhoek had his microscope. Darwin had his idea.

Charles Darwin is the most written-about scientist in history. The reader may well ask, in picking up the present Companion, whether we need more: do we need more, even as part of a centennial celebration? The answer is yes! Light continues to be thrown by evolutionary thought on more and more subjects. The human self-image continues to grow in depth and clarity as a result. All this is worth an ever-evolving commentary. The history, provenance, and impact of the *Origin* and Darwin's other great books deserve repeated rounds of assessment.

INTRODUCTION

In 1859, the English naturalist Charles Robert Darwin published his major work, the *Origin of Species*. In this work, he argued that all organisms living and dead are the end result of a long, slow, natural process of development from forms far simpler and that indeed all life, by reason of its descent from but a few ancestors, is related. He also proposed a mechanism, natural selection, meaning that only a few survive and reproduce and that success in this process is on average a function of the distinctive features of organisms – over time, this leads to change, change that is in the direction of adaptation. Eyes, ears, noses, leaves, trunks, flowers, flippers, fins – these are the things that are produced by evolution, and these are the things crucial for survival and reproduction.

At once, it was recognized that the *Origin* was a major work of science. Indeed, it was seen as a major event in the history of Western civilization. As Copernicus had expanded space, so Darwin expanded time. Moreover, this was something that impinges on human beings. The *Origin* is not directly about humans. The only explicit reference is an almost throwaway passage at the end of the book. "Light will be thrown on the origin of man and his history." But no one was fooled. Humans may be important, but our importance must be tempered by our shared links with the rest of life. What perhaps could not have been seen back in 1859 was the extent to which Darwin and his book would be subjects of intense interest and controversy at the beginning of the new millennium, here in the first decade of the twenty-first century. Thinking of other major events in the years of the *Origin*, clearly the Crimean War and the Indian Mutiny had far-reaching effects – the former if only for medical care in battle and the latter eventually for the independence of the subcontinent. But today these are matters of historical interest and not of contemporary controversy. Not so the *Origin*. In part because of the controversy about creationism, in part because the mechanism of natural selection is debated vigorously in the science community, in part because Darwin's thinking does strike so fundamentally at the visions that we all have of ourselves, the *Origin* itself is still reprinted, reread, discussed and debated.

This Cambridge Companion has been written, edited, and published precisely because the *Origin* is still today a work of vital significance and interest. It is not a substitute for reading the *Origin* itself. We suggest that you have a copy of the *Origin* beside you as you open this volume – for reasons that will be made clear, we suggest that you use the first edition of the *Origin*, and it will be easiest to use the readily available facsimile, for it is to this that the contributors in this volume refer. The order of the contributions to this volume is straightforward. After an initial introduction, the topics of the *Origin* are introduced and discussed in order. Then there are several essays on issues that come out of the *Origin*. We stress that this is a work on the *Origin* and not on Darwin or evolution generally. As always with Cambridge Companions, the intent is to introduce pertinent ideas to readers new to the issues, but at the same time to try to offer thoughts that the more experienced and professional will find relevant and challenging.

The volume opens with a piece by one of the editors, Michael Ruse, that gives some background to the *Origin*, why it was written and when. There is an ongoing debate about the long interval of time between the moments at which Darwin thought up his ideas and the date of eventual publication. Why was there this gap, and indeed is it even right to speak in terms of a "gap"? At the same time, the piece gives an overview of the work, suggesting that – as Darwin himself said – it is "one long argument" and not simply a bunch of ideas thrown together. It is suggested that the work is skillfully constructed in the light of the dictates of the leading methodologists of science active when Darwin was a young man, but also that in judging the style one should take note of the sponsors of the naturalist, and why he would be trying to meet their approval and gain their sympathy. Also discussed are a number of issues arising from the *Origin*, for instance, the extent to which one can truly say that

Darwin was working in a British tradition as opposed to the style and methods of various continental thinkers.

The *Origin* opens with an extended discussion of the work of animal and plant breeders and the uses that they make of artificial selection. This is the topic of Mark A. Largent's paper. After a careful discussion of the information offered by Darwin, including a recognition of the ways in which Darwin separated out the differing intentions of and effects produced by the breeders – including a discussion of something Darwin called "unconscious selection" that he thought very important – Largent takes up a topic about artificial selection that is much discussed, namely, its role in the *Origin*. Is it just a heuristic device to get the reader ready for natural selection, or does it have a deeper role, that of making natural selection and the evolution consequent upon its action more evidentially plausible? Largent argues we can answer questions like these only by considering the extended work that Darwin himself did on breeding such organisms as pigeons.

Robert Olby tackles the important question of variation. In order for natural selection to work, there must be a renewable supply of variation. If there is not, then selection will soon have used up everything and ongoing change will grind to a halt. Darwin had no theory of heredity – what we call "genetics" – and although he did later come up with (what we now consider) a misguided theory, it was never introduced into the *Origin*. Olby looks at the data that Darwin did introduce into the *Origin*, and at the kinds of speculations that he made about it. Olby also looks at the changes that Darwin made through various editions of the *Origin*, in the light of his own work and the criticisms of others. Finally, Olby tackles the much-debated question of whether Darwin could and would have put his theory on a much firmer basis had he read the work of the Moravian monk Gregor Mendel, the man who is today regarded as the father of modern genetics. Olby suggests that an adequate answer to this question is far from obvious.

We have next a discussion of the linchpin of Darwin's thinking, natural selection. Robert J. Richards (the other editor of this volume) tackles this topic, taking us through the discovery of natural selection as a force for change, together with a secondary mechanism that Darwin called "sexual selection," and on to the eventual discussion of selection in the *Origin*. Richards's discussion uncovers a major difference between two camps of contemporary Darwin scholars (a difference that separates the two editors of this Companion!) – between those, like Ruse, who think that Darwin was the quintessential English thinker and that natural selection was a mechanism very much in the mold of the industrialists of the eighteenth century, and those, like Richards, who see Darwin much more in the continental Romantic tradition and who therefore see selection as something contributing to this tradition, being focused essentially on humans and their interests, their moral concerns, and their unique, high place in the world, which they owe to the world-wide progressive force of nature.

One still sometimes hears people say that although Darwin's book was titled the *Origin of Species*, in fact the topic of species is virtually absent from the book. This is simply not true, for there is much said in the *Origin* both on the nature of species in a world of evolution and on the reasons for their coming into being – the next chapter shows the latter in some detail. It is true, however, that one gets no explicit formal discussion of the topic of species, and it is Phillip R. Sloan's task to pull together what Darwin has to say about species in the *Origin* and to put the discussion in a historical context. This Sloan does by going back and looking at earlier (especially eighteenth-century) discussions of classification (systematics) and then at how Darwin tackled the topic in various writings up to and including the *Origin*. An important part of Sloan's discussion focuses on the debate after publication of the *Origin* and how Darwin's critics and friends tried to incorporate his thinking into their discussions and scientific forays.

Following on Sloan's discussion, David Kohn tackles a very important topic in the *Origin* (the word "keystone" in the title to the chapter is a word that Darwin himself used about the topic). How is it that organisms diverge in kind, and more importantly, why do they do so? Why are not all organisms identical? If you look at Darwin's notebooks, you can see that from the beginning he is wrestling with this problem, but it was not until the 1850s that he got on top of it, seeing that it is through divergence and diversification that many more organisms can exist and be supported than otherwise. Kohn points to the very great significance of metaphor for Darwin – in this case, the industrialist's metaphor of the division of labor. The creation of organisms that are different and adapted for different ends means that more life can flourish than would happen otherwise. As Kohn points out, this is the basis for another of Darwin's celebrated metaphors, the tree of life.

A. J. Lustig takes on two linking chapters in the *Origin*, first that dealing with organs of great perfection and second that dealing with instinct. The former basically completes the case for evolution through selection. Lustig shows how Darwin cleverly turns the argument on its head, or at least on its side, arguing that although we cannot see the evolution of perfection through time, we can in space, by setting up a series of existent organisms from the simplest to the most complex. Lustig also notes how here (as elsewhere) there are theological concerns lying beneath the most secular of discussions. The latter chapter, on instinct, starts to take us to the second part of the *Origin*, where Darwin applies his theory across a wide range of topics and problems. The discussion is important both for the extent to which it shows that behavior is as much a subject for selection and evolution as are physical features, and also for the ways in which Darwin had to wrestle with what today is known as the units-of-selection problem. In social organisms, like the ants and bees, was selection working on the individual insect or on the hive as a whole, and why does this matter?

When the layperson thinks of evolution, at once the fossil record comes to mind. Indeed, how often has one heard someone say that they believe in evolution because of the fossils – or, alternatively, that they do not believe in evolution because of the fossils! As it happens, however, although the fossil record is clearly crucial for the belief in evolution, when it comes to mechanisms there is often debate – who, after all, could see selection working on the trilobites or the dinosaurs? Darwin realized this to the full, and his extensive discussion in the *Origin* of paleontology is concerned in major part with answering problems and objections. However, as Sandra Herbert and David Norman show in their contribution, there is more to the story than this, and in the *Origin* Darwin is also concerned to make the positive case for evolution and selection as based on the fossil record. Note how some of the things that Darwin has to say about the record – for instance, about the status of higher organisms, especially humans – clearly bear on topics mentioned earlier. (Although Darwin makes explicit mention of humans only at the end of the *Origin*, throughout the work there are hints about and references to our species.)

Next, Peter Bowler takes up Darwin's discussion of geographical distribution. It is well known that this was an important topic for Darwin, because as a young man, when he was the ship's naturalist on HMS *Beagle*, he had been tremendously impressed by the distributions of the birds and reptiles on the Galápagos Archipelago. Why do you have different finches and mockingbirds and tortoises on islands within sight of each other, when on the mainland there is far less biodiversity? And why are the animals of the Galápagos like the animals of the South American continent and not like those of Africa, whereas the animals of the Canary Islands are like those of Africa and not like those of South America? These are facts that Darwin uses to bolster his case for evolution against the religious who want everything to be the function of creative miracle. Bowler shows how Darwin fits his discussion into a pattern that mirrors that of the earlier discussion of paleontology – one dealing with space and the other with time.

The penultimate chapter of the *Origin* (the final is recapitulatory) is something of a grab-bag discussion, as Darwin starts to clean up the topics he has so far left undiscussed. The first is that of classification or systematics, and Richard A. Richards's discussion of this subject is a nice complement to the earlier discussion of species by Phillip Sloan. Richards also puts things in historical context, talking both of the system of the Swedish taxonomist Linnaeus in the eighteenth century and of classificatory suggestions of the early nineteenth century, including the rather odd quinary system of William MacLeay. Richards also shows how very crucial was the thinking about homology and analogy by the anatomist Richard Owen, a man who was to become much hated by the Darwinian party. This is all a springboard for a discussion of classification in the *Origin*, where Darwin shows that his theory explains the possibility and nature of organic classification and how in turn this possibility and nature (in a kind of feedback way) make plausible Darwin's theory. Richards also goes on to show how Darwin's thinking was to impinge on the thinking of classifiers after the *Origin*.

Darwin was very proud of his discussion of embryology in the *Origin*. The big puzzling question was why exactly it is that, although the adults of different species may be very different, their embryos are often nigh identical – the naked eye cannot distinguish a human embryo from that of a dog. Darwin saw that this all followed from the different selective pressures on organisms at different times of their lives. Embryos are basically under the same conditions, whereas adults are not, and hence they get driven apart. But there was more to embryology than this, and Lynn K. Nyhart's contribution also focuses on these extra facts and problems, particularly the extent to which Darwin shared the views of the continental embryologists who thought that there were parallels between the development of the individual organism and the history (whether evolutionary or not) of the group or race. Obviously these issues take us right to the heart of the debate about the extent(s) to which Darwin can be considered a naturalist in the English tradition or a more Romantic thinker in the continental line.

We now start on a slightly different track. The main arguments of the *Origin* have been covered. The contributions from now on deal with particular issues arising from the *Origin*, issues that merit discussion in their own right. (For this reason, there is no longer any binding reason behind the ordering of the contributions, and the reader should feel free to read them in any order.) Vassiliki Betty Smocovitis writes about Darwin's botany in the *Origin*. Although we tend to think of Darwin as an animal man – finches, pigeons, tortoises, bees, and so forth – the study of plants was always something of prime importance for him, and indeed after the *Origin* most of his own empirical research was on plants, starting with a sprightly little book on orchids. Smocovitis shows how plants do in fact play a large role in the *Origin*, especially in the early chapters. There is discussion of them in the world of breeders, as a source for variation and for the argument that varieties can turn into species, for refining the meaning of the struggle for existence, as well as for examples of fantastic adaptive abilities. Smocovitis argues that in many respects Darwin was way ahead of his time and that his evolutionary speculations about plants were fully appreciated only in the middle of the last century, almost a century after the *Origin* first appeared.

David J. Depew raises topics that have been gaining increasing scrutiny from scholars recently, topics that more traditional scientific and philosophical approaches tend to miss. He is concerned with Darwin's style and his argumentative strategy. One very interesting point that Depew makes speaks to an issue that has often puzzled readers of the *Origin*, namely, who exactly does Darwin take to be his opponents? He often speaks of religiously based opposing ideas, but it is not easy to see exactly who the people holding such ideas might be. Depew suggests that this fuzziness might have been deliberate. Darwin did not want to make too strong and clear a case for the opposition because readers might be swung to this opposition! Depew also raises a topic that has been discussed since Aristotle, namely, the role of metaphor in thought and especially in scientific thought. There is much use of metaphor in the *Origin*, and Depew's question is about its use and dispensability. Could one have Darwin's theory without all of the flowery language?

What about religion and the *Origin*? Fortunately, as John Hedley Brooke shows in his contribution, we know a lot about Darwin's religious beliefs, starting with a fairly literal Anglicanism when at college, which on the *Beagle* voyage turned into a kind of deism – God as unmoved mover, Who works through unbroken law – and finally towards the end of his life turns again into a form of agnosticism, so favored by many leading Victorian intellectuals. We know also that it was never really science that changed Darwin's thinking about religion, but other generally more theological issues like hellfire and damnation. Brooke focuses much of his discussion on the religion of the *Origin*, arguing that Darwin at that time was no atheist or even agnostic, but trying to work out a form of deism, which frees God from the details but puts Him behind the overall working and excellence of the living world. Brooke also shows that the religious response to the *Origin* was by no means uniform, and that reactions varied from enthusiastic acceptance to outright hostility. Matters were also complicated by the fact that the Anglican Church in Britain had other more pressing issues to deal with, especially the trends in continental theology ("higher criticism"), which was throwing much uncomfortable light on the literal claims of Holy Scripture. Whatever else, Brooke shows that the relationship between the *Origin* and religion was complex and multifaceted.

As a young man, Darwin loved music and literature and the arts generally. As an old man, he regretted that, under the pressure of a life of science, his feeling for the aesthetic side of human existence had withered and gone. At most, he wanted a novel with a good story, a happy ending, and preferably a pretty heroine. In fact, he sold himself short, for it is clear that as a writer he always showed the influence of great works of creative art, novels and poetry, as well as related volumes of travel and the like. The *Origin* certainly exhibits this. In her contribution, Gillian Beer shows that the relationship between evolution, especially Darwin's evolution, and literature – novels and poetry – has always been much closer than many realize and that Darwin's thinking has been a rich source of inspiration for authors, from the creators of the very greatest works of fiction, notably George Eliot's *Middlemarch*, down to novelists and poets writing today. Beer shows also how this influence has extended from the most sober and widely admired works to what can only be described as light fiction of the frothiest kind. *Tarzan of the Apes* is influenced by the *Origin* just as much as Thomas Hardy's *Jude the Obscure*. Through Beer's essay, we start to grasp how the *Origin* has been as much a cultural influence as a simple and pure work of science.

Naomi Beck takes up the question of the influence of the *Origin of Species* on political thinking. This is a vast subject, and she distills the discussion down to three important responses to the *Origin*. First, Beck looks at the reactions of the man who in that day was even more closely identified with evolution than Darwin, namely, Herbert Spencer. She shows that although Spencer welcomed the *Origin* with enthusiasm, in fact what he did (perhaps typically) was to read Darwin through his own progress-tinted spectacles, and that the Darwin who found his way into Spencer's writings owed little to the efforts of the author of the *Origin*. Clémence Royer, the first translator of the *Origin* into French, was an ardent progressionist and advocate of forward-looking movements, from republicanism to feminism. The way she introduced the *Origin* may have given her and her admirers great satisfaction, but it made the rather staid author of the book itself very uncomfortable. Finally, there are the fathers of Marxism, Karl Marx himself and his great supporter and collaborator Friedrich Engels. They too praised Darwin – Marx particularly was much in the camp of those who saw Darwin as a very English thinker – but eventually their own thoughts and conclusions were little related to the ideas of the *Origin*. Finishing Beck's piece, one might conclude that the *Origin* is a little bit like tofu – it can take on flavorings of many kinds, according to the wishes of the chef.

Darwin was not a philosopher, and the *Origin* is not a work of philosophy. Nevertheless, Darwin was an educated Englishman; he had read philosophy as part of this education – Plato, the British empiricists, some continental thinkers (his later work *The Descent of Man* shows a keen appreciation of Kant), and most particularly the methodologists of science of his day. Tim Lewens teases out these themes, particularly the influence of the methodologists. Lewens has much to say about what today is regarded as a particularly important kind of scientific strategy, namely, the use of "inferences to the best explanation," where one gathers up the information and tries then to infer the best overall explanation of what is empirically at stake. Lewens argues that this is an important part of the argumentation of Darwin and that the chief influence here was the astronomer and philosopher of science John Herschel. Seeing how today we are still embroiled in disputes about Darwin and his *Origin*, the reader might compare Lewens's account of Darwin's methodology to the rather different one given in the introductory essay of Ruse.

Finally, Michèle and Chris Kohler talk about the *Origin* as a physical object – what sort of book it was, how it was published and distributed, what it cost, and all of those sorts of matters. They also trace the publishing history of the *Origin*, first in English and then in translation. Particularly interesting are their comparisons of the *Origin*'s fate to that of other well-known works on evolution, notably Robert Chambers's anonymously published *Vestiges of the Natural History of Creation*. The *Origin* holds its own, but not much more than that. However, do note the interesting point that the Kohlers make about the purchase of the *Origin* by lending libraries. The actual readership of the *Origin*, in whole or in part, may have been large. Among other interesting facts that the Kohlers have unearthed is the way in which the first edition of the *Origin* has shot up in value in recent years. The book may never sell in its lifetime what a book on wizards sells in its first twenty-four hours, but the original now costs what no scholar could ever afford. Let us hope that the tax laws incline rich collectors to give to research libraries, so that all can continue to enjoy the great work by a great scientist.

It remains now only for us as editors to thank Cambridge University Press for letting us produce *The Cambridge Companion to the "Origin of Species,"* to thank our contributors for writing such

splendid pieces, to thank the world's most distinguished living evolutionary biologist, Edward O. Wilson of Harvard University, for
writing a Foreword, and to thank the William and Lucyle T. Werkmeister Fund of Florida State University for supporting a conference
where early versions of these contributions could be presented and
discussed by all.

1   The Origin of the *Origin*

Charles Robert Darwin was born in 1809. His great book, *the Origin of Species*, was published in 1859, when he was fifty. He was to live another twenty-plus years, dying in 1882, by which time the *Origin* had gone through six editions and been extensively revised and rewritten. It used to be the case that it was the sixth edition of 1872 that was most frequently reproduced, but more recently scholars have insisted that the first edition is the really important one – we not only see Darwin's thinking in its original form but the revisions today are often judged to have been made for less than worthy reasons (in the sense that the criticisms now no longer seem so forceful). It is therefore the first edition that will be the focus of this piece, and my question opening this volume is about its genesis, and the implications that this had for the actual book that Darwin produced. While I do not think that the *Origin* is a particularly mysterious book, I believe that there are aspects to it that are not quite as obvious as we today often assume.

THE ROUTE TO DISCOVERY

Undistinguished at school, Darwin went first to the University of Edinburgh to study medicine and then (after that proved not to be to his liking) to the University of Cambridge to prepare for the life of an Anglican clergyman. (Janet Browne's [1995, 2002] biography is definitive.) We know now that, although Darwin had no formal training as a biologist, by the time he graduated (in 1831) he not

The late Sydney Smith once wrote a paper with the same title, and in using it again I show how much I owe to his friendliness and scholarship.

only was showing an aptitude for science but also was long versed in the ways of empirical study and research. From early years, Charles and his older brother Erasmus had played with chemical ideas and experiments, and at both universities he had immersed himself in the active groups of naturalists and empirical inquirers. At the end of 1831, Darwin joined HMS *Beagle*, about to start what proved to be a five-year trip mapping the coast of South America and then going on around the world before returning home. Darwin started as a kind of gentleman companion to the captain, Robert Fitzroy, but soon became the de facto ship's naturalist, in which role his earlier scientific activities and training served him very well. The notebooks that he kept show that he was serious and competent right from the start. (Sandra Herbert [2005] is very insightful on Darwin's move into serious science.)

The time on the *Beagle* was important for many reasons, not the least of which was that, being away from his Cambridge mentors, Darwin was forced to think independently. This was shown particularly in geology, the science that was most important to him in these early years. Darwin became enthused with the uniformitarian thinking of Charles Lyell in his *Principles of Geology* (1830–33) and broke with the catastrophism of people like Adam Sedgwick (1831), a professor of geology at Cambridge and the man who had taken Darwin on a crash course in Wales in the summer of 1831. In religion, the trip was important because Darwin's rather literalistic Christianity started to fade and he became something of a deist, believing in God as unmoved mover and that the greatest signs of His powers are the workings of unbroken law rather than signs of miraculous intervention.

Most significantly, perhaps because he was now thinking of God as someone Whose greatness is evidenced by unbroken law rather than by miracle, Darwin started on the path to evolution. It is generally agreed that Darwin (who knew about evolutionary ideas from reading *Zoonomia*, an evolution-favoring book by his grandfather Erasmus Darwin, as well as from encounters at Edinburgh with the future London professor of anatomy Robert Grant, and from Lyell's discussion of the thinking of Jean Baptiste de Lamarck) did not actually become an evolutionist on the voyage. But his encounter with the different reptiles and birds on the Galápagos Archipelago shocked him. How could one have different-but-similar forms on islands only a few miles apart? When, on his return to England, Darwin learned that the birds were undoubtedly of different species, this was enough to tip the balance. In the spring of 1837, Charles Darwin slipped over to transmutationism.

For eighteen months, until the end of September 1838, Darwin worked hard looking for a cause of evolution. One suspects that it was the ideal of Newton – much praised by the day's scientific methodological gurus, especially John Herschel (1830) and William Whewell (1837) – that spurred Darwin here. He wanted to find a force for evolution akin to Newton's force of gravitational attraction. For all that we have Darwin's detailed notebooks – perhaps because the notebooks are so detailed – there has been debate about the exact course of Darwin's thinking. Darwin himself always claimed that he started with artificial selection, realizing that this was the way in which breeders change their animals and plants. Then he started to look for a natural equivalent, and this he found at the end of September 1838 after he had read Thomas Robert Malthus's (1826) treatise on population. More organisms are born than can survive and reproduce. Those that get through will, on average, be different from those that do not. And it is these differences – shaggier coats, stronger legs, sharper eyes – that are crucial. Given enough time, there will be overall change – descent with modification (what we call "evolution") – and, moreover, this will be in the direction of adaptive advantage. Shaggier coats keep sheep warm; stronger legs let the wolf catch the deer; sharper eyes mean that the eagle can spot the rabbit.

Through a careful reading of the notebooks that Darwin kept while he was searching for his mechanism – a mechanism that, when discovered, he clearly did think was akin to a Newtonian force – some scholars have concluded that, although in his various sketches and published versions of his theory Darwin does use artificial selection to lead into natural selection, it is unlikely that he really did have the analogy in mind on his way to natural selection (Barrett et al. 1987; Herbert 1971; Limoges 1970). He never really thought that artificial selection could do the job, or at least that a natural equivalent would be sufficiently powerful to get full-blooded change. Whether this interpretation is correct is something that has been argued for some time now. My own feeling, looking at some of the material that Darwin read during the crucial discovery months – some material, incidentally, that not only drew attention to artificial selection but also showed that one might expect a natural equivalent, some material that Darwin highlighted particularly[1] – is that he probably did have the analogy in mind. But I would agree that he was more hesitant at the time than his confident later recollections suggest (Ruse 1979).

Darwin did not at once write things up in any formal way. Indeed, we have to work rather carefully through the notebooks to see that he did appreciate the full worth of natural selection. (He did. Jottings later in 1838 about human mental evolution put this fact beyond doubt.) Moreover, it was to be another four years before he actually wrote out what was a thirty-five-page, penciled *Sketch* (as we now call it) of his ideas (Darwin and Wallace 1958). This was then extended in 1844 to a 230-page *Essay*, which Darwin had fair-copied by the local schoolmaster. It should be added that in his *Autobiography* and elsewhere Darwin referred to 1838 as the point at which he first thought up his species theory, and this may well be true, although there seems to be no written record (nor indeed should there necessarily be).

THE LONG DELAY

Darwin then put things on hold, and having written a letter to his wife asking that in the event of his death she arrange that some competent biologist bring the *Essay* to publication, he turned to a massive eight-year-long study of barnacles (Darwin 1851a, b, 1854a, b). It was not until around 1854 that he turned back to his evolutionary theory. It is clear that, by this time, word was starting to get out that Darwin was an evolutionist – and he was in the habit of showing bits of his writings to some of the young men he was encouraging around him. His friends urged him to get back to the job and to go public, lest he be scooped. Darwin therefore started to write a massive book about his theory. This was interrupted by the arrival, in the early summer of 1858, of the essay by Alfred Russel Wallace,

---

[1] Like most people who actually take seriously the task of uncovering Darwin's thought processes, rather than triumphantly holding up something as evidence that he was both unoriginal and a plagiarist, I do not in any sense suggest that Darwin pinched natural selection from someone else, or that someone else should get the real credit. None of his precursors were seeing natural selection as a mechanism of evolutionary change, and some indeed denied that it could be.

a naturalist and collector in the Malay Archipelago – the essay in which Wallace captured almost exactly the ideas that Darwin had discovered twenty years before.

Extracts of Darwin's writings along with Wallace's essay were at once read at the next meeting of the Linnaean Society and published. Despite stories about the ideas being disregarded, there was immediate interest. Later in the summer, in his presidential address to the British Association for the Advancement of Science, quite favourable notice was made of the papers by Richard Owen (for all that he later was cast as the Darth Vader of the *Darwin Wars*). By now, Darwin had launched frenetically into the writing of what he wanted to call an "abstract" of his thinking – a qualification that his publisher, John Murray, wisely declined to accept for a work that in print extended to 490 pages – and so finally *On the Origin of Species by Means of Natural Selection, or the Preservation of the Favoured Races in the Struggle for Life*, by Charles Darwin, M.A., appeared in November 1859.

There has been and still is considerable controversy over the reasons why Darwin took so long to bring his theory into print. Recently it has been suggested that this is a bit of a pseudo-problem, because the delay was not really that long and because Darwin was, after all, working away for much of the time on matters evolutionary (Van Wyhe 2007). Those who apparently do not consider it a pseudo-problem have included Thomas Henry Huxley (1893), who praised Darwin for spending so much time on the barnacles and turning himself into a real zoologist before he published; the late Dov Ospovat (1981), who thought that Darwin moved from an ultra-natural theological stance of seeing all adaptation as perfect to seeing it as relative; and Robert J. Richards (1987), who thinks that Darwin was so worried about the sterility of the hymenoptera (seemingly a counterexample to a process that stresses reproduction) that it was not until he had seen (in the early 1850s) that breeders could get desired traits possessed by animals that do not breed (steers killed for food, for instance) by going back to the family stock, that he realized that something comparable could happen in nature and so felt free to get cracking again on his theory.

For the record, I have been marked as one who thinks there was a genuine delay, and I continue to think so (Ruse 1979). I am not too bothered by the jump between 1839 and 1842 or between 1842 and 1844. Darwin was working flat out on other projects, the geology in particular. Most notably, making him into a household name, Darwin wrote up his account of the trip around the world on the *Beagle*, and what started as a formal report for the Admiralty turned into one of the most popular of travel books at a time when society just loved stories of exploration in distant and strange lands (Darwin 1839). Darwin was also newly married, moving to the house in Kent (and having it extended), starting a family, and feeling sick. He had more than enough on his plate at that time.

It is the gap between 1844 and 1858 that fascinates me. I am happy to accept the bits and pieces of new information that come into Darwin's thinking between 1844 and 1859. I have always been impressed by the way that the barnacle work so convinced Darwin of the variation that exists in all natural populations, something that was crucial for a mechanism like natural selection. And let us not forget the "principle of divergence," tied to the tree-of-life metaphor, where Darwin saw that divergence is the way in which selection maximizes the use that organisms can make of resources. Although I think that in fact there are hints of it even in his notebooks of the late 1830s, I accept fully that Darwin did not really realize the problem and the solution until much later.

However, I have to say that none of this alone or in conjunction really convinces me that this yields the solution. Two things always strike me. First, Charles Darwin was always so ambitious. Never let the friendly, warm, almost-casual man and his style deceive you. At the beginning of her biography, Janet Browne speaks of the sliver of ice in the heart of Charles Darwin. I have always thought that this is so. He was not a nasty man in any way, but he did want to make his mark as a scientist, and nothing was going to stand in his way. The sickness was genuine, but he used it to advantage to avoid boring jobs and people. His massive letter writing was sincere, but again and again it was a medium through which Darwin could get others to do jobs for him. And above all, he was going to get into print. Just after the *Beagle* voyage, Darwin dashed up to Glen Roy in Scotland to look at the parallel roads around the sides of the valley, arguing that they were of marine origin. Unfortunately, the subsequent paper in the *Transactions of the Royal Society* turned out to be a bit of a disaster. Louis Agassiz was soon to point out that the parallel roads were produced by a lake dammed by a glacier, not by the now-receded sea. But the drive to get a paper on a hot topic into a prestigious journal certainly fits the pattern: a young man on the make. I am not criticizing Darwin for this. If I did, I would have to extend my comments to every successful scientist I have ever met – in science as in love, being reticent gets you nowhere. So I just cannot see Darwin, a man who knows he has solved the mystery of organic origins, sitting on his hands for fifteen years.

My second point is that truly I cannot find all of that much difference between the *Essay* of 1844 and the *Origin* of 1859. I have long argued – and continue to argue – that Darwin's theory is a very skilful piece of work. It is, as he truly said, one long argument, not simply one damn thing after another. I am convinced that the men influencing him on matters of methodology – William Whewell particularly, but also John F. W. Herschel – taught Darwin that he had to find a *vera causa* if he was to solve the organic origins problem. Darwin knew that natural selection could do the trick. It was a force-like phenomenon that explained adaptation, something that both scientists and theologians (often one and the same person) were trumpeting. But selection had to be set in the right justificatory framework. Satisfying Herschel, who as an empiricist demanded direct or analogical evidence, Darwin made much of the analogy between artificial and natural selection – this is so whatever the role of artificial selection in finding natural selection. Satisfying Whewell, who as a rationalist demanded that one's cause be at the apex of a consilience of inductions (Whewell 1840, Herschel 1841) – the cause explains the phenomena, the phenomena make reasonable the cause – the whole of the second part of his theory is a trip through the sub-branches of biology (paleontology, biogeography, systematics, anatomy, embryology) as Darwin shows that selection provides explanations in such areas and in turn is justified by such areas. The point I make here is that this structure is in the *Sketch*, the *Essay*, and the *Origin* – identical in form and presentation – and much of the evidence is just the same. Even the sub-bits, like the introduction of sexual selection along with natural selection, are the same.

ANSWERING THE QUESTION

So I still have the question of the delay. Why did Darwin not publish the *Essay* back in 1844? My answer is twofold. First, he was scared. Not of his wife or anything like that; and I doubt that being labeled a materialist much bothered him. He came from a family and a set (particularly connected with his brother Erasmus) where that was not much of a taunt. In any case, Darwin was not a materialist. He was a deist, and the various writings up to and including the *Origin* make that very clear. (He even added additional references to the Creator in later editions.) It was precisely the leaders of his scientific set – those very men who had nurtured him and made his early career possible – whom Darwin feared offending. 1844 was the year in which the notorious evolutionary work the *Vestiges of the Natural History of Creation* was published, and the set went after the work with a vengeance. Adam Sedgwick raged against it in the *Edinburgh Review* – it was so vile it must have been written by a woman, but surely no woman could pen such filthy muck. David Brewster (physicist, biographer of Newton, and the inspiration for the flowery passage with which Darwin ends the *Origin*) declaimed against it in the *North British Review*. And Whewell thought it so disgusting that he did not write against it but merely collected selected passages from earlier writings for a little book – *Indications of the Creator*. The first edition did not even mention the *Vestiges* by name. I realize that the reception of *Vestiges* was by no means uniformly negative – Tennyson, for instance, was to use its ideas to finish *In Memoriam* – but for Darwin's group it was anathema. So he knew that he had better stay silent.

The second reason is simply, as many have noted, that Darwin just did not expect the delay to be so long. He set out on his barnacle work thinking that it would take but a year, and it kept stretching on and on as he worked obsessively on the project. One year stretched to eight. The species book – which in the light of the reactions would need very careful documentation – did not get written. I should say that I see here, balancing the ambition, the other side of Darwin's character. He was selfish – call it self-centered if you like – because, as a rich man who had been favoured in his youth, he was accustomed to doing what he liked. He became obsessed with a project, and nothing was going to stop him. To put the matter in modern terms, he did not have to write research grants to show that his work would cure cancer. He could just amuse himself, although perhaps "amuse" is not the right word for someone who did work so hard. I see this pattern again after the *Origin*. Why did Darwin not set up a selection lab at Down? He had the money, and there were those who wanted to join him in doing just that. He could see that selection was being downgraded by people like Huxley, but he did not really fight back. Although Darwin did write the *Variation* as an extension of the *Origin* and was sufficiently threatened by Wallace's apostasy (arguing that human evolution demanded divine intervention) that he felt compelled to write the *Descent*, scientifically Darwin went on doing what he had always done, namely, working away on projects – orchids (1862), climbing plants (1880), earthworms (1881) – that caught his fancy.

A final comment. I see Darwin's sharing his evolutionary ideas with others as part and parcel of this picture. He was not about to share them with Sedgwick and Whewell – still the people who really controlled science – but the younger members of the set had long been discussing origins in a potentially naturalistic way. As soon as he came back from the *Beagle* voyage, Darwin and Owen began chewing the fat over such things. (Pertinently, Owen, the best scientist of them all at this time, was probably well on the way to some kind of Germanic evolutionism, but dared not publish his work because he was so dependent on the established powers. He did not dare accept a knighthood lest he appear too uppity.) Darwin knew full well that when he did publish he would need supporters. So it was quite natural to talk about these things with those who were potential supporters and who, although they may have been cowed by people like Whewell, certainly did not necessarily agree with them.

THE *ORIGIN* AS ANACHRONISTIC

In a way, talking about the long delay is a bit like speculating on whether Queen Victoria had sex with John Brown or whether the heir to the throne was Jack the Ripper. Fun to do, but not really that important, and probably ultimately futile. I would truly query only whether it was not that important. If the *Origin* is fundamentally different from the earlier versions, then it should be judged on its own terms. I would hate, for instance, for someone to judge my present taste in food and drink on my convictions as a small child in postwar England. It would be baked beans and sliced white bread all the way, washed down by a rather revolting fizzy concoction called *Vimto*. But if the *Origin* is more a product of the late 1830s and early 1840s, then we should judge it on those terms.

Let me make five points showing that such an approach pays explanatory dividends. First, take the book's topic. Of course, Charles Darwin was not the first to ask about organic origins. His grandfather Erasmus had done so, for one. And in the 1840s and 1850s people went on asking about the topic – Chambers (1844) in the first decade and Herbert Spencer (1852) in the second. However, I wonder if this was something on the front burner of the top professional biologists. Huxley was happy to get on board when the time came, although it took through the 1860s for him to accept that the fossil record showed evolution, and he never taught the topic in his classes (Ruse 1996). For him, it was indeed the materialism and like elements that were attractive. In the 1830s, however, Darwin's set did rather obsess about the topic – usually very negatively! It was described as the "mystery of mysteries" in a letter from Herschel to Lyell – a letter that became very public thanks to its being reprinted in Charles Babbage's *Ninth Bridgewater Treatise* (1838). My sense is that Darwin brought the issues back into discussion – incidentally, just at the time when Pasteur was showing the impossibility of spontaneous generation, and so in a way making the whole question of origins a bit iffy.

Second, consider the style of the *Origin*. From the beginning, everyone recognized that it was a remarkably easy read, especially for a work that was doing so much and claiming to be scientific. Richard Owen (1860) in his review in the *Quarterly* was quite nasty about this, congratulating Darwin for writing in a way that we have come to expect from the author of travel books and the like – the implication being that, written as it was, this could not be a serious work. Darwin was certainly capable of writing stuff that could be read only by the expert, if at all. Look at the barnacle monographs, for example. But we must think of Darwin's patrons. He may not have had to work, but there were those whose approbation he sought, namely his father and his Uncle Josh (later his father-in-law). Darwin had a rather rocky start – second-rate at school, and dropping out of medicine – and his father was rightly skeptical of his abilities and his willingness to get down to things.

After the *Beagle* voyage – even during the *Beagle* voyage – all changed. The good reports came in, and then the wonderful travel book showed what an all-around family credit he had become. This was just the time when the *Origin* was being conceived, and, right the way through, Darwin wrote for those who paid the bills, even though by 1859 both father and uncle were dead. Owen was right. It was an odd book for a professional scientist, but then Darwin was an odd professional scientist, always with one foot in another world. We see this in what I have referred to as his dual sides of ambition and selfishness (or self-centeredness).

Third, there is the book's structure. People like Huxley, dominating biology by the time the *Origin* was published, could not have cared less about that kind of methodology. Do your anatomy, draw your analogies, trace your phylogenies. It was really all a bit noncausal in its way – certainly noncausal in the sense of forces. Darwin of the 1830s thought otherwise. This was just the time when men like Whewell were trying to define what it is to be a scientist, and that is a major reason why they wrote their books showing the nature of good science. It had to be causal, in a Newtonian sort of way, and hence the debate over Newton's rather mysterious notion of a *vera causa*. This was no abstract debate, nor was the difference insignificant between Herschel and his call for experience and Whewell and his call for consilience. There was debate over geology, with Herschel (1830) thinking Lyell did the right kind of stuff and Whewell (1837) thinking that the catastrophists did the right kind of stuff. Darwin, trained by catastrophists, converted to Lyell, was very sensitive to these issues.

The same is true of another matter, namely, the debate over the theory of light. By the 1830s, it was clear (thanks to the work of people like Fresnel) that the particle theory of Newton was inferior to the wave (or undulatory) theory of Huygens. Herschel (1827) and Whewell (1837, 1840) battled over this, since no one actually sees either particles or waves. Herschel tied himself into knots with analogies to sound – tuning forks with sealing wax stuck on them and that sort of thing. Whewell, somewhat smugly, simply identified the waves as the center of a consilience. Case closed. Darwin was sensitive to this and in later life, likewise postulating an unseen cause, came to use the wave theory as a good example for his own theory (Darwin 1868). The point is that all of this went back to the 1830s, when Darwin was at his creative peak.

Fourth, take the obvious matter of function and final cause or teleology (Ruse 2003). People like Huxley were just not interested in these sorts of issues. For them, it was structure and form all the way. In Darwin's language (which he took from others), they were interested in Unity of Type, not Conditions of Existence. Huxley declared openly that he was no naturalist. What turned him on were the wonderful structures that were revealed through his work at the dissecting table. The ends that features served were at best irrelevancies and at worst hindrances to finding true relationships. After the *Origin* also, Huxley and his various scientific friends showed much less interest in adaptation and more in formal issues. It was really not until the twentieth century, after the work of the population geneticists, that function really began to ride again. This was because it was only then that natural selection began to take its place as the dominant mechanism of evolutionary change. And why was there this connection between function and selection? Precisely because, as we have seen, natural selection is a mechanism expressly intended to speak to function, and the reason for this is that Darwin thought that (what John Maynard Smith called) "organized complexity" is the dominant feature of the living world. He thought this because it was precisely what his teachers and senior friends, like Sedgwick and Whewell, were telling him. In a way, by the time the *Origin* appeared, it was already old-fashioned.

Fifth and finally, something to irritate my coeditor, there is the matter of progress. Pretty much every Darwin scholar now agrees that he accepted some form of biological progress. The question is: what form of progress? I would argue that it is a form that makes selection central, even though many think (with good reason) that the relativity of selection drives a stake through the heart of progress (Ruse 1996). Darwin saw organisms engaged in what today we call "arms races," with adaptations getting better because of the competition with other lines of organisms. Ultimately, this all leads to intelligence and the evolution of upper-class Englishmen. What makes Darwin's treatment of progress so difficult to discern is that he often seems opposed to the very notion. But what he is against is not the very notion, but a kind of Germanic notion of inevitable upward change for the better – a kind of progress through momentum, which many morphologists saw as analogous to the development of the embryo from blob to fully complete organism.

Darwin was dead set against this. And why? Once again, because such transcendentalism – such *Naturphilosoph* thinking – was hated by Darwin's group in the 1830s. So he had to stay away from it. It is this that makes the *Origin* very different from any of those evolutionary (or quasi-evolutionary) tracts written in the Germanic mode, whether it be Richard Owen's *On the Nature of Limbs* (1849), published ten years before the *Origin*, or Ernst Haeckel's *Generelle Morphologie* (1866), published nearly ten years after the *Origin*.[2]

CONCLUSION

The *Origin of Species* is a very great book and a very important book. As I have tried to show, in respects it is also a very puzzling book. I shall be disappointed if the contributors coming after me do not challenge just about every substantive claim that I have made. That is what makes the *Origin*, to this day, a very exciting book.

[2] Robert J. Richards could not disagree more with this understanding of Darwin and progress. For now, I will simply direct you to his writings where he makes his case – Richards (1992, 2002, and 2004).

2 Darwin's Analogy between Artificial and Natural Selection in the *Origin of Species*

Darwin opened the *Origin of Species* precisely where he claimed that his recognition of the engine of evolutionary change itself originated: with the transforming effects that breeders' selection of preferred characters had on their plants and animals. In the first chapter of *Origin*, titled "Variation under Domestication," Darwin lays the groundwork for his analogy between artificial and natural selection by asserting that "the great power of this principle of selection is not hypothetical," and that through selection, either methodological or unconscious, breeders fundamentally and permanently alter the domesticated plants and animals in their charge (*Origin*, 30). Motivated by their interest in enabling organisms that possess desired traits to produce offspring, which would presumably share their parents' preferred characteristics, breeders select favored individuals for reproduction and destroy or otherwise discourage the perpetuation of those that lack desired qualities. Darwin argued that the conditions of life in the wild allow for a struggle for existence within and between species that likewise tends to select for those characteristics that permit survival and reproduction and against those that tend to hinder organisms' opportunity to survive long enough to reproduce and pass along their particular qualities. In later chapters – in particular, Chapter 4, "Natural Selection" – these arguments coalesce into his assertion that just as breeders could use selection to change the characters of a particular breed, nature too could select and thus modify a wild species.

This chapter will explore the specific role that the analogy between artificial and natural selection played in Darwin's *Origin of Species*. Was his analysis of artificial selection as significant for the original formulation of his theory of evolution by natural selection as he maintained in a letter to Alfred Russel Wallace in 1859 and again in his autobiography almost three decades later (*More Letters,* 1: 108; *Letters,* 67–8)? Or did Darwin come to appreciate the analogy between artificial and natural selection only after he had conceived of natural selection, which might lead one to believe that the analogy served only as a means to convince the reader of the reality of evolution and natural selection? I will begin with a detailed description of Darwin's arguments in the first chapter of the *Origin* about the efficacy of selection and the relationship between domesticated and wild species, placing them in the context of claims made by his contemporaries. I will then examine how he employs the analogy between artificial and natural selection to support his argument about the reality of evolution and his theory of natural selection. In offering the analogy, Darwin had to demonstrate the ability of artificial selection to alter domesticated organisms, and he had to demonstrate a method by which nature could select for preferred characteristics. I will conclude by addressing the question of how breeders' work itself shaped the development of Darwin's theory. While he later depicted the emergence of his theory and of the analogy itself as a singular event, analyses over the last several decades suggest that it arose slowly through a reciprocal relationship between his theorizing and his evidence-gathering activities. Darwin and many of the scholars who have described the genesis of his theory have depicted it as springing to life between late September and early October of 1838. I will ultimately argue that fully to appreciate the role of the analogy between artificial and natural selection in the production and presentation of Darwin's theory, we must consider the two decades between the conception and reception of his theory as one long period, condensed and presented in the *Origin* as simply "one long argument" in a manner that reflects both the norms of the philosophy of science of his day and the context of the book's author and its audience (*Origin*, 459).

THE FIRST CHAPTER OF THE *ORIGIN*

Beginning with the empirical evidence available to him and to his readers, Darwin explains that plants and animals that have been under domestication for long periods of time show a broader range of differences among varieties and subvarieties than do similar organisms in the wild. Whereas various authors have disputed the possible sources of these variations, Darwin explains that he is "strongly inclined to suspect" that in both domesticated and wild species variations originate with changes in the parents' "reproductive elements" that occur because of environmental forces (*Origin*, 8). This, combined with his assertion that confinement somehow affects the reproductive system, explains why domesticated organisms demonstrate greater variation, while animals in the wild show less variation. The wider range of variation in domesticated organisms is also evident in the production of a sport, which is "a new and sometimes very different character from that of the rest of the plant." Sports, Darwin explains, are more commonly found under domestication than in nature, where they are "extremely rare" (*Origin*, 9–10).

The conditions of life, Darwin asserts, are of little importance in producing variability in organisms directly. Plants are apparently more directly influenced by conditions than are animals, but both are capable of being slightly altered by their environments. The various habits of individuals seem to have some influence in shaping individuals' characters. For example, Darwin describes how the bones of the domesticated duck's wings and legs have thickened as compared to their wild counterparts. In similar fashion, the udders of cows and goats that have been regularly milked for generations have increased in size, and the ears of every domesticated animal have grown droopy because of "the disuse of the muscles of the ears, from the animals not being much alarmed by danger" (*Origin*, 11). These sorts of examples, however, are relatively uncommon, so he concludes that direct environmental effects are not potent enough to explain the tremendous range of variation that he has recorded among various individual members of a given species. Whereas Lamarck had seen in the conditions of life the source for evolutionary change, Darwin rejects the claim that environmental sources are directly capable of producing significant new variations in an organism and concludes that they must emerge through conditions affecting the sexual organs of parents.

Darwin was not the first to address the potential analogy between the manipulation of species by plant and animal breeders and the transmutation of species in nature. Several earlier authors, including Charles Lyell, John Sebright, and William Whewell, had discussed the potential relevance of evidence collected from domesticated organisms to the debate over the mutability of species. A consensus had emerged among these authors that breeding shed no light on the potential mutability of wild plants and animals (Cornell, 307). Most significant was Lyell's treatment of the subject in the second volume of *Principles of Geology*, which Darwin had received when HMS *Beagle* had reached South America. Lyell had argued that one could admit "a considerable degree of variability in the specific character" of a breed without accepting the mutability of species, which allowed him to explain how breeders could temporarily alter or enhance preexisting characters in the organisms they bred (Lyell, 2: 36). In the *Origin*, Darwin indirectly counters Lyell's claim by arguing that alterations to domesticated organisms, which have occurred for "an almost infinite number of generations," indicate that large modifications can become permanent, despite the occasional reversion (*Origin* 15). He then introduces a different kind of selection, natural selection, to explain how new traits could be made permanent in the wild.

The basis for Darwin's analogy between artificial and natural selection was predicated on two related assertions: first, that breeders did permanently alter breeds though selection and, second, that they brought about modifications that had not existed in the domesticated organisms' wild ancestors. For Darwin to support his eventual argument about evolution by natural selection, he would have to demonstrate the truth of both of these claims.

Is Selection Efficacious?

Darwin relied on the tremendous variety found among domesticated pigeons, more than any other domesticated plant or animal, to demonstrate breeders' ability to produce distinct varieties by manipulating the variations within a given species. In letters to his cousin, William Darwin Fox, he indicated that he had found Fox's "careful work at pigeons really invaluable." It had allowed him to "trace the gradual changes in the breeds of pigeons" and was "extraordinarily useful" (*Letters*, 2: 84; Secord, 183). Domesticated pigeons revealed to Darwin that the diversity of breeds humans had produced from limited wild stock was "something astonishing" (*Origin*, 19). Darwin himself experimented with varietal crosses in order to show that offspring in the third generation often reverted to colors characteristic of the rock pigeon. This and other kinds of evidence indicated to Darwin that domestic selection could introduce new varieties that were permanent modifications of an aboriginal stock. For the oppositional claim that domestic breeds descended from similar wild organisms to hold true, he explained, the supposed aboriginal stocks must either still have existed in the countries where they were originally domesticated and not yet have been found by ornithologists, or they must have gone extinct in the wild. Neither possibility seemed likely.

Darwin formally introduces the concept of selection in the *Origin* by offering an explanation of how breeders could have produced the current domestic varieties from one or several allied wild species. He states that one of the most remarkable features of domesticated plants and animals is that their adaptations are not for their own good, but rather for man's use or fancy. How do humans manipulate wild species in order to produce the desired domesticated plants and animals? Darwin explains, "The key is man's power of accumulative selection: nature gives successive variations; man adds them up in certain directions useful to him. In this sense he may be said to make for himself useful breeds" (*Origin*, 30). The differences from which the breeder selects are often "absolutely unappreciable" to the "uneducated eye," so a breeder has to be gifted with a natural capacity and years of practice to identify such minute differences (*Origin*, 32). Darwin's analogy between artificial selection and natural selection is also an analogy between the practiced eye of the breeder and nature itself. As James Secord explains: "By demonstrating the existence of the smallest differences in creatures that looked identical to the untrained observer, he hoped to show his readers that wild nature could be seen with the practiced eye of the pigeon fancier" (Secord, 164).

Darwin suggests that one might object to his claims about the efficacy of breeders' selection, which he terms "methodological selection," by arguing that it had been practiced only for about seventy-five years. However, he responds, the potential effects of selection are hardly a modern discovery; "the inheritance of good and bad qualities is so obvious" that it most assuredly influences the actions of even the "lowest savages" (*Origin*, 34). Nonetheless, there is arguably a difference between the work of modern method-ological selection and earlier breeding efforts, so Darwin introduces the concept of unconscious selection, which results from "every one trying to possess and breed from the best individual animals" (*Origin*, 34). Using the example of the many varieties of dogs that have descended from the spaniel, including setters and the English pointer, he argues that changes in the animals have "been effected unconsciously and gradually, and yet so effectually that, though the old Spanish pointer certainly came from Spain," there is no native dog in Spain that is anything like the English pointer (*Origin*, 35). Darwin makes similar arguments using horses, pigeons, and sheep as examples. Unconscious selection could also occur, he later argues, when "many men, without intending to collectively alter a breed toward a particular ideal, have a nearly common standard of perfec-tion, and all try to get and breed from the best animals" (*Origin*, 102).

Darwin's distinction between methodological and unconscious selection includes a relatively low estimation of the apparent differ-ence in efficacy between the two. This may have been an argument against those who would have dismissed the power of selection by claiming that humans had practiced it rigorously for less than a century. One might also argue, as some scholars have, that Darwin underestimated the differences between methodological and uncon-scious selection in order to avoid the accusation that he was pro-moting the concept of design in nature by asserting that nature, like the breeder, exerted a will in selecting for particular character-istics. When he develops the analogy between artificial and natural selection later in the *Origin*, Darwin downplays the overall effects of man's selection as compared to the selective influence of the struggle for existence, and judges that nature is ultimately more successful in selecting organisms: "How fleeting are the wishes and efforts of man! how short his time! and consequently how poor will his products be, compared with those accumulated by nature during whole geolog-ical periods." Darwin likewise describes how organisms subjected to natural selection are, as a result, "truer" in character and thus plainly bear "the stamp of far higher workmanship" (*Origin*, 84). John Angus Campbell has argued that Darwin did this in order to take advantage of mid-Victorian readers' affinity for the language of design (Campbell, 361). Take, for example his explanation of how nature selects: "It may be said that natural selection is daily and hourly scrutinizing, throughout the world, every variation, even the slightest; rejecting that which is bad, preserving and adding up all that is good; silently and insensibly working" (*Origin*, 84).

An even stronger argument to explain Darwin's motive for his assertion that methodological selection is only slightly more efficacious than unconscious selection contends that Darwin discounted their differences in order to effectively claim that unconscious selection is merely a type of natural selection. We see evidence for this when, foreshadowing his ultimate claim that nature itself selects, Darwin describes a type of unconscious selection that occurs in times of great stress: Even "savages so barbarous as never to think of the inherited character of the offspring of their domesticated animals" still preserve their most useful animals even in times of famine (*Origin*, 36–7). Darwin also introduces the concept of the struggle for survival, which readers would soon learn is vital to his theory of evolution by natural selection, by discussing the special pressures on the domesticated animals kept by uncivilized man. Unlike the modern breeders' animals, those raised by uncivilized man would have had to struggle for their own food, at least during certain seasons. This struggle would lead to a kind of "natural selection," which Darwin places inside quotation marks (*Origin*, 38). He views the animals kept by uncivilized man as existing in conditions somewhere between the wild and those offered by modern breeders. In this interpretation, human intention played no essential role; to convince his readers of the efficacy of unconscious selection would be to get them simultaneously to admit to the reality of natural selection.

Did Domesticated Characters Exist in Wild Ancestors?

In the case of those species that have been domesticated for the longest periods of time, Darwin does not believe it is possible to come to any definite conclusion about whether they descended from one or from several parent species. He nonetheless argues that the broad range of domesticated varieties has derived from a very few wild ancestors, which have demonstrated far less variation than did their domesticated progeny. Darwin claims to have once believed, as did most every breeder and many naturalists of his day, that there were multiple original breeds of wild animals that exhibited the same range of variation found in current domesticated animals. He had "never met a pigeon, or poultry, or duck, or rabbit fancier, who was not fully convinced that each main breed was descended from a distinct species." It was sensible for them to believe this, he explains, as "from long-continued study they are strongly impressed with the differences between the several races." But they refuse to "sum up in their minds slight differences accumulated during many successive generations" (*Origin*, 29).

In drawing an analogy between the alleged efficacy of the breeders' selection and selection in nature, Darwin confronted the well-established claims that breeders could do little more than sharpen or emphasize characters already in existence among wild species and that they could not make permanent any alterations they achieved through selective breeding. For example, in the second volume of *Principles of Geology*, Lyell had agreed with Frederic Cuvier that domesticated animals' "domestic qualities" were "modifications of instincts...implanted in them in a state of nature" (Lyell, 2: 41). In cases in which these animals appeared to have acquired "attainments foreign to their natural habits and faculties," he asserted that such habits were never transmitted from one generation to the next (Lyell, 2: 42). The apparently rapid changes that breeders were able to achieve among their animals occurred only in the first generation or two, and they were lost once animals returned to a wild state. "Every husbandman and gardener," he wrote, was aware that experiments to acclimatize species substantially and permanently to environments to which they were not suited were doomed to failure (Lyell, 2: 38). Likewise, among the breeders Darwin consulted during the 1850s, most believed that the current variety of domesticated plants and animals mirrored a preexisting range of variation of original wild stocks (Eaton 1851; Eaton 1858; Ruse 1975a, 344–349; Secord, 169). Darwin went to great pains to undermine the breeders' notion that domestic stocks had multiple, similar wild ancestors and to encourage readers to accept the idea that fundamentally new variations could be derived from existing stocks. As Secord concluded, "Darwin's discussion of the pigeon was almost entirely an attempt to prove conclusively...that the incredible diversity of the domestic pigeons was indeed countered by the unity of their origin" (Secord, 179).

If existing domesticated varieties were mere reflections of their wild ancestors, Darwin's analogy between artificial and natural selection would have failed because selection would not have brought about fundamentally new characters. It was therefore vital for Darwin to demonstrate the contrary position, and he argues that his interpretation of the origin of apparently unique characters found in domesticated pigeons is a better explanation than other potential accounts because it is simpler. He explains that it is rare to find in nature the sort of structures that exist in the domesticated species of pigeons, and in order for this to occur, "half-civilized man" would have had to have thoroughly domesticated several species of wild pigeons and, by chance or by intention, have picked several extraordinary abnormal species, all of which then went extinct. "So many strange contingencies," he concludes, "seem to me improbable in the highest degree" (*Origin*, 24). Likewise, taking into consideration the coloring of pigeons, he finds incredible diversity. If all the domestic breeds descended from similarly diverse wild species, it would be easy to assume that humans had had nothing to do with producing the diversity of coloring. However, doing so leaves us with two "highly improbable suppositions:" first, that all the "several imagined aboriginal stocks were coloured and marked like the rock pigeon," and second, that domesticated species had simply reverted to the very same colors and markings. Or that each of the domesticated species has been crossed with the rock pigeon and reverted to the characters of the non–rock pigeon parent, and Darwin explains that there is no evidence that offspring reverted to the characters of only one parent (*Origin*, 25–6). All of the competing explanations that Darwin offers require extraordinary events or circumstances, so his claim that new characters arose via selection appears the most viable given William Whewell's claims that the simpler and more harmonious explanations tend to be true. In his *Philosophy of the Inductive Sciences*, Whewell had argued that the ability to explain a wide range of phenomena with fewer and simpler arguments was the mark of a good theory. In true theories, he explained, "all the additional suppositions *tend to simplicity* and harmony; the new suppositions resolve themselves into the old ones, or at least require only some easy modification of the hypothesis first assumed." In false theories, by contrast, "the new suppositions are something altogether additional" (Whewell, 2: 68–9). Darwin was well aware of Whewell's work and genuinely enthusiastic about it (Ruse 1975c, 166).

Darwin concludes the first chapter of the *Origin* by identifying a number of factors that aid or hinder the power of breeders' selective efforts to alter breeds. A high degree of variability is favorable, and large numbers of kept individuals increase the chance that preferred qualities will appear, while low numbers encourage breeders to perpetuate every individual and effectively prevent selection. Those plants and animals that are especially useful compel breeders to give each individual the closest attention as they search for the slightest favorable deviation in qualities or structure. Preventing crosses between emerging varieties encourages divergence, so geographical separation of varieties by the enclosure of land is a significant factor in allowing the selection of traits to generate fundamentally new breeds. He also explains why, in contrast to organisms like pigeons, certain breeds of domestic animals have not developed new varieties. In the case of cats, he asserts, there is great difficulty in controlling their pairing. With donkeys, too few have been kept and those by poor people, so little attention has been paid to their breeding. Peacocks are not easily reared and large stocks are not kept, and geese are valuable for only two purposes, food and feathers, so there is little compulsion to produce new varieties and thus little attention is paid to their selection. Darwin concludes: "Over all these causes of Change I am convinced that the accumulative action of Selection, whether applied methodically and more quickly, or unconsciously and more slowly, but more efficiently, is by far the predominant Power" (*Origin*, 42–3).

Darwin's discussion in the first chapter of the *Origin* of the efficacy of selection in the hands of plant and animal breeders provides the basis for the analogy he draws between artificial and natural selection in the fourth chapter. He begins Chapter 4, titled "Natural Selection," by pointedly asking, "Can the principle of selection, which we have seen is so potent in the hands of man, apply in nature?" (*Origin*, 80). Keeping in mind the apparently endless number of potential variations in all organisms, domesticated or wild, the apparent plasticity of organisms, and the powerful effects of the struggle for life, he asserts that the realities and particularities of life in the state of nature differentially affect plants and animals. Just as certain variations in domesticated organisms are useful in some way to humans, some traits must be "useful in some way to each being in the great and complex battle of life" (*Origin*, 80). He concludes, "it would be a most extraordinary fact, if no variation ever had occurred useful to each being's welfare," given the fact that humans have found so many variations in plants and animals useful to them (*Origin*, 127).

THE ROLE OF THE ANALOGY BETWEEN ARTIFICIAL AND NATURAL SELECTION IN DARWIN'S THEORY

From the examination of Darwin's notebooks, letters, and the various editions of his *Origin of Species* by science studies scholars, we have learned a great deal about the process by which Darwin developed his theory and the manner in which he presented it to professional and public audiences. Of particular interest to many scholars has been the question of the role that his analogy between artificial and natural selection played in the actual development of his theory. Darwin claimed that he had conceived of the mechanism of natural selection only after coming to appreciate the significance of selection in the work of plant and animal breeders. Both in an 1859 letter to Wallace and in his autobiography, Darwin affirmed this scenario: "I came to the conclusion that selection was the principle cause of change from the study of domesticated productions; and then, reading Malthus, I saw at once how to apply this principle" (*More Letters*, 1: 108; *Autography*, 67–8). However, on examining his private notebooks, some have argued that in fact Darwin adopted the potential analogy only after he had conceived of natural selection (Young, 109–45; Vorzimmer, 225–59). Several scholars have argued that, contrary to the actual structure of his argument in *Origin* as well as his later claims, Darwin had not fully developed his understanding of the power of breeders' selective activities until after he had established in his mind his explanation of the process of natural selection. Sandra Herbert, for example, asserts that Darwin does not seem to have held a "sufficiently unambiguous notion of artificial selection" that would have enabled him to anticipate finding "a similar process at work in untended nature." She concludes, "the discovery of natural selection made the domestic analogy much more clear to Darwin than it had been before" (Herbert, 212–13). A recognition of the vital role that Darwin's description of pigeon breeding played in the first chapter combined with Secord's description of how Darwin immersed himself in the world of English pigeon fanciers from 1855 until he began writing the *Origin* in the summer of 1858 likewise suggests that his reliance on the analogy between artificial and natural selection emerged in concert with, rather than subsequent to, his description of the mechanism of natural selection (Secord, 163–4). Even taking into account earlier statements by Darwin, such as the passage in his 1844 letter to J. D. Hooker in which he discounts proposed evolutionary mechanisms because their authors have not "approached the subject on the side of variation under domestication," it appears that Darwin increasingly explored and emphasized the analogy after his 1838 reading of Malthus (*Letters*, 2: 29; Evans, 114). Other scholars, such as Limoges (1970) and Kohn (1980), likewise dispute Darwin's later claim that his full appreciation of the efficacy of artificial selection led him to perceive in Malthus's passages a comparable mode of selection in nature.

Further evidence that undermines Darwin's claim about the priority of artificial selection in the formulation of his theory of natural selection can be found in a document he published several months after reading the *Principles of Population*. The "printed enquiries" that Darwin mentions in his autobiography that he used to collect "facts on a wholesale scale" consisted of an eight-page quarto pamphlet titled "Questions about the Breeding of Animals," which contained twenty-one numbered paragraphs containing forty-eight questions (Vorzimmer, 269–81). Darwin self-published the pamphlet in the spring of 1839 and presumably sent it to breeders to collect information about their methods. His inquiry, as Peter Vorzimmer explains, consisted almost exclusively of leading questions. "That is, they are framed in such a way as to elicit specific responses which may or may not confirm an existing hypothesis. This is borne out by the conditional interrogative ("*If* this ... *will* this ... ?") syntactic arrangement of nearly all the queries" (Vorzimmer, 281). It is therefore apparent that Darwin had in mind a specific hypothesis when he published the pamphlet. Knowing the specific questions Darwin asked breeders, we can see how he integrated their answers into the first chapter of the *Origin*. For example, the answers apparently given to question #19 – "Can you give the history of the production in any country of any new but now permanent variety, in quadrupeds or birds, which was not simply intermediate between two established kinds?" – were borne out in Darwin's assertion in the first chapter of the *Origin* that "the possibility of making distinct races by crossing has been greatly exaggerated" and that "a breed intermediate between *two very distinct* breeds could not be got without extreme care and long-continued selection; nor can I find a single case on record of a permanent race having been thus formed" (*Origin*, 20).

Michael Ruse (1975a) has summarized and commented upon claims, principally those by Herbert, that Darwin misrepresented the significance of his examination of domesticated plants and animals in the development of his theory of evolution by natural selection. After presenting Darwin's assertions about the significance of his study of domesticated plants and animals to the development of his theory of natural selection, Ruse reviews the claims made by more recent commentators and initially concludes that they are essentially correct. "The artificial selection analogy did not have a crucial role in Darwin's discovery of natural selection as a cause of evolutionary change. It was only after he had grasped this later concept that he came to stress the analogy" (Ruse 1975a, 344). However, he likewise identifies one significant problem with this conclusion: it ignores the influence of two pamphlets written by breeders, works that Darwin had carefully examined about six months before he read Malthus's *Essay on Population*. These pamphlets were John Wilkinson's "Remarks on the Improvement of Cattle" (1820) and John Sebright's "The Art of Improving the Breeds of Domestic Animals" (1809).

Ultimately, Ruse seeks to reconcile Darwin's claims about the development of his theory with scholarship that disputes the role of artificial selection in the evolution of Darwin's thought by weaving "a fabric which makes better sense of what Darwin himself said about his path of discovery, while not denying the actual things Darwin wrote in his notebooks at the time of the discovery" (Ruse 1975a). He does this by depicting the emergence of Darwin's theory as the result of his simultaneous appreciation of both the struggle for existence and the analogy between artificial and natural selection, all of which came about when he "read for amusement" Malthus's *Essay on Population* (*Autobiography*, 68). Darwin had come across depictions of what he eventually called the "struggle for existence" as well as an analogy between the actions of breeders and those of nature in his earlier readings, the former in Lyell's *Principles of Geology* and the latter both in Sebright's 1809 "The Art of Improving the Breeds of Domestic Animals" and in Lamarck's work, which Lyell had disputed. Later scholars who have examined the genesis of Darwin's analogy between artificial and natural selection, such as L. T. Evans, have described how these works "stimulated him to seek a mechanism in nature equivalent to the sustained gradual picking employed by breeders" (Evans, 124).

How are we to answer the question of the relationship between Darwin's use of the analogy between natural selection and selection as it was performed by breeders (and thoroughly described in the first chapter of the *Origin*) and the actual impact of his study of the work of breeders on his own development of the theory of evolution by natural selection? Did Darwin, either intentionally or unintentionally, misrepresent the path he had followed in generating his theory of evolution by natural selection? Are the historians and philosophers who, over a century later, dispute his recollection of the emergence of his theory correct in their assertions that he did not first comprehend the efficacy of artificial selection and then discover the engine of natural selection in the struggle for existence? I conclude that while both Darwin and his analysts are prone to depict the theory of evolution by natural selection as having sprung to life one evening in the fall of 1838, it was in fact merely conceived the evening that Darwin amused himself by reading Malthus's *Essay on Population*. It took twenty-one years of gestation – which included at least four years of labor during which Darwin "entered with enthusiasm on forays into the world of the English pigeon fanciers through printed inquiries and the study of breeders' writings," the inhibiting influence of Chambers' 1844 *Vestiges of the Natural History of Creation*, and a final push motivated in part by a letter from Alfred Russel

Wallace – before Darwin's theory of evolution by natural selection finally emerged into the public's view (Secord, 164; Chambers; *More Letters*, 1: 118). Imagining Darwin's theory as coming into existence with his instant appreciation of Malthus's description of the struggle for existence in the preexisting context of his awareness of the efficacy of artificial selection leads us to debate awkward and often uninformative questions about chronology. Shifting our frame of reference from one evening in 1838 to nearly two decades of intellectual development and evidence gathering allows Darwin the time to find support for his theory, recognize patterns, and develop strategies with which to present his argument to both professional and lay audiences.

If, in fact, the emergence of Darwin's theory of evolution by natural selection was a process that took many years, why did he describe it in letters to Wallace and in his autobiography as having come to him all at once upon reading Malthus? The answer may well rest not, as philosophers have assumed, within Darwin's argument itself, but rather within the larger context of his audience. That is, while I claim that Darwin's production of his theory was an iterative process during which he took years to consider evidence and sharpen the argument, his presentation of the theory in the *Origin* was necessarily a single, linear explanation of it. "One long argument," as he called it at the end of the book and as subsequent scholars have emphasized (*Origin*, 459; Mayr). He thus framed his argument in the form of what Campbell has depicted as a set of stairs, with the first four chapters leading readers from artificial selection, to variation in nature, to the struggle for existence, to natural selection (Campbell 2003, 203–7). Doing so led his readers from the apparent effects of breeders' selection, via analogy, to nature's selective influence. It reflected Bacon's advice to scientific authors that one ought to recapitulate the order of discovery in order to give readers a sense of discovering the argument's points independently (Campbell 2003, 361). Both the *Origin* and Darwin's later depiction of the emergence of his theory condensed its development into a series of linear events. Thus, the book began precisely where Darwin would later assert that his recognition of the engine of evolutionary change originated: with the apparent effects that breeders' selection of preferable characters had on their plants and animals, and its possible analog in nature.

ACKNOWLEDGMENTS

The author wishes to thank Michael Ruse, Robert Richards, Paul Farber, John P. Jackson, David Depew, and Chris Foley for their comments and suggestions.

3   Variation and Inheritance

Darwin signaled the preeminent importance of natural selection in both the title and subtitle of his book the *Origin of Species*. But without variation as the raw material upon which selection can act and inheritance as the means for preserving favorable variations in the future, natural selection would not lead to the genesis of new species. How, then, did Darwin present these two topics in the *Origin*? Answering this question will require textual analysis of the first edition of the *Origin*, including some reference to changes in later editions, and a little indulgence in some more speculative discussion of the subject. The chapter will address the following questions:

1. Why did Darwin privilege variation in 1859?
2. Why did he hold that changes in the conditions of life provide the chief raw material for evolution?
3. Had Darwin known of Mendel's critique of his work, how might the sixth edition of the *Origin* have differed from what was published?

INTRODUCTION

To the contemporary reader with some knowledge of modern biology, the crucial importance of heritable variation to the theory of species transmutation has no need of justification. It is obvious. Yet Darwin's manner of presenting it and the theoretical stance he adopted must seem strange, even extraordinary.

As to his manner, Darwin assumes a strong inductive style: he repeatedly refers to facts, "copious details" of them (*Origin*, 8), "a long list" on this, "numerous instances" of that (*Origin*, 9), "a long catalogue of facts" on something else (*Origin*, 131), none of which he judged he had the space to detail. But repeatedly he tells us how these facts have convinced him of their portent. Many data he finds "perplexing," others "doubtful." There are doctrines that he "cannot doubt" (*Origin*, 17), others that he cannot "believe" (*Origin*, 18), yet others that drive us to "conclude . . ." (*Origin*, 7). This very personal style is also couched in slippery language with double negatives, and qualifiers like: "It seems to me not improbable that . . ." (*Origin*, 15). As the physicist William Hopkins commented, in reviewing the *Origin*, where Newton demonstrated, Darwin asserted (Hull 1973, 239–40). The zoologist Richard Owen likewise remarked that to be served up an "expression of a belief, where one looks for a demonstration, is simply provoking" (Hull 1973, 209). Today we can look in vain for a rigorously controlled experiment on variation or heredity yielding numerical data sufficient to convince us. We have to be satisfied with the occasional crucial experiment that he suggests but does not carry out. Little words like "perhaps," "probable," and "possible" repeated again and again may carry conviction. But the critic Fleeming Jenkin complained about "all these maybe's happening on an enormous scale, in order that we may believe the final Darwinian 'maybe', as to the origin of species." And he concluded: "There is little direct evidence that any of these maybe's actually *have been*" (Hull 1973, 339). Over a century later, Gertrude Himmelfarb took up this theme, remarking: "As possibilities were promoted into probabilities, and probabilities into certainties, so ignorance itself was raised to a position only once removed from certain knowledge" (Himmelfarb, 335, cited in Gale, 157).

We know, of course, that the absence of fuller empirical data was due to the rush with which the book was prepared following Alfred Russel Wallace's letter of June 1858. The manuscript of *Natural Selection*, already more than 225,000 words long, had to be shaved down. Only one half of the text in the *Origin* can be identified with passages in *Natural Selection*. The two chapters on Variation under Domestication in *Natural Selection* were reduced to one in the *Origin* (*Species Book*, 10–14).

The tone, bordering on pleading for the reader's confidence in his claims regarding variation, had much to do, no doubt, with this enforced absence of so much of his data, but also, more importantly, with his prior knowledge of the disinclination of a number of his friends (not to mention his critics) to accept his claims regarding variation – Joseph Hooker, Asa Gray, Thomas Henry Huxley, and Charles Lyell, in particular.

PRIORITIZING VARIATION

Darwin devoted Chapters 1, 2, and 5 of the *Origin* to variation. "Variation under Domestication" comes first, followed by "Variation under Nature." After moving on to the struggle for existence (p. 60) and natural selection (p. 80), we are back to variation under "Laws of Variation" (p. 131). Inheritance is not given a chapter to itself, nor is the subject ever discussed fully in its own right elsewhere in the *Origin*, but the inheritance of variations is briefly discussed in all three of the variation chapters[1]. No other topic in the *Origin* was given as much stand-alone coverage as variation. Inheritance, by contrast, did not receive separate treatment until 1868 in his great two-volume work *The Variation of Animals and Plants under Domestication*. Inheritance never achieved that status in later editions of the *Origin*.

Darwin explained in his introduction to the *Origin* how, when he began his transmutation studies and was seeking for the "means of modification" of species, he judged that "a careful study of domesticated animals and cultivated plants would offer the best chance of making out this obscure problem" (*Origin*, 4). In addition to the analogy he intended to draw between the breeder's practice of selection and nature's, such a study would show that "a large amount of hereditary modification is at least possible" (*Origin*, 4). As Jon Hodge has pointed out, the variation chapters illustrate Darwin's attempt to establish the existence, competence, and responsibility of variation as a *vera causa* (true cause) in his theory of the origin of species (Hodge 1977).

Inheritance was at the time both a popular and a problematic subject. Henry Thomas Buckle's *History of Civilisation in England*,

---

[1] It was somewhat misleading for Vorzimmer to state that inheritance "was confined to a single chapter" in the *Origin* (Vorzimmer, 21).

volume 1, came out in 1857. There Buckle castigated the many authors writing on the subject of man's hereditary qualities:

The way in which they are commonly proved is in the highest degree illogical; the usual course being for writers to collect instances of some mental peculiarity found in a parent and in his child, and then to infer that the peculiarity was bequeathed. By this mode of reasoning we might demonstrate any proposition. (Buckle, 1: 159)

He was attacking the medical profession, the source of claims for an hereditary "diathesis" or constitutional predisposition of people to gout, asthma, adverse reaction to coffee, and so many other afflictions and aberrant behaviors. Darwin needed inheritance, and he bought into the diathesis craze, but he also had need of a disrupting force that would overcome inheritance to yield variations. This brings us to the conditions of life.

CHANGES IN THE CONDITIONS OF LIFE

Already on the opening page of Chapter 1 we find the following:

When we reflect on the vast diversity of the plants and animals which have been cultivated, and which have varied during all ages under the most different climates and treatments, I think we are driven to conclude that this greater variability is simply due to our domestic productions having been raised under conditions of life not so uniform as, and somewhat different from, those to which the parent-species have been exposed under nature. (*Origin*, 7)

Brought up in Shropshire, and since 1841 living in Kent, Darwin was well situated to appreciate this remarkable variability. For England, despite its industrial development, was still a landed society where the wealthy could easily live in a rural setting. It was, too, the foremost country in agriculture and horticulture, its plant and animal breeders having won international renown.

The circumstances that he judged significant for variation were those of changes in the "conditions of life." This choice among the possible causes of variation resulted neither from his intense study of the works of the German hybridists Joseph Kölreuter and Carl von Gärtner, nor from his study of the British hybridists – the Hon. and

Rev. Dean Herbert, Thomas Andrew Knight, and others – but from pondering the "favourable circumstances" for variation, namely, the domestication of animals and the cultivation of plants.

Alongside the variability of species under domestication, Darwin was also impressed by the evidence of decreased fertility, often amounting to complete sterility. A frequent visitor to the Zoological Gardens in London, Darwin requested data on the reproductive success achieved by the more exotic inmates (*Correspondence*, 3: 404–5; Desmond 1985). The many failures reported to him convinced him that "[n]othing is more easy than to tame an animal, and few things more difficult than to get it to breed freely under confinement, even in the many cases when the male and female unite." This was not merely a matter of "vitiated instincts," he wrote, for "how many cultivated plants display the utmost vigour and yet rarely or never seed!" (*Origin*, 8). Surely it was the changed conditions of domestication and cultivation that were disturbing the reproductive systems of these organisms, yielding degrees of reduced fertility, even utter sterility.

Professor Richard Owen, in his review of the *Origin*, was quite unconvinced. He appealed to the recent experience at London's Zoological Gardens, where the young pair of giraffes introduced in 1838 had, since attaining maturity, produced nine offspring. Other success stories included the breeding pair of hippopotami at the Jardin des Plantes in Paris (Hull, 208). Indeed, Darwin himself admitted in the *Origin* that tropical carnivores, as well as rabbits and ferrets kept in most unnatural conditions (i.e., in hutches), reproduce freely (*Origin*, 9). Here, he admitted, their reproductive systems were not affected. He cited further striking examples of domestic fecundity in 1868 (Darwin 1868, 2: 135).

Among animals subjected to confinement, Darwin noted that the reproductive systems of some still functioned, yet they acted "not quite regularly." So they produced "offspring not perfectly like their parents," that is, they showed variability (*Origin*, 9). This reminded him of the discovery by the French comparative anatomist Geoffroy St. Hilaire that "unnatural treatment" of embryos caused the production of "monstrosities." These, Darwin held, "cannot be distinguished by any clear line from mere variations" (*Origin*, 8). Maybe domestication is a very mild form of "unnatural treatment." That being the case, he "strongly inclined to suspect that the most frequent cause of variability may be attributed to the male and female reproductive elements having been affected prior to the act of conception" (*Origin*, 8). Thus arose Darwin's hypothesis of the "indirect action of the environment" – the assumption that in this case the environment did not cause variation in the exposed parental generation. Instead, it was acting on the parental reproductive systems, yielding variability, first manifested in the subsequent offspring. Moreover, such variability bore no necessary relation to the nature of the environmental influences, and the variants were not all the same – the variation was "indefinite." Clearly this is not a Lamarckian conception. Here is where Darwin believed were to be found the numerous small individual differences upon which natural selection could act in a gradual but cumulative fashion.

Just how un-Lamarckian, then, is the first edition of the *Origin*? Lamarckian variation is there, under the terms "direct action of the conditions of life" and the inheritance of "habit," and hence "use and disuse." "Direct" because in this case the variation is produced directly in the parent and then reproduced in the offspring. It is also adaptive, as in the case of thick fur under cold climates. Since all exposed individuals are affected in the same manner, Darwin called this variation "definite." He followed his mentor in physiology, Johannes Müller, in doubting many cases of this kind. "Some slight amount of change" was as far as he would go (*Origin*, 11). On habit, his own experiments comparing the skeletons of domestic and wild ducks were designed to test this effect. The results revealed a contrast that he attributed to "the domestic duck flying much less, and walking more, than its wild parent." He also mentioned the drooping ears of many domestic animals, due probably to the disuse of the ear muscles "from the animals not being much alarmed by danger" (ibid.).

Summing up Chapter 1, Darwin declared;

I believe that the conditions of life, from their action on the reproductive system, are so far of the highest importance as causing variability. I do not believe that variability is an inherent and necessary contingency, under all circumstances, with all organic beings, as some authors have thought. (*Origin*, 43)

His close friend and scientific correspondent Joseph Hooker disagreed and advised him to start by adopting instead variation as a "fundamental" and "innate principle; & afterwards [make] a few remarks, showing that hereafter perhaps this principle would be explicable" (Darwin paraphrasing Hooker's suggestion, *Correspondence*, 10: 135).

BREEDS OF THE DOMESTIC PIGEON

Center stage in Darwin's treatment of variation in the *Origin* is occupied by the breeds of pigeons. One quarter of Chapter 1 is given to them. This was for two reasons:

1. Because they displayed such a remarkable diversity of characteristics. Consider the tumblers, Jacobins, pouters, runts, barbs, carriers, trumpeters, fantails, laughers, and more, each looking as distinctive as the several species of a genus. Yet all produced viable offspring when crossed with one another, suggesting their descent from the same aboriginal form.
2. In 1855 he discovered that two Frenchmen, Boitard and Corbié, had reported characteristics of the wild rock dove in the progeny of hybrids between different domestic breeds (Boitard and Corbié 1824) – slate-blue plumage, two black wing bars, white rump, and a terminal dark bar to the tail, its feathers being white at the edges (*Origin*, 25). This could provide stunning evidence for the origin of all these breeds from a single species. Darwin set to work, and by November 1855 he was inviting Charles Lyell to visit and see his pigeons (*Correspondence*, 5: 492).

Most breeders thought otherwise. They were convinced that in former times a distinct progenitor of each of their breeds had existed. And indeed Boitard and Corbié thought their work indicated that the breeds in question were distinct species. To Darwin neither of these conclusions was plausible. As for the numerous alleged originating species, there was no sign of their former presence. Even if hybridization between aboriginal forms were also involved, he reckoned, "at least seven or eight aboriginal stocks" would still be required to yield all the varieties in existence. He tried crossing different breeds himself, and noted the uniformity of the [F_1] generation. But in succeeding generations bred from them [F_2 . . .], "hardly two of them will be alike, and then the extreme difficulty, or rather utter hopelessness, of the task becomes apparent" (*Origin*, 20). From this point on in the *Origin*, Darwin's message to the reader on variation is firmly against cross-breeding as a significant source of new forms.

Darwin's discussion of variation under nature in Chapter 2 is dominated by the results of his enquiry into the comparison of the number of varieties presented by species belonging to large genera compared to those presented by those belonging to smaller genera. The tables he used to assemble the quantitative data did not appear in the *Origin* (*Species Book*, 149–54), but their message was clear to him when he placed the larger genera on one side of a table and the smaller on the other. He found that

it has invariably proved to be the case that a large proportion of the species on the side of the larger genera present varieties, than on the side of the smaller genera. Moreover, the species of the larger genera which present any varieties, invariably present a larger average number of varieties than do the species of the smaller genera. . . . These facts are of plain signification on the view that species are only strongly marked and permanent varieties; for wherever many species of the same genus have been formed, or where, if we may use the expression, the manufactory of species has been active, we ought generally to find the manufactory still in action, more especially as we have every reason to believe the process of manufacturing a new species to be a slow one. (*Origin*, 55–6)

Darwin's most trenchant critic, Fleeming Jenkin, was to complain about the approach adopted here, for it is an argument by default. Darwin wanted his explanation to be accepted because it was better than the creationists' explanation. But, objected Jenkin, "our inability to account for certain phenomena, in any way but one, is no proof of the truth of the explanation given, but simply is a confession of our ignorance" (Hull 1973, 340).

HYBRIDIZATION AND CROSS-BREEDING

In Chapter 1, Darwin declared that the creation of distinct races by crossing "has been greatly exaggerated." Indeed, he could "hardly believe" that in this manner "a race could be obtained nearly intermediate between two extremely different races or species" (*Origin*, 20). He repeated this claim at the end of the chapter, using the same phrase, "greatly exaggerated" (*Origin*, 43). The full treatment of hybridism comes in Chapter 8, but here neither the laws of heredity nor those of variation are on the agenda. Instead, the discussion of hybridism serves to explore the nature of sterility in order to show that it is not "a specially acquired or endowed quality, but is incidental on other acquired differences" (*Origin*, 245). In this way Darwin sought to undermine the claim that sterility is the Creator's way of preserving the fixity of species. That claim had been supported by the assumed sterility of hybrids (species crosses) in contrast to the fertility of mongrels (variety crosses). Accordingly, his theme for the chapter is the degrees of fertility and sterility of hybrids and mongrels, relying chiefly on the experimental studies of Kölreuter and Gärtner. Darwin's aim was to construct a continuum between hybrids and mongrels regarding this phenomenon. Cautiously, he concluded the chapter by remarking that "the facts briefly given in this chapter do not seem to me opposed to, but even rather to support the view, that there is no fundamental distinction between species and varieties" (*Origin*, 278).

Darwin did not consider the act of hybridization itself to be the source of the resulting variation. Instead, he subsumed it under what for him was the more fundamental process, namely, the action of changed conditions of life. Hence his "double parallel" between the effects of changed conditions and hybridism.

The sterility of hybrids, which have their reproductive systems imperfect, and which have had this system and their whole organization disturbed by being compounded of two distinct species, seems closely allied to that sterility which so frequently affects pure species, when their natural conditions of life have been disturbed. This view is supported by a parallelism of another kind; – namely, that the crossing of forms only slightly different is favourable to the vigour and fertility of their offspring; and that slight changes in the conditions of life are apparently favourable to the vigour and fertility of all organic beings. (*Origin*, 277)

Nine years later he was to ask: "Can this parallelism be accidental? Does it not rather indicate some real bond of connection?" (Darwin 1868, 2: 177) Another eight years on he answered:

The most important conclusion at which I have arrived is that the mere act of crossing by itself does no good. The good depends on the individuals which are being crossed differing slightly in constitution, owing to their progenitors having been subjected during several generations to slightly different conditions, or to what we call in our ignorance spontaneous variation. (Darwin 1878, 27)

What irony that the one who had revealed the way in which so many hermaphrodite organisms are designed to prevent self-fertilization was the same author who belittled the role of crossing in the generation of variability! Was this not due to his concern to provide his theory of natural selection with the appropriate kind of variations – small differences – appropriate, that is, for their gradual, stepwise accumulation under natural selection? For how else could adaptive change be achieved and natural selection's creative role be assured?

INHERITANCE

Admitting that the "laws governing inheritance are quite unknown" (*Origin*, 13), Darwin singled out one rule: "at whatever period of life a peculiarity appears, it tends to appear in the offspring at a corresponding age, though sometimes earlier" (ibid.). The time of onset of hereditary diseases, especially those known as diatheses, offered him striking examples. This rule, he judged, was of the "highest importance in explaining the laws of embryology" (*Origin*, 14). It seems he was suggesting that the cause of the sequential development of organs (etc.) was to be understood in terms of the particular nature of the hereditary process involved. Of course, he was considering inheritance as part of that broader nineteenth-century concept of "generation," which included both regeneration and embryological development. Nevertheless, like his contemporaries, he referred to what it is that governs inherited traits by the term "constitution."Thus in the quotation given earlier, he refers to individuals "differing slightly in constitution." Then again, discussing the phenomena of acclimatization, he regards "adaptation to any special climate as a quality readily grafted on an innate wide flexibility of constitution" (*Origin*, 141). The term "constitution" appears at least two dozen times in the book. It was widely used in the medical literature on diatheses (Olby 1993, 412ff.).

When he considered the faithful hereditary transmission of rare abnormalities – "albinism, prickly skin (ichtheosis), hairy bodies

&c," he reasoned that "less strange and commoner deviations may be freely admitted to be inheritable. Perhaps the correct way of viewing the whole subject, would be, to look at the inheritance of every character whatever as the rule, and non-inheritance as the anomaly." (*Origin*, 13)

However, he did draw the line at the inheritance of mutilations (*Origin*, 135). Nor did he judge that variations acquired rapidly would be inherited. Curiously, he also expected that offspring born before the parent has reached full maturity will lack those characters not yet developed in the parent – for example, the horns of horned cattle.

Another feature of inheritance that he highlighted was the "correlation of growth." That is: "The whole organization is so tied together during its growth and development, that when slight variations in any one part occur, and are accumulated through natural selection, other parts become modified" (*Origin*, 143). Some of these correlations he found "quite obscure" – for instance, blue eyes and deafness in cats, feathered feet and skin between the outer toes in pigeons. Were these, he queried, due to growth, or to two quite separate modifications of the inherited constitution? (*Origin*, 146)

The feature of inheritance that Darwin discussed at some length is reversion. He recognized the tendency of hybrid offspring to return to the character of their originating species as a problem for the generation of novel descendents, their characteristics intermediate between those of the two originating species. But he denied that there was any good evidence that domesticated varieties, when "run wild, gradually but certainly revert in character to their aboriginal stocks" (*Origin*, 14). He suggested experiments that could be carried out to test the claim (*Origin*, 19).

Reversion was for him a form of ancestral inheritance. That meant that he included under the term any returns to a distant ancestor, usually termed "atavism." While reversion was considered the barrier that kept hybrids from becoming new constant forms, atavism was the phenomenon Darwin used to assert the common descent of domesticated breeds from a single source – famously, pigeons from the rock dove (*Origin*, 27ff.) and horses from some striped animal like a zebra or quagga (*Origin*, 167). His discussion of the means by which these cases of distant reversion could occur reveals his assumption of what we may call the fractional theory of inheritance, that is, assuming that bisexual reproduction involves a blending of the constitutions of two individuals, then the hereditary contribution of each ancestor will be halved in each generation. Darwin then calculated that:

After twelve generations, the proportion of blood, to use a common expression, of any one ancestor, is only 1 in 2048; and yet, as we see, it is generally believed that a tendency to reversion is retained by this very small proportion of foreign blood.... When a character which has been lost in a breed, reappears after a great number of generations, the most probable hypothesis is, not that the offspring suddenly takes after an ancestor some hundred generations distant, but that in each successive generation there has been a tendency to reproduce the character in question, which at last, under unknown favourable conditions, gains an ascendancy. (*Origin*, 160–1)

This notion of latent tendencies can perhaps be accommodated in a model of hereditary determination by powers, but hardly by a fractionating process. Interestingly, Darwin did not appeal to such "latencies" when, ten years later, he was confronted by the Scottish engineer Fleeming Jenkin's calculation of the "swamping" effect of an outbreeding population on rare variants (Hull 1973, 303ff.). This calculation appeared in Jenkin's (anonymous) review of the fourth edition of the *Origin*. After reading it, Darwin confessed to Hooker: "Fleeming Jenkins [sic] has given me much trouble, but has been of more real use to me than any other essay" (*More Letters*, 2: 379). Assuming blending inheritance, Jenkin claimed that in an outbreeding population, 'single' variations, in spite of being advantageous, will soon be swamped by the hereditary contributions of the individuals with whom they will breed.

Jenkin's critique was directed at the imaginary illustration Darwin had given, in the first four editions of the *Origin*, of wolves preying on deer. Darwin had asserted that the slimmest and fleetest wolves would have an advantage over the slower and heavier and that their descendants would increase, spreading through the population. Then he illustrated the case of a single wolf as follows:

Now if any slight innate change of habit or of structure benefited an individual wolf, it would have the best chance of surviving and of leaving progeny. Some of its young would probably inherit the same habits or structure, and by the repetition of this process, a new variety might be formed.... (*Origin*, 91)

In the fifth edition (1869) of the *Origin*, Darwin altered this passage by substituting "the slimmest individual wolves" for a "single wolf," explaining that he had not appreciated "how rarely single variations, whether slight or strongly marked, could be perpetuated" until he had read Jenkin's review (*Variorum*, 178). This revision was minor. He did not go further into Jenkin's critique on this subject. But Jenkin had gone on to point out that if the variant were a nonblending "sport,"[2] then

a great number of offspring will retain in full vigour the peculiarity constituting the favourable sport.... [L]et all his descendants retain his peculiarity in an eminent degree, however little of the first ancestor's blood be in them, then it follows, from mere mathematics, that the descendants of our gifted beast will probably exterminate the descendants of his inferior brethren.... What is this but stating that, from time to time, a new species is created? (Hull 1973, 317–18)

Jenkin was clearly seeking to force Darwin to take a position that his critics could interpret as a form of continuous creation. Variation would thus take away from natural selection its most important and creative role – accounting for adaptation. Darwin preferred to hope that by relying on individual differences and assuming their great prevalence, he could leave the nonblending "sports" on one side. Accordingly, he altered two passages in the *Origin* where he had emphasized the rarity of useful adaptations. The first edition passage "variations useful to each being... should sometimes occur in the course of thousands of generations" was changed in the sixth edition to "in the course of many successive generations" (*Origin*, 80; *Variorum*, 38). And the passage in the first edition, "Nothing can be effected, unless favourable variations occur, and variation itself is apparently always a very slow process," was changed in the fifth edition to: "Although all the individuals of the same species differ more or less from each other, differences of the right nature, better adapted to the then existing conditions, may not soon occur" (*Origin*, 108; *Variorum*, 202). Darwin clearly aimed to make natural selection, rather than the availability of variations, the chief limiting factor on the rate of evolutionary change. This debate over the

---

[2] "Sport" here means an individual that deviates singularly and spontaneously from the rest of the population.

"swamping effect" continued to the end of the nineteenth century and has been discussed elsewhere (Bulmer 2004).

AN IMAGINARY SIXTH EDITION

Anyone in 1859 reading the chapters of the *Origin* discussed here was made well aware that Darwin was forcing his way through fresh territory. "Our ignorance of the laws of variation is profound," he acknowledged (*Origin*, 167). "Nevertheless, we can here and there dimly catch a faint ray of light, and we may feel sure that there must be some cause for each deviation of structure, however slight" (*Origin*, 132). That being the case, why did he so consistently push to one side the resource of variation provided by cross-breeding in the form of what we refer to as genetic recombination? Of course, recombination on its own would not have been enough. He needed some source of spontaneous change as well. But he could have utilized both genetic recombination and innate variability, however caused. The suggestion here is that changed conditions of life offered him the foundation stone for his theory of variation. Domestication of animals and cultivation of plants were less important for the opportunities they gave for cross-breeding than they were for the opportunities they afforded for changed conditions to act and for humankind to select the resulting variations. Cross-breeding, insofar as it brought together the effects of different conditions of life in the hybrid constitution, was icing on the cake.

Appealing to innate variability of unknown cause had no attraction for Darwin. That would open the door to creationist influences – St. George Mivart provides a case in point. Those, like Pallas, who gave a central role to hybridization obviously needed a substantial stock of species to begin the process. Darwin repeatedly pointed to the limits of such a mechanism.

It is therefore instructive to see how Gregor Mendel, in the closing passages of his 1865 paper, challenged the very essence of Darwin's conception of variation, declaring:

The opinion has often been expressed that the stability of the species is greatly disturbed or entirely upset by cultivation. . . . No one will seriously maintain that in the open country the development of plants is ruled by other laws than in the garden bed. . . . Were the change in the conditions the sole cause of variability we might expect that those cultivated plants which are grown for centuries under almost identical conditions would again attain constancy. That, as is well known, is not the case. . . . Our cultivated plants are members of various hybrid series whose further development in conformity with law is varied and interrupted by frequent crossing inter se. (Mendel 1865, section 10)

Let us imagine what might have happened had Darwin seen Mendel's paper. We know that when he read Charles Naudin's paper of 1865 he judged that the French hybridist's notion of germinal segregation in hybrids would not account for "distant reversion," as Darwin had found it in pigeons. But Mendel's paper was far more impressive than Naudin's. Moreover, as cited earlier, Mendel confronted Darwin head-on over his claims for the difference between the conditions of life under cultivation and in the wild. If Darwin took due note of Fleeming Jenkin's critique, would he not have been impressed by Mendel?

To accept Mendel's critique would surely have overturned the argument for the role of the conditions of life. But it would not have undermined the case for some unknown cause of innate variation – call it mutation, as did the Dutch botanist Hugo de Vries in 1901. The external environment could still have been implicated, although – as Mendel, his teacher Franz Unger, and Mendel's contemporary Kerner von Marilaun had all demonstrated, simply transporting a species from one location to another or from the wild into the garden did not yield lasting variations. Furthermore, cross-breeding does produce individual differences, does create novelties, as had been proven again and again by the very breeders Darwin had consulted.

But Mendel's pea characteristics were all strongly marked ones. They were not the stuff of what Darwin considered to be individual differences. Would Darwin have launched a major breeding program along the lines of Mendel's experiments, had he known of them? Can we picture the multitude of pigeon houses that would be needed, crowded around Down House in order to reproduce Mendel's results in an animal? Darwin's pigeon research was unpleasant enough with just a few birds, but with thousands of birds, what would it have been like?

Mendel's results were established with strongly marked (non-blending) characters. Such were not, in Darwin's opinion, relevant to evolution. Just such an objection was voiced by British evolutionists in the early years of the twentieth century after Mendel's work had been rediscovered. Thus, in his critical survey of 1908, Alfred Russel Wallace belittled Mendel's achievement and quoted Darwin to the effect that "hybridization...had no place whatever in the natural process of species-formation"; and that, he added, "was the reason why Darwin did not prosecute the research further." Wallace preferred Darwin's text to "any amount of study of the complex diagrams and tabular statements which the Mendelians are for ever putting before us with great flourish of trumpets and reiterated assertions of their importance." Wallace's anger grew as he wrote. For the Mendelians to "set upon a pinnacle this mere side-issue of biological research" was to invite ridicule. Their claims were, he declared, "monstrous" (Wallace 1908). The distinguished Oxford Darwinian Sir Edward Poulton reported that all the eminent zoologists to whom he explained his grounds for indicting the Mendelian writings as "injurious to Biological Science, and a hindrance in the attempt to solve the problem of evolution" had agreed with him (Poulton 1908). Reginald Punnett, a colleague of the Mendelian William Bateson, replied: "The Sacred College has convened and orthodoxy has spoken through its chosen mouthpiece" (Punnett 1909, 107).

The great German evolutionist August Weismann was cautious on Mendel in public, but writing to his translator, William Parker, he expressed his dislike of mathematics and his expectation that reduction division in animals introduces complications that are not present in plants (Churchill 1999, 1: 375). Famous for his denial of the inheritance of acquired characters and his reliance on cross-breeding for the emergence of variation, Weismann had long claimed that for the original source of variation one had to turn to lower forms of life, where there is no separation between germ cells and body cells. In the course of evolution, he explained, higher forms, by cross-breeding, expose these variations, combining them in myriad ways. But in Weismann's great book, *The Germ-Plasm*, the arch-critic of variation due to changes in the conditions of life made a 180° turn. Now he accepted the "slight inequalities of nutrition in the germ-plasm" (Weismann 1893, 431) as the cause of individual variations, as had Darwin in 1859. If Weismann at the close of the century felt he needed external conditions in order to produce variation, surely the Darwinians would too? And if Darwin had known and accepted

Mendel's critique, his supporters would surely have drowned out his admission!

Has the analysis in this chapter any relevance to current concerns over the adequacy of evidence for evolution by natural selection? Clearly not. Instead, it serves to remind us forcibly how much the evidence has strengthened since 1859. Furthermore, it should serve to underline Darwin's courage in publishing his theory at a time when much of the data on variation and inheritance was confused. Hopefully, it will also prove an antidote to any tendency to assume that when Darwin pondered and wrote about variation and heredity, he was always thinking along the same lines as that do today.

4 Darwin's Theory of Natural Selection and Its Moral Purpose

Thomas Henry Huxley recalled that after he had read Darwin's *Origin of Species*, he had exclaimed to himself: "How extremely stupid not to have thought of that!" (Huxley 1900, 1: 183). It is a famous but puzzling remark. In his contribution to Francis Darwin's *Life and Letters of Charles Darwin*, Huxley rehearsed the history of his engagement with the idea of transmutation of species. He mentioned the views of Robert Grant, an advocate of Lamarck, and Robert Chambers, the anonymous author of *Vestiges of the Natural History of Creation* (1844), which advanced a crude idea of transmutation. He also recounted his rejection of Agassiz's belief that species were progressively replaced by the divine hand. He neglected altogether his friend Herbert Spencer's early Lamarckian ideas about species development, which were also part of the long history of his encounters with the theory of descent. None of these sources moved him to adopt any version of the transmutation hypothesis.

Huxley was clear about what finally led him to abandon his longstanding belief in species stability:

The facts of variability, of the struggle for existence, of adaptation to conditions, were notorious enough; but none of us had suspected that the road to the heart of the species problem lay through them, until Darwin and Wallace dispelled the darkness, and the beacon-fire of the "Origin" guided the benighted. (Huxley 1900, 1: 179–83)

The elements that Huxley indicated – variability, struggle for existence, adaptation – form core features of Darwin's conception of natural selection. Thus what Huxley admonished himself for not immediately comprehending was not the fact, as it might be called, of species change but the cause of that change. Huxley's exclamation suggests – and it has usually been interpreted to affirm – that the idea of natural selection was really quite simple and that when the few elements composing it were held before the mind's eye, the principle and its significance would flash out. The elements, it is supposed, fall together in the following way: species members vary in their heritable traits; more individuals are produced than the resources of the environment can sustain; those that by chance have traits that better fit them to circumstances compared to others of their kind will more likely survive to pass on those traits to offspring; consequently, the structural character of the species will continue to alter over generations until individuals come to be specifically different from their ancestors.

Yet, if the idea of natural selection were as simple and fundamental as Huxley suggested and as countless scholars have maintained, why did it take so long for the theory to be published after Darwin supposedly discovered it? And why did it then require a very long book to make its truth obvious? In this chapter, I will try to answer these questions. I will do so by showing that the principle of natural selection is not simple but complex and that it took shape only gradually in Darwin's mind. In what follows, I will refer to the "principle" or "device" of natural selection, never to the "mechanism" of selection. Though the phrase "mechanism of natural selection" comes trippingly to our lips, it never came to Darwin's in the *Origin*; and I will explain why. I will also use the term "evolution" to describe the idea of species descent with modification. Somehow the notion has gained currency that Darwin avoided the term because it suggested progressive development. This assumption has no warrant for two reasons. First, the term is obviously present, in its participial form, as the very last word of the *Origin*, as well as being freely used as a noun in the last edition of the *Origin* (1872), in the *Variation of Animals and Plants under Domestication* (1868), and throughout *The Descent of Man* (1871) and *The Expression of the Emotions in Man and Animals* (1872). But the second reason for rejecting the assumption is that Darwin's theory is indeed progressivist, and his device of natural selection was designed to produce evolutionary progress.

DARWIN'S EARLY EFFORTS TO EXPLAIN TRANSFORMATION

Shortly after he returned from his voyage on HMS *Beagle* (1831–36), Darwin began seriously to entertain the hypothesis of species change over time. He had been introduced to the idea, when a teenager, through reading his grandfather Erasmus Darwin's *Zoonomia* (1794–96), which included speculations about species development; and, while at medical school in Edinburgh (1825–27), he had studied Lamarck's *Système des animaux sans vertèbres* (1801) under the tutelage of Robert Grant, a convinced evolutionist. On the voyage, he took with him Lamarck's *Histoire naturelle des animaux sans vertèbres* (1815–22), in which the idea of evolutionary change was prominent. He got another large dose of the Frenchman's ideas during his time off the coast of South America, where he received by merchant ship the second volume of Charles Lyell's *Principles of Geology* (1831–33), which contained a searching discussion and negative critique of the fanciful supposition of an "evolution of one species out of another" (Lyell 1987, 2: 60). Undoubtedly the rejection of Lamarck by Lyell and most British naturalists gave Darwin pause; but after his return to England, while sorting and cataloguing his specimens from the Galápagos, he came to understand that his materials supplied compelling evidence for the suspect theory.

In his various early notebooks (January 1837 to June 1838), Darwin began to work out different possible ways to explain species change (Richards 1987, 85–98). Initially, he supposed that a species might be "created for a definite time," so that when its span of years was exhausted, it went extinct and another, affiliated species took its place (*Notebooks*, 12, 62). He rather quickly abandoned the idea of species senescence, and began to think in terms of Lamarck's notion of the direct effects of the environment, especially the possible impact of the imponderable fluids of heat and electricity (*Notebooks*, 175). If the device of environmental impact were to meet what seemed to be the empirical requirement – as evidenced by the pattern of fossil deposits, going from simple shells at the deepest levels to complex vertebrate remains at higher levels – then it had to produce progressive development. If species resembled ideas, then progressive change would seem to be a natural result, or so

Darwin speculated: "Each species changes. Does it progress. Man gains ideas. The simplest cannot help. – becoming more complicated; & if we look to first origin there must be progress" (*Notebooks*, 175). Being the conservative thinker that he was, Darwin retained in the *Origin* the idea that some species, under special conditions, might alter through direct environmental impact as well as the conviction that modifications would be progressive.

Darwin seems to have soon recognized that the direct influence of surroundings on an organism could not account for its more complex adaptations, and so he began constructing another causal device. He had been stimulated by an essay of Frédéric Cuvier, which suggested that animals might acquire heritable traits through exercise in response to particular circumstances. He rather quickly concluded that "all structures either direct effect of habit, or hereditary <& combined> effect of habit" (*Notebooks*, 259).[1] Darwin thus assumed that new habits, if practiced by the population over long periods of time, would turn into instincts; and these latter would eventually modify anatomical structures, thus altering the species. Use inheritance was, of course, a principal mode of species transformation for Lamarck.

In developing his own theory of use inheritance, Darwin carefully distinguished his ideas from those of his discredited predecessor – or at least he convinced himself that their ideas were quite different. He attempted to distance himself from the French naturalist by proposing that habits introduced into a population would first gradually become instinctual before they altered anatomy. And instinct – innate patterns of behavior – would be expressed automatically, without the intervention of conscious will power, the presumptive Lamarckian mode (*Notebooks*, 292). By early summer of 1838, Darwin thus had two means by which to explain descent of species with modification: the direct effects of the environment and his habit-instinct device.

ELEMENTS OF THE THEORY OF NATURAL SELECTION

At the end of September 1838, Darwin paged through Thomas Malthus's *Essay on the Principle of Population*. As he later recalled

---

[1] Single wedges indicate erasure; double wedges indicate addition.

in his *Autobiography*, this happy event changed everything for his developing conceptions:

I soon perceived that selection was the keystone of man's success in making useful races of animals and plants. But how selection could be applied to organisms living in a state of nature remained for some time a mystery to me.

In October 1838, that is, fifteen months after I had begun my systematic enquiry, I happened to read for amusement Malthus on *Population*, and being well prepared to appreciate the struggle for existence which everywhere goes on from long-continued observation of the habits of animals and plants, it at once struck me that under these circumstances favourable variations would tend to be preserved, and unfavourable ones to be destroyed. The result of this would be the formation of new species. Here, then, I had at last got a theory by which to work. (*Autobiography*, 119–20)

Darwin's description provides the classic account of his discovery, and it does capture a moment of that discovery, though not the complete character or full scope of his mature conception. The account in the *Autobiography* needs to be placed against the notebooks, essays, and various editions of the *Origin* and *The Descent of Man*. These comparisons will reveal many moments of discovery, and a gradual development of his theory of natural selection from 1838 through the next four decades.

In the *Autobiography*, Darwin mentioned two considerations that had readied him to detect in Malthus a new possibility for the explanation of species development: the power of artificial selection and the role of struggle. Lamarck had suggested domestic breeding as the model for what occurred in nature. Undeterred by Lyell's objection that domestic animals and plants were specially created for man (Lyell 1987, 2: 41), Darwin began reading in breeders' manuals, such as those by John Sebright (1809) and John Wilkinson (1820). This literature brought him to understand the power of domestic "selection" (Sebright's term), but he was initially puzzled, as his *Autobiography* suggests, about what might play the role of the natural selector or "picker." In midsummer of 1838, he observed:

The Varieties of the domesticated animals must be most complicated, because they are partly local & then the local ones are taken to fresh country

& breed confined, to certain best individuals. – scarcely any breed but what some individuals are picked out. – in a really natural breed, not one is picked out.... (*Notebooks*, 337)

In this passage, he appears to have been wondering how selecting could occur in nature when no agent was picking the few "best individuals" to breed.

In the *Autobiography*, Darwin indicated that the second idea that prepared the way for him to divine the significance of Malthus's *Essay* was that of the struggle for existence. Lyell, in the *Principles of Geology*, had mentioned de Candolle's observation that all the plants of a country "are at war with one another" (Lyell 1987, 2: 131). This kind of struggle, Lyell believed, would be the cause of "mortality" of species, of which fossils gave abundant evidence (Lyell 1987, 2: 130). In his own reading of Lyell, Darwin took to heart the implied admonition to "study the wars of organic being" (*Notebooks*, 262).

These antecedent notions gleaned from Lamarck, Lyell, and the breeders led Darwin to the brink of a stable conception that would begin to take more explicit form after his reading of Malthus's *Essay* in late September 1838. In spring of 1837, for instance, he considered how a multitude of varieties might yield creatures better adapted to circumstances: "whether every animal produces in course of ages ten thousand varieties, (influenced itself perhaps by circumstances) & those alone preserved which are well adapted" (*Notebooks*, 193). Here Darwin mentioned in passing a central element of his principle of natural selection without, apparently, detecting its significance. And a year later something like both natural and sexual selection spilled onto the pages of his Notebook C: "Whether species may not be made by a little more vigour being given to the chance offspring who have any slight peculiarity of structure. <<hence seals take victorious seals, hence deer victorious deer, hence males armed & pugnacious all orders; cocks all war-like)>>" (*Notebooks*, 258; likely a gloss on Sebright 1809, 15–16). It is fair to say, nonetheless, that the foundations for Darwin's device of natural selection were laid on the ground of Malthus's *Essay*. His reading of that book caused those earlier presentiments to settle into a firm platform for further development.

THE MALTHUS EPISODE

Malthus had composed his book in order to investigate two questions: What has kept humankind from steadily advancing in happiness? Can the impediments to happiness be removed? Famously, he argued that the chief barrier to the progress of civil society was that population increase would always outstrip the growth in the food supply, thus causing periodic misery and famine. What caught Darwin's eye in the opening sections of Malthus's *Essay*, as suggested by scorings in his copy of the book, was the notion of population pressure through geometric increase:

In the northern states of America, where the means of subsistence have been more ample... the population has been found to double itself, for above a century and half successively, in less than twenty-five years.... It may safely be pronounced, therefore, that population, when unchecked, goes on doubling itself every twenty-five years, or increases in a geometrical ratio.... But the food to support the increase from the greater number will by no means be obtained with the same facility. Man is necessarily confined in room. (Malthus 1914, 5–7)

Darwin found in these passages from Malthus a propulsive force that had two effects: it would cause the death of the vast number in the population by reason of the better adapted pushing out the weaker, and thus it would sort out, or transform, the population. On September 28, 1838, Darwin phrased it this way in his Notebook D:

Even the energetic language of <Malthus> <<Decandoelle>> does not convey the warring of the species as inference from Malthus.... population in increase at geometrical ratio in FAR SHORTER time than 25 years – yet until the one sentence of Malthus no one clearly perceived the great check amongst men.... One may say there is a force like a hundred thousand wedges trying force <into> every kind of adapted structure into the gaps <of> in the oeconomy of Nature, or rather forming gaps by thrusting out weaker ones. <<The final cause of all this wedging, must be to sort out proper structure & adapt it to change. (*Notebooks*, 375–6)

All the "wedging" caused by population pressure would have the effect, according to Darwin, of filtering out all but the most fit organisms and thus adapting the latter (actually, leaving them pre-adapted) to their circumstances.

Though natural selection is the linchpin of Darwin's theory of evolution, his notebooks indicate only the slow emergence of its ramifying features. He reflected on his burgeoning notions through the first week of October 1838, but then turned to other matters. Through the next few months, here and there, the implications became more prominent in his thought. In early December, for instance, he explicitly drew for the first time the analogy between natural selection and domestic selection: "It is a beautiful part of my theory, that <<domesticated>> races... are made by percisely [sic] same means as species" (*Notebooks*, 416). But the most interesting reflections, which belie the standard assumptions about Darwin's theory, were directed to the *final cause* or *purpose* of evolution. This teleological framework would help to organize several other elements constituting his developing notion.

THE PURPOSE OF PROGRESSIVE EVOLUTION: HUMAN BEINGS AND MORALITY

The great peroration at the very end of the *Origin of Species* asserts a long-standing and permanent conviction of Darwin, namely, that the "object," or purpose, of the "war of nature" is "the production of the higher animals" (*Origin*, 490). And the unspoken, but clearly intended, higher animals are human beings with their moral sentiments. Darwin imbedded his developing theory of natural selection in a decidedly progessivist and teleological framework, a framework quite obvious when one examines the initial construction of his theory.

At the end of October 1838, he focused on the newly formulated device:

My theory gives great final cause <<I do not wish to say only cause, but one great final cause...>> of sexes...for otherwise there would be as many species, as individuals, &...few only social...hence not social instincts, which as I hope to show is <<probably>> the foundation of all that is most beautiful in the moral sentiments of the animated beings. (*Notebooks*, 409)

In this intricate cascade of ideas, Darwin traced a path from sexual generation to its consequences: the establishment of stable species, then the appearance of social species, and finally the ultimate purpose of the process, the production of human beings with their moral sentiments. This trajectory needs further explication.

When Darwin opened his first transmutation notebook in the spring of 1837, he began with his grandfather's reflections on the special value of sexual generation over asexual kinds of reproduction. The grandson supposed that sexually produced offspring would, during gestation, recapitulate the forms of ancestor species. As he initially put the principle of recapitulation: "The ordinary kind [i.e., sexual reproduction], which is a longer process, the new individual passing through several stages (typical, <of the> or shortened repetition of what the original molecule has done)" (*Notebooks*, 170). Darwin retained the principle of embryological recapitulation right through the several editions of the *Origin* (Nyhart, this volume). Recapitulation produced an individual that gathered in itself all the progressive adaptations of its ancestors. But the key to progressive adaptation was the variability that came with sexual reproduction (*Notebooks*, 171). In the spring of 1837, he still did not understand exactly how variability might function in adaptation; he yet perceived that variable offspring could adjust to a changing environment in ways that clonally reproducing plants and animals could not. Moreover, in variable offspring accidental injuries would not accumulate as they would in continuously reproducing asexual organisms. Hence stable species would result from sexual generation. For "without sexual crossing, there would be endless changes . . . & hence there could not be *improvement*. <<& hence not <<be>> higher animals" (*Notebooks*, 410). But once stable species obtained, social behavior and ultimately moral behavior might ensue.

Just at the time Darwin considered the "great final cause" of sexual generation – namely, the production of higher animals with their moral traits – he opened his Notebook N, in which he began to compose an account of the moral sentiments. He worked out the kernel of his conception, which would later flower in *The Descent of Man*, in a fanciful example. He imagined the case of a dog with incipient moral instincts:

Dog obeying instincts of running hare is stopped by fleas, also by greater temptation as bitch. . . . Now if dogs mind were so framed that he constantly compared his impressions, & wished he had done so & so for his interest, & found he disobeyed a wish which was part of his system, & constant, for a wish which was only short & might otherwise have been relieved, he would be sorry or have troubled conscience – therefore I say grant reason to any animal with social & sexual instincts <<& yet with passions he *must* have conscience – this is capital view. – Dogs conscience would not have been same with mans because original instinct different. (*Notebooks*, 563–4)

Darwin believed that the moral instincts were essentially persistent social instincts that might continue to urge cooperative action even after being interrupted by a more powerful, self-directed impulse. As he suggested to himself at this time: "May not moral sense arise from our enlarged capacity <acting> <<yet being obscurely guided>> or strong instinctive sexual, parental & social instincts give rise 'do unto others as yourself', 'love thy neighbour as thyself'. Analyse this out" (*Notebooks*, 558). He would, indeed, continue to analyze out his theory; for at this point in its development, he did not see how other-directed, social instincts, which gave no benefit to their carrier, could be produced by selection. This difficulty seems to have led him to retain the device of inherited habit to explain the origin of the social instincts. Thus in late spring of 1839, he formulated what he called the "law of utility" – derived from Paley – which supposed that social utility would lead the whole species to adopt certain habits that, through dint of exercise, would become instinctive: "*On Law of* Utility Nothing but that which has beneficial tendency through many ages [i.e., necessary social habits] would be acquired.... It is probable that becomes instinctive which is repeated under many generations" (*Notebooks*, 623). While Darwin never gave up the idea that habits could be inherited, he would solve the problem of the natural selection of social instincts only in the final throes of composing the *Origin*.

At the very end of October 1838, Darwin gave an analytic summary of his developing idea, a neat set of virtually axiomatic principles constituting his device:

Three principles, will account for all

    (1) Grandchildren. like grandfathers
    (2) Tendency to small change...
    (3) Great fertility in proportion to support of parents. (*Notebooks*, 412–13)

These factors may be interpreted as: traits of organisms are heritable (with occasional reversions); these traits vary slightly from generation to generation; and reproduction outstrips food resources (the Malthusian factor). These principles seem very much like those "necessary and sufficient" axioms advanced by contemporary evolutionary theorists: variation, heritability, and differential survival (Lewontin 1978). Such analytic reduction does appear to render evolution by natural selection a quite simple concept, as Huxley supposed. However, these bare principles do not identify a causal force that might scrutinize the traits of organisms to pick out just those that could provide an advantage and thus be preserved. Darwin would shortly construct that force as both a moral and an intelligent agent, and the structure of that conception would sink deeply into the language of the *Origin*.

NATURAL SELECTION AS AN INTELLIGENT AND MORAL FORCE

In 1842, Darwin roughly sketched the outlines of his theory, and two years later he enlarged the essay to compose a more complete and systematic version. In the first section of both essays, as in the first chapter of the *Origin*, Darwin discussed artificial selection. He suggested that variations in traits of plants and animals occurred as the result of the effects of the environment in two different ways: directly, by the environment's affecting features of the malleable body of the young progeny; but also indirectly, by the environment's affecting the sexual organs of the parents (*Foundations*, 1–2). Typically, a breeder would examine variations in plant or animal offspring, and if any captured his fancy, he would breed only from those suitable varieties and prevent back-crosses to the general stock. Back-crosses, of course, would damp out any advantages the selected organisms might possess.

In the next section of the essays, Darwin inquired whether variation and selection could be found in nature. Variations in the wild, he thought, would occur much as they did in domestic stocks. But the crucial, two-pronged issue was: "is there any means of selecting those offspring which vary in the same manner, crossing them and keeping their offspring separate and thus producing selected races" (*Foundations*, 5)? The first of these problems might be called the problem of selection, the second that of swamping out. In beginning to deal with these difficulties (and more to come), Darwin proposed to himself a certain model against which he would construct his device of natural selection. This model would control his language and the concepts deployed in the *Origin*. In the 1844 essay, he described the model this way:

Let us now suppose a Being with penetration sufficient to perceive the differences in the outer and innermost organization quite imperceptible to man, and with forethought extending over future centuries to watch with unerring care and select for any object the offspring of an organism produced under the foregoing circumstances; I can see no conceivable reason why he could not form a new race (or several were he to separate the stock of the original organism and work on several islands) adapted to new ends. As we assume his discrimination, and his forethought, and his steadiness of object, to be incomparably greater than those qualities in man, so we may suppose the beauty and complications of the adaptations of the new races and their differences from the original stock to be greater than in the domestic races produced by man's agency. (*Foundations*, 85)

The model Darwin had chosen to explain to himself the process of selection in nature was that of a powerfully intelligent being, one that had foresight and that selected animals to produce beautiful and intricate structures. This prescient being made choices that were "infinitely wise compared to those of man" (*Foundations*, 21). As a wise breeder, this being would prevent back-crosses of his flocks. Nature, the analog of this being, was thus conceived not as a machine but as a supremely intelligent force.

In the succeeding sections of both essays, Darwin began specifying the analogs for the model, that is, those features of nature that operated in a fashion comparable to the actions of the imaginary being. He stipulated, for instance, that variations in nature would be very slight and intermittent due to the actions of a slowly changing environment. But, looking to his model, he supposed that nature would compensate for very gradually appearing variations by acting in a way "far more rigid and scrutinizing" (*Foundations*, 9). He then brought to bear the Malthusian idea of geometrical increase of offspring, and the consequent struggle for existence that would cull all but those having the most beneficial traits.

Many difficulties in the theory of natural selection were yet unsolved in the essays. Darwin had not really dealt with the problem of swamping. Nor had he succeeded in working out how nature might select social, or altruistic, instincts, the ultimate goal of evolution. And as he considered the operations of natural selection, it seemed improbable that it could produce organs of great perfection, such as the vertebrate eye. His strategy for solving this last problem, however, did seem ready to hand – namely, to find a graduation of structures in various different species that might illustrate how organs like the eye might have evolved over long periods of time. Moreover, if natural selection had virtually preternatural discernment, it could operate on exquisitely small variations to produce something as intricate as an eye.

DARWIN'S BIG SPECIES BOOK: GROUP SELECTION AND THE MORALITY OF NATURE

In September 1854, Darwin noted in his pocket diary, "Began sorting notes for Species theory." His friends had urged him not to delay in publishing his theory, lest someone else beat him to the goal. His diary records on May 14, 1856: "Began by Lyell's advice *writing* species sketch."[2] By the following fall, the sketch grew far beyond his initial intention. His expanding composition was to be called *Natural Selection*, though in his notes he referred to it affectionately as "my Big Species Book." And big it would have been: his efforts would have yielded a very large work, perhaps extending to two or three fat volumes. But the writing was interrupted when Lyell's prophecy about someone else forestalling him came true. In mid-June 1858, he received the famous letter from Wallace, then in Malaya, in which that naturalist included an essay that could have been purloined from Darwin's notebooks. After reassurances from friends that honor did not require him to toss his manuscript into the flames, Darwin compressed that part of the composition already completed and quickly wrote out the remaining chapters of what became the *Origin of Species*.

At the beginning of March 1858, a few months before he was to receive Wallace's letter, Darwin had finished a chapter in his manuscript entitled "Mental Powers and Instincts of Animals." In that chapter he solved a problem about which he had been worrying for almost a decade. In his study of the social insects – especially

---

[2] Charles Darwin, Personal Journal MS 34, Cambridge University Library, DAR 158.1–76.

ants and bees – he had recognized that the workers formed different castes with peculiar anatomies and instincts. Yet the workers were sterile, and so natural selection could not act on the individuals to preserve in their offspring any useful habits. How, then, had these features of the social insects evolved? In a loose note, dated June 1848, in which he sketched out the problem, he remarked, "I must get up this subject – it is the greatest *special* difficulty I have met with."[3]

Though Darwin had identified the problem many years before, it was only in the actual writing of his *Big Species Book* that he arrived at a solution. He took his cue from William Youatt's *Cattle: Their Breeds, Management, and Disease* (1834). When breeders wished to produce a herd with desirable characteristics, they would choose animals from several groups and slaughter them. If one or another had, say, desired marbling, they would breed from the family of the animal with that characteristic.[4] In the *Species Book*, Darwin rendered the discovery this way:

This principle of selection, namely not of the individual which cannot breed, but of the family which produced such individual, has I believe been followed by nature in regard to the neuters amongst social insects; the selected characters being attached exclusively not only to one sex, which is a circumstance of the commonest occurrences, but to a peculiar & sterile state of one sex. (*Species Book*, 370)

Darwin thus came to understand that natural selection could operate not only on individuals but also on whole families, hives, or tribes. This insight and the expansion of his theory of natural selection would have two important dividends: first, he could exclude a Lamarckian explanation of the wonderful instincts of the social insects – since no acquired habits could be passed to offspring – and simultaneously he could overcome a potentially fatal objection to his theory (see Lustig, this volume). But second, this theory of family selection (or community selection, as he came to call it) would enable him to solve the like problem in human evolution, namely, the origin of the altruistic instincts. In *The Descent of Man*, Darwin

[3] Charles Darwin, Cambridge University Library, DAR 73.1–4. I have discussed the problem of the social insects in Richards 1987, 142–52.
[4] Charles Darwin, Cambridge University Library, DAR 73.1–4.

would mobilize the model of the social insects precisely in order to construct a theory of human moral behavior that contained a core of pure, unselfish altruism – that is, acts that benefited others at a cost to self, something that could not occur under individual selection (Richards 1987, 206–19). Hence the final goal of evolution, as he originally conceived its telic purpose, could be realized: the production of the higher animals having moral sentiments. Yet not only did Darwin construe natural selection as producing moral creatures, he conceived of natural selection itself as a moral and intelligent agent.

The model of an intelligent and moral selector, which Darwin cultivated in the earlier essays, makes an appearance in the *Big Species Book*. In the chapter "On Natural Selection," he contrasted man's selection with nature's. The human breeder did not allow "each being to struggle for life"; rather, he protected animals "from all enemies." Further, man judged animals only on surface characteristics and often picked countervailing traits. The human breeder also allowed crosses that reduced the power of selection. And finally, man acted selfishly, choosing only that property that "pleases or is useful to him." Nature acted quite differently:

> She cares not for mere external appearances; she may be said to scrutinize with a severe eye, every nerve, vessel & muscle; every habit, instinct, shade of constitution, – the whole machinery of the organization. There will be here no caprice, no favouring: the good will be preserve & the bad rigidly destroyed. (*Species Book*, 224)

Nature thus acted steadily, justly, and with divine discernment, separating the good from the bad. Nature, in this conception, was God's surrogate, which Darwin signaled by penciling in his manuscript above the passage just quoted: "By nature, I mean the laws ordained by God to govern the Universe" (*Species Book*, 224; see also Brooke, this volume). As Darwin pared away the overgrowth of the *Big Species Book*, the intelligent and moral character of natural selection stood out even more boldly in the précis, that is, in the *Origin of Species*.

NATURAL SELECTION IN THE *ORIGIN OF SPECIES*

In the first edition of the *Origin*, Darwin approached natural selection from two distinct perspectives, conveyed in two chapters whose titles suggest the distinction: "Struggle for Existence" and "Natural Selection" (Chapters 3 and 4). Though their considerations overlap, the first focuses on the details of the operations of selection and the second contains the more highly personified re-conceptualization of its activities. In Chapter 3, Darwin proposed that small variations in organisms would give some an advantage in the struggle for life. He then defined natural selection:

Owing to this struggle for life, any variation, however slight and from whatever cause proceeding, if it be in any degree profitable to an individual of any species, . . . will tend to the preservation of that individual, and will generally be inherited by its offspring. The offspring, also, will thus have a better chance of surviving. . . . I have called this principle, by which each slight variation, if useful, is preserved by the term Natural Selection. (*Origin*, 61)

Darwin would explain what he meant by "struggle" a bit later in the chapter, and I will discuss that in a moment. Here, I would like to note several revealing features of his definition. First, selection is supposed to operate on all variations, even those produced by the inheritance of acquired characters, and not just on those that arise accidentally from the environment's acting on the sex organs of parents. Second, Darwin believed that virtually all traits, useful or not, would be heritable – what he called the "strong principle of inheritance" (*Origin*, 5). Third, though the initial part of the definition indicates that it is the individual that is preserved, in the second part it is the slight variation that is preserved – which latter is the meaning of the phrase "natural selection" (*Origin*, 61, 81). The passage draws out the "chicken and egg" problem for Darwin: a trait gives an individual an advantage in its struggle, so that the individual is preserved; the individual, in turn, preserves the trait by passing it on to offspring. Finally, the definition looks to the future, when useful traits will be sifted out and the nonuseful extinguished, along with their carriers. In the short run, individuals are preserved; in the long run, it is their morphologies that are both perpetuated and slowly changed as the result of continued selection.

"We behold," Darwin observed (using a recurring metaphor), "the face of nature bright with gladness"; but we do not see the struggle that occurs beneath her beaming countenance (*Origin*, 62). But what does "struggle" mean, and who are the antagonists in a struggle for existence? Darwin said he meant "struggle" in a "large and metaphorical sense," which, as he spun out his meandering notion, would cover three or four distinct meanings (*Origin*, 62–3). First, an animal preyed upon will struggle with its aggressor. But as well, two canine animals will "struggle with one another to get food and live." The image Darwin seems to have had in mind was that of two dogs struggling over a piece of meat. Furthermore, struggle can be used to characterize a plant at the edge of the desert: it struggles "for life against the drought." In addition, one can say that plants struggle with other plants of the same and different species for their seeds to occupy fertile ground. These different kinds of struggle, in Darwin's estimation, can be aligned according to a sliding scale of severity. Accordingly, the struggle will move from most to least intense: between individuals of the same variety of a species; between individuals of different varieties of the same species; between individuals of different species of the same genus; between species members of quite different types; and finally, between individuals and climate. These various and divergent meanings of struggle seem to have come from the two different sources for Darwin's concept: de Candolle, who proclaimed that all of nature was at war, and Malthus, who emphasized the consequences of dearth to whole populations. Today, we would say that struggle – granted its metaphorical sense – properly occurs only between members of the same species to leave progeny. Adopting de Candolle's emphasis on the warlike aspects of struggle may have led Darwin to distinguish natural selection from sexual selection, which latter concerns not a death struggle for existence but males' struggling for mating opportunities.

In the chapter "Natural Selection" in the *Origin*, Darwin characterized his device in this way, pulling phrases from his earlier essays and *Big Species Book* but rendering them with a biblical inflexion:

Man can act only on external and visible characters: nature cares nothing for appearances, except in so far as they may be useful to any being. She can act on every internal organ, on every shade of constitutional difference, on the whole machinery of life. Man selects only for his own good; Nature only for that of the being which she tends. . . . Can we wonder, then, that nature's productions should be far "truer" in character than man's productions; that they should be infinitely better adapted to the most complex conditions of life, and should plainly bear the stamp of far higher workmanship? (*Origin*, 83–4).

The biblical coloring of Darwin's text is condign for a nature that is the divine surrogate and that acts only altruistically for the welfare of creatures. The attribution of benevolence to nature becomes explicit in Darwin's attempt to mitigate what might seem the harsh language of struggle. He concludes his chapter "Struggle for Existence" with the solace: "When we reflect on this struggle, we may console ourselves with the full belief, that the war of nature is not incessant, that no fear is felt, that death is generally prompt, and that the vigorous, the healthy, and the happy survive and multiply" (*Origin*, 79). Darwin's model of moral agency mitigated the force of Malthusian pitilessness.

CONCLUSION

I have argued that Darwin did not come to his conception of natural selection in a flash that yielded a fully formed theory. What appears as the intuitive clarity of his device is, I believe, quite deceptive. I have tried to show that his notions about the parameters of natural selection, what it operates on and its mode of operation, gradually took shape in Darwin's mind, and hardly came to final form even with the publication of the first edition of the *Origin of Species*. In this gradual evolution of a concept – actually a set of concepts – I have emphasized the way Darwin characterized selection as a moral and intelligent agent. Most contemporary scholars have described Darwinian nature as mechanical, even amoral in its ruthlessness. To be sure, when Wallace and others pointed out what seemed the misleading implications of the device, Darwin protested that, of course, he did not mean to argue that natural selection was actually an intelligent or moral agent. But even Darwin recognized, if dimly, that his original formulation of the device and the cognitively laden language of his writing carried certain consequences with which he did not wish to dispense – and, indeed, could not dispense with without altering his deeper conception of the character and goal of evolution. Darwin's language and metaphorical mode of thought gave his theory a meaning resistant to any mechanistic interpretation and unyielding even to his later, more cautious reflections.

Let me spell out some of those consequences to make clear how markedly Darwin's original notion of evolution by natural selection differs from what is usually attributed to him. Natural selection, in Darwin's view, moved very slowly and gradually, operating at a stately Lyellian pace (perhaps seizing on useful variations that might occur only after thousands of generations; *Origin*, 80, 82). It compensated for meager variability by daily and hourly scrutinizing every individual for even the slightest and most obscure variation, selecting just those that gave the organism an advantage. A nineteenth-century machine could not be calibrated to operate on such small variations or on features that might escape human notice. If natural selection clanked along like a Manchester spinning loom, one would not have fine damask – only a skillful and intelligent hand could spin that – or the fabric of the eye.

Second, Darwin frequently remarked in the *Origin* that selection operated more efficiently on species with a large number of individuals in an extensive, open area (*Origin*, 41, 70, 102, 105, 125, 177, 179). He presumed that, as in the case of the human breeder, a large number of individual animals or plants would produce more favorable variations upon which selection might act. The greater quantities would also create Malthusian pressure. Yet in the wild, this scenario for selection could occur only if the watchful eye of an intelligent selector somehow gathered the favored varieties together and isolated them so as to prevent back-crossing into the rest of the stock. After Fleeming Jenkin, in his review of the *Origin*, pointed out the problem of swamping of single variations, Darwin suggested in the fifth edition that groups of individuals would all vary in the same way due to the impact of the local environment (*Variorum*, 179). Thus when the implications of his model of intelligent nature were recognized, Darwin had to invoke as analog a Lamarckian scenario. Today, we assume that small breeding groups isolated by physical barriers would more likely furnish the requisite conditions for natural selection.

Third, a wise selector that has the good of creatures at heart would produce a progressive evolution, one that created ever-more-improved organization, which Darwin certainly thought to be the case. He believed that more recent creatures had accumulated progressive traits and would triumph over more ancient creatures in comparable environments (*Origin*, 205–6, 336–7). He summed up his view in the last section of the *Origin*: "And as natural selection works solely by and for the good of each being, all corporeal and mental endowments will tend to progress towards perfection"

(*Origin*, 489). This passage, which remains unchanged through the several editions of the *Origin*, is an index both of Darwin's moral conception of nature and of its progressive intent. The moral overlay of the passage has blotted out the winnowing force of selection, which hardly works to the benefit of every creature. And, as Darwin made clear in the third edition, the "improvements" wrought by selection will "inevitably lead to the gradual advancement of the organization of the greater number of living beings throughout the world" (*Variorum*, 221).

Fourth, such an intelligent agency would select not merely for each creature's good, but also for that of the community. Darwin, in the fifth and sixth editions of the *Origin*, extended his model of family selection to one that operated simply on a community: "In social animals it [natural selection] will adapt the structure of each individual for the benefit of the community; if this in consequence profits by the selected change" (*Variorum*, 172).

Finally, the intelligent and moral character of natural selection would produce the goal that Darwin had sighted early in his notebooks, namely, the production of the higher animals with their moral sentiments. Darwin thus concluded his volume with the Miltonic and salvific vision that he had harbored from his earliest days:

Thus, from the war of nature, from famine and death, the most exalted object which we are capable of conceiving, namely, the production of the higher animals, directly follows. There is grandeur in this view of life, with its several powers, having been originally breathed into a few forms or into one; and that, whilst this planet has gone cycling on according to the fixed laws of gravity, from so simple a beginning endless forms most beautiful and most wonderful have been, and are being, evolved. (*Origin*, 490)

Darwin's vision of the process of natural selection was anything but mechanical and brutal. Nature, while it may have sacrificed a multitude of its creatures, did so for the higher "object," or purpose, of creating beings with a moral spine – out of death came life more abundant. We humans, Darwin believed, were the goal of evolution by natural selection.

5 Originating Species
Darwin on the Species Problem

Darwin's revolutionary arguments for the transformation of species emerged against the backdrop of a century-long transnational debate over the nature of organic species. The aim of this chapter is to clarify this context and relate it to Darwin's *Origin*.

Following a brief summary of the "species problem" in the period before 1850, I analyze some key aspects of Darwin's species concept in the so-called *Big Species Book* (1856–58) (*Species Book*) that directly underlies the published *Origin*. This will be followed by a discussion of Darwin's public presentation in the *Origin*, followed by some brief remarks on the divergent interpretations by Darwin's successors.

I. THE SPECIES QUESTION BEFORE 1850

Available literature on the history of the species concept (Bachmann 1906; Uhlmann 1923; Stamos 2003 and references therein) only partially clarifies the nature of the debate over organic species in the period before the *Origin*. Anglophone discussions typically take as their reference framework British, or to a lesser degree French, natural history, assumed to be the main relevant background against which to assess Darwin's own reflections. This chapter seeks to complete some of these lines of discussion by consideration of developments in German-language natural history that were known to

I wish to acknowledge my appreciation for valuable comments on earlier versions of this chapter from David Depew, James Barham, James Lennox, David Stamos, Robert O'Hara, and the editors of this volume.

Darwin and that played an important role in his mature reflections on the species question.

A pre-*Origin* debate over the nature of organic species can be dated from the 1750s; it followed upon the novel reflections on the issue by Georges-Louis le Clerc, comte de Buffon (1707–1788) in his *Histoire naturelle, générale et particulière* (1749–67 with supplements). The novelty of Buffon's innovation, first introduced in 1749, revolved around the epistemological and metaphysical shift that Buffon introduced into the discussion of species in a form that affected subsequent reflections in a profound way (Farber 1972; Sloan 1986, 2002, 2006).

Buffon's arguments can be positioned against a heritage of discussion that reached back to Aristotelian and Scholastic sources and the remnants of the so-called universals problem inherited from the medieval period (Stamos 2003, Chapter 2 and references therein). One aspect of this heritage was the dual reference of the term *species* or *eidos* in Greek. One employment of this term applied to the universal in thought and language, such as the universal noun "man" or "horse." The other referred to the individualized substantial form in the particular existent thing, for example, the specific principle in "Socrates" or "Dobbin" that with matter constituted the existent thing in Aristotelian metaphysics. Although these two concepts of *eidos* or *species* were related within Aristotelian metaphysics and epistemology, they were nonetheless distinct concepts and referred to different domains – one to thought, language, and logic, the other to physical reality. The use of the same Greek, and later Latin, term – *eidos* or *species* – was not, however, accidental, nor was it the product of conceptual imprecision. The tie between universal and particular, ultimately linking conceptual thought to the existent world of real substances, formed a critical component of the Aristotelian theory of how mind and world connect. "Man" and the specific principle in "Socrates" are different but still intimately connected. Scholarly research has shown that it is the meaning of *eidos* as form, and not in its sense as universal, that is the primary meaning encountered in the Aristotelian biological treatises, a point that is important for the subsequent discussion (Balme 1962).

The positions developed as criticisms of the Aristotelian theory of universals took the form of either "nominalism" – the claim that universals are only "bare names" that do not denote any shared form or essence in things – or "conceptualism," developed particularly in the early modern period by the English philosopher John Locke (1632–1704) (Stamos 2003, Chapter 2 and references therein).

The development of modern rational systematics, commonly attributed to Carl von Linné (Linnaeus) (1707–1778), involved the development of a stabilized hierarchy of classes in the now-familiar sequence of Kingdom, Class, Order, Genus, Species, and Variety. Furthermore, with some important exceptions that cannot be explored here, Linnean taxonomic philosophy was built upon implicit Aristotelian Realist foundations (Larson 1971). Within the Linnean hiearchy, only taxa at the next-to-lowest category level were to be designated as a "Species." Divisions below this – groups forming the Linnean category "Variety" – were only accidental variations within the group sharing a single form defined by the species definition.

There were no fundamental difficulties, except practical ones, in applying this same logic to the classification of inorganic entities, and Linnaeus himself classified minerals in the *Systema naturae* by the same principles. Linnean classification can be generally analyzed in terms of class logic and set theory (Buck and Hull 1966). In the following discussion, I will understand the Linnean "classificatory" or "logical" conception of a natural species as a class of particulars that satisfy the defining necessary and sufficient criteria for class membership at that level of the Linnean hierarchy. Such groups will be denoted as **species**$_L$, including within this designation similar classificatory and logical senses that might be given to taxon or group names at the levels of Variety, Genus, Family, and so on.

The novelty of Buffon's new program in natural history involved a creative reinterpretation of the by-then established Linnean system. This involved his explicit separation of the two fundamental meanings of *species* – universal concept and individualized substantial form – that had been connected in the Realist tradition just outlined. Furthermore, this separation of meanings was posed by Buffon in the form of an *opposition* between these two traditional meanings of *species*.

Buffon's primary definition of an organic species was remarkably close to the conception of *eidos* as encountered in the biological treatises of Aristotle, and may in fact have been derived from

Aristotle's biological treatises through Buffon's reading of William Harvey (Sloan 1986).[1] In this sense, *species* refers to the individualized substantial form perpetuated through a lineage of particular individuals forming an ancestor-and-descendant relationship, extended in time such that it renders the species enduring and even eternal (e.g., Aristotle, *De anima*, II.iv. 415b 1–10). This conception of natural species Buffon explicitly contrasted with the meaning of species as universals, or species$_L$, which he considered only "artificial" and "arbitrary." Buffon's "real and physical" species were, unlike Linnean species, grounded in the *material* connectedness of individuals through reproduction, either contemporaneously or as extended in time.

We can see the novel ingredients in Buffon's species concept in his most extended statement of 1753, subsequently given wide currency by its verbatim republication in 1755 in the influential article "Espèce: histoire naturelle" in Diderot's and D'Alembert's *Encyclopédie ou dictionnaire raisonné*:

An individual is a creature by itself, isolated and detached, which has nothing in common with other beings, except in that which it resembles or differs from them. All the similar individuals which exist on the surface of the globe are regarded as composing the species of these individuals....

However, it is neither the number nor the collection of similar individuals which forms the species. It is the constant succession and uninterrupted renewal of these individuals which comprises it. Because a being which would last forever would not be a species, no more than would a thousand similar beings which would last forever. The species is thus an abstract and general term, for which the thing exists only in considering Nature in the succession of time, and in the constant destruction and renewal of creatures....

It is thus in the characteristic diversity of species that the intervals in the gradations of nature are most sensible and best marked.... The species being nothing else than a constant succession of similar individuals that reproduce themselves, it is clear that this denomination must be applied only to animals and plants, and that it is an abuse of terms or ideas that

---

[1] Some textual evidence supports the claim that Buffon and Daubenton recognized these different usages in Aristotle's biological works and consciously exploited this difference (Sloan 2002, 252 n. 25 and references therein). Appeal to recent scholarly readings, such as Balme's (1962), is therefore warranted.

the classifiers [*nomenclateurs*] have used it to designate different kinds of minerals. (Buffon 1753 quoted in Piveteau 1954, 355–6)

As Buffon developed these concepts in the course of his *Natural History*, his "physical" species had both synchronic aspects – membership in such a species was defined by relationships of interfertility – and diachronic dimensions – the individual organisms comprising a physical species were connected in time by descent from a common source by physical reproduction. In both dimensions, organisms are related by some kind of material continuity. The identity of the species over time is maintained by the conservation of the underlying "internal mold" that underlies the reproduction of like by like (Farber 1972). Much of this sounds like Aristotle's conception of *eidos* in the metaphysical sense employed in the biological treatises, even to the degree that Buffon's "internal mold" functions much like Aristotle's substantial form in this relation.

I shall designate such "physical" species in Buffon's sense as **species**$_H$ in the following discussion, with the subscript intended to denote the historical and material connectedness implied in his designation. I also include in this designation the broader conceptions of *genre* and *famille* in the "real and physical" sense in which Buffon employed these genealogical meanings in subsequent volumes of his *Natural History*, broadening his original conception of species fixity.

The confusions in the subsequent tradition created by Buffon's arguments were substantial. Many of his contemporaries either read him as some kind of species nominalist, or failed to see the point of his distinction between traditional species$_L$ and his new emphasis on the ontological reality of species$_H$. But this was not the case in the Germanies. Through the reworking of some of Buffon's arguments by Immanuel Kant in influential discussions initiated in his anthropology lectures in 1775, a formal and epistemically grounded systematization of the distinction between species$_L$ and species$_H$ was introduced that reverberated through German-language life science in the early nineteenth century (Sloan 2006 and references therein). This was an outcome of Kant's efforts to systematize the distinction between Linnean and Buffonian projects in natural history. One form of "natural history" – primarily the Linnean – was conceptualized as a synchronic descriptive project, the "description of nature"

(*Naturbeschreibung, Physiographie*), which included biogeography and the geographical distribution of organisms. Classification within the "description" of nature was concerned with synchronic relationship rather than vertical historical genesis. To a large degree this mapped onto the tradition of species$_L$ discussions.

Kant located the historical lineage dimensions of Buffonian species$_H$ within a historical and genetic understanding of nature (*Naturgeschichte, Physiogonie*). Furthermore, the distinction between these two programs was not only relevant to the conception of groups at the species level; it also extended to the distinction between the Linnean conception of a Variety and the newly emerging conception of Race (*Rasse*), given a new theoretical status by Kant's distinctions. For Kant, a "Race" designates the group of related and interfertile but semipermanent lineages descending from a common source. This is a concept that is valid within the "history," but not the "description," of nature. A Linnean "Variety" lacks these historical aspects, even if it is validly employed in Linnean fashion in the synchronic classification of forms. From these distinctions we find a parallel terminology developing in the German biological literature that differentiated concepts valid within the "history" from those valid in the "description" of nature. The importance of this will emerge later.

In spite of Kant's attempts to systematize and clarify the distinctions between the referents of species$_L$ and species$_H$ and related concepts, the historical interaction, and indeed the inevitable confusion, that ensued in the early decades of the nineteenth century among the various referents of the term "species" is evident from the literature of natural history of this period. In this we perceive a rich and complex interplay of these alternative species concepts, with one body of literature continuing to interpret species in the species$_L$ tradition, and a second body of literature, developed particularly within German philosophy of nature (*Naturphilosophie*), exploiting the concept of species$_H$ (Spring 1838). This later meaning was also picked up by Hegel and given a further philosophical foundation in his notion of the concrete universal (Stamos 2003, Chapter 4).

This German tradition of discussion emphasized the notion of species taxa as holistic and historical lineages constituted by reproductive relationships, descending from a common ancestral source, and extended in time and space. Such species and their sub-divisions or conglomerations (*Arten, Abarten, Unterarten, Rassen, Gattungen*) in the living world were to be distinguished from classes of minerals or other nonbiological entities. Such species were also, in the tradition of Buffon, designated by adjectives like "real," "physical," and "historical" in opposition to the "arbitrary" group-ings of logical arrangement. The complex framework of these early nineteenth-century debates within theoretical natural history, rang-ing across metaphysical, epistemological, and empirical questions, provides a complex background against which we can address Dar-win's mature discussion of these questions in the *Origin*.

II. THE *BIG SPECIES BOOK* FOUNDATIONS

Darwin's omnivorous readings in the twenty-year period following his return to England in October of 1836 and the commencement of the *Big Species Book* manuscript in 1856 involved a wide sampling of sources drawn from several national traditions. Well known is the close reading and annotation of Charles Lyell's discussion of the "species question" in the fifth (not the first) edition of the *Principles of Geology*, published after Darwin's return from the *Beagle* voyage. Lyell's long discussion of the species question, presented in the con-text of a critique of Lamarck's transformism, defined much of the "species problem" for British natural history. His discussions also gave a specific meaning to the "reality of species" question that cen-tered it around issues bearing primarily on what I have termed the species$_L$ problems. For Lyell, the issue of the "reality" of species con-cerned the evidence for the existence of a stable "specific character" in spite of variations related to geography and geological changes (Lyell 1832, 2: 18). Lyell defended the "reality" of species in his exhaustive examination of the evidence for and against variation beyond the limits of a fixed specific type. Darwin also encountered some further analysis and reworking of Lyell's arguments through his reading of William Whewell's discussion in the *History of the Inductive Sciences* of 1837 (Whewell 1837, 3: 573–80).

Less well understood is the impact of Darwin's reading of other sources, particularly those drawing on the species$_H$ tradition. Commencing in 1845, following the completion of the 230–page rough draft laying out the primary argument of the future *Origin*

(*Foundations*), Darwin undertook an intensive period of correspondence with his closest confidant, Joseph Dalton Hooker, to whom he had shown his 1844 manuscript. These readings also took place in the wake of the sensation created by the anonymous publication in 1844 of the *Vestiges of the Natural History of Creation* by the Scottish publisher Robert Chambers, who made a sweeping case for species transformism. As a botanist, Hooker referred Darwin to several sources dealing with the species question as it was encountered in botany. These readings also transpired shortly before Darwin began his intensive study of barnacles in September of 1846.

Primarily socialized as a zoologist and geologist, Darwin was now immediately engaged with the problems of botanical classification, issues he would then confront in a similar form in the zoological world in his massive study of the barnacles. This sequence of mid-1840s readings can be followed in Darwin's Reading Notebooks (DAR 119), in his correspondence, and also through his annotations of key texts. These readings included several papers by the botanist Hewlett C. Watson (Watson 1843, 1845), issues of the *London Journal of Botany* and *The Phytologist*, a long review article in French on the species question by Frédéric Gérard, an article on geographical botany by Alphonse De Candolle, and a monograph on the species question by the German botanist Johann Jacob Bernhardi. In these readings we can follow an interplay between the species$_L$ and species$_H$ concepts we have outlined earlier. Darwin's direct encounter with species$_H$ discussions in these 1840s readings added a new dimension to his reflections, and they allowed him to draw into a creative synthesis issues from these two separable domains of discourse. Through these readings in the botanical literature, Darwin engaged the complex issues of criteria and group definition that beset any practical application of the species$_L$ concept in botany. Similar issues were then engaged in his barnacle studies, which involved him in a messy worldwide revision of a difficult invertebrate group (*Correspondence*, 5: 155–56; Darwin 1975, 101). Darwin's readings in the mid-1840s laid foundations, through annotations and notes, to which he returned in 1856 in composing the *Big Species Book* manuscript. To these 1840s readings were also added new readings in works that had appeared during the intervening period, such as De Candolle's *Geographical Botany* (1855) and papers by Edward Blyth (Stamos 2007, 140–1). The result of this engagement can be discerned from a close examination of the *Species Book* discussion.

The manuscript discussion opens with a commentary on the confusion over the various definitions of "species" found in the literature. Much of this literature was summarized for Darwin in De Candolle's review (Candolle 1855, 2: 1071–8). Of all the options summarized by Candolle, he sees common descent as the preferred basis for species definition (*Species Book*, 96). Elaborating on this point, Darwin cites empirical problems created by the fact that many organisms – specifically, in this case, the primrose and cowslip – fit all the morphological criteria for distinct species but are "produced from the same stock," even though "they cannot be said to resemble each other as much as *analogous* plants do, which we positively & habitually know to have descended from a common source. Hence I conclude, that descent is a prominent idea under the word species as commonly accepted" (*Species Book*, 96). This leads him to conclude that "the idea of descent inevitably leads the mind to the first parent, & consequently to its first appearance, or creation" (ibid.).

Darwin then returns to a text he had read carefully and annotated in the mid-1840s, *Concerning the Concept of the Plant Species and its Application*, by the German botanist Johann Jacob Bernhardi (1774–1850). Bernhardi's short (sixty-eight-page) text displays some implicit familiarity with the post-Kantian distinctions between the "descriptive" and "historical" analyses of nature and the concepts relevant to each we have outlined here. At least some contemporaries placed Bernhardi in the same tradition as the *Naturphilosoph* Lorenz Oken. This context is not, however, made explicit in Bernhardi's text, and the content of his essay is concerned with practical classification rather than with issues of biological theory.

His text opens with an introductory discussion of the concept of species, which Darwin annotated in several places. This begins with a distinction between the conception of species in logic – a "sum of individuals which agree in certain characters" – and other meanings of the term. The "logical" meaning is a "concept, which lies at the basis of the classification of organic beings," but this definition "remains too indeterminate" because of its inapplicability in practical natural history and its inability to capture the "extraordinary changes according to the age, the time of birth, the location, climate, etc." necessary to discriminate between a species [*Art*] and a variation [*Abart*]." Required also are concepts applicable to "organic creation and specifically to the plants"(Bernhardi 1834, 1).

In answering this difficulty, Bernhardi develops an extended argument that interweaves several issues, interfacing these with the practical problems of distinguishing between species and varieties in the botanical world. However, these difficulties are not focused on the standard problems of character overlap and variation, as we might encounter in species$_L$ discussions. The difficulties instead are those presented by the different degrees of genealogical descent. Ideally, the general rule for defining a species would be the communality of genealogical descent. In a passage marked by Darwin, Bernhardi offers such a primary species definition: "One unites all individuals into one species which have been generated up to the present from an original stem-parent from seeds or germs" (ibid., 2).

But Bernhardi then raises several problems with this definition. Even if descent is the desirable means of defining natural species, problems presented by cryptogams and the difficulty of determining sources of descent require other supplementary considerations. To accomplish this, Bernhardi proposes consideration of the "total conformity in organization which must have been generated under the same circumstances" as a needed addition (ibid., 3).

Because Bernhardi sees a range of difficulties created by different degrees of genealogical relationship, he proposes a set of terminological distinctions that are intended to capture these descent relations, which he distinguishes from "logical" usages. This requires a new terminology. In place of the category name "Variety" he offers "Subspecies" (*Unterart*), designating groups defined by fertile interbreeding and common descent (ibid., 4). Further degrees of genealogical relationship are designated by such terms as *Abart, Bastarde, Spielart*, and *Abanderung.* The distinctions among such subgroupings are not clear-cut, but for reasons other than character overlap. As Bernhardi comments in a passage underlined (as indicated here) by Darwin:

As it appears essential to adopt these concepts generally, <u>it is still not to be denied that in many cases it can be doubtful</u> whether one is dealing with a Species [*Art*], a Subspecies [*Unterart*], a Variation [*Abart*] or a Hybrid [*Bastarde*]. (Ibid., 5)

Bernhardi's distinctions are then directly translated into more familiar Linnean terms by Darwin. His annotations on this discussion also make clearer than do his own published texts the differences between these descent-concepts and those of Linnean classification. In a marginal note on Bernhardi's definition of *Abarten*, he writes "'Abarten' a variety which does not tend to go back to parent form." Similarly, alongside *Spielarten* he writes: "'Spielarten', those that go back one or more generations" (Bernhardi 1834; DiGregorio and Gill 1990, 55). And in a longer annotation at the bottom of the next page he writes:

Unterart subspecies = doubtful races or the close species/ Abarten – hereditary = race (or variety in animals)/ 'Spielarten' – which are herditary [sic] for few generns [sic] – variety of DeCandolle/ Abanderung – which are not at all hereditary – allied to Monstrosities. (Ibid., 55)

By understanding the interplay of these issues in Darwin's thought, we can unpack several otherwise puzzling issues in his discussion of species, and particularly the fluidity in his concept of "variety" (Stamos 2007, Chapter 7). On the one hand, the issues of morphological variation and character overlap, something made strikingly evident at the practical level to Darwin through his barnacle work, which displayed how taxonomic distinctions of taxa at the levels of varieties$_L$ and species$_L$ were rendered difficult, and even to a large degree arbitrary, by the overlap produced by character variation. Such variation blurred the boundaries and introduced the need for "expert" skill and consensus among "competent" authorities as the only means of deciding on these limits. On the other hand, as we pursue these issues within the framework developed earlier, we see that Darwin was drawing upon more than one tradition of discourse. The problems created by variation for species could be *reinterpreted* by the use of species$_H$ concepts. This combination of traditions is strikingly apparent in the following quotation from the *Big Species* manuscript:

The term 'Variety' is applied to forms often offering considerable differences, & which can be securely propagated by buds, grafts, cuttings, suckers &c, but which are believed not to be inheritable by seed. This class nearly corresponds with "abanderungen" in Bernhardi's classification in which the form is not hereditary or only so in certain soils; & likewise in a lesser degree with his "Spielarten" in which the form tends to go back in one or more generations to the parent type.... Lastly we have the class "Race", corresponding with "Abarten" of Bernhardi & with subspecies of some authors, in which the form is strictly inherited, often even under changed conditions; of this class we know there are plenty under domestication, some known, & more suspected in a state of nature, as in the geographical races of some Zoologists. But the term subspecies is used by some authors, to define (& corresponds in this sense with "unterart" of Bernhardi) very close species, in which they cannot determine whether to consider them as species or varieties.... (*Species Book*, 98–9)

This interweaving of issues of taxonomic classification and genealogical descent, conflating different referents for the terms "species," "varieties," "subspecies," and "races," underlies the conceptual fluidity of Darwin's discussions that we can follow directly into the published *Origin*.

III. GOING PUBLIC

Presented initially as an "abstract" of his intended book, the argument of the *Origin* involved a drastic stripping out of the detailed references and supporting materials from the *Natural Selection* manuscript. This created a condensed, and ultimately more popularly accessible, form of presentation. Arguments developed in the manuscript with reference to specific issues or groups were rendered universal claims by this condensation. For his critics, this also meant that supporting documentation and detailed references to sources were typically not present.

On the issue of species, this condensed form of presentation submerged the important details in the genesis of his novel views we have examined in the previous section. Instead, we find a compressed discussion that is difficult to clarify. To understand Darwin's arguments, I have examined the various usages of the 1,489 occurrences of the term "species" in the first edition of the *Origin* (Barrett et al. 1981). This review of word usage shows an interweaving of several diverse meanings. Judged from these usages alone, it is difficult to defend a systematic "rhetorical strategy" thesis advanced by some scholars – the thesis that Darwin consciously utilized different meanings of "species" to win over adherents to his arguments – although this may be argued on additional evidence from other texts (Beatty 1985; Stamos 2007, Chapter 8). In some places, Darwin's usage is consistent with species$_L$ concepts, and in this context he even affirms a kind of species realism consistent with this framework, as when he speaks of "good species" or of species as "tolerably well-defined objects" that "do not at any one period present an inextricable chaos of varying and intermediate links," meaning primarily that they can be discriminated by nonoverlapping characters (*Origin*, 177). We also find this context in play when he describes how in many organisms, the issues of character variation tend to break down such clear definitions, resulting in groups that display "intermediate gradations" that make distinctions between species and varieties endlessly difficult and even arbitrary (ibid., 47).

But in other places, species$_H$ meanings predominate, and it is remarkable to see the way he has interwoven issues that arise from these considerations with those of traditional logical classification. Here Darwin interprets species and varieties as genealogical lineages that display varying degrees of historical relationship. In such contexts Darwin speaks, often interchangeably, of "subspecies," "races," and "varieties," in a diachronic and genealogical sense similar to the way he integrated Bernhardi's discussions in the *Species Book* manuscript. As we read through Darwin's central discussion of these issues in the second chapter of the published *Origin*, it is striking to see the way in which he has interwoven these different usages in critical passages.

For example, in the central discussion of species in Chapter 2, Darwin introduces his analysis with a description of the traditional practical problems created by character variation that face any naturalist working on preserved specimens. But remarkable in this discussion, and a claim that was startling to some of Darwin's readers (see, e.g., Owen 1860 in Hull 1973, 211), is that this practical problem facing species$_L$ concepts is then integrated, without detailed argument, with the gradation in degrees of genealogical relationship at issue in species$_H$ distinctions. We see how this is being done in the following passages:

Certainly no clear line of demarcation has as yet been drawn between species and sub-species – that is, the forms which in the opinion of some naturalists come very near to, but do not quite arrive at the rank of species; or, again, between sub-species and well-marked varieties, or between lesser varieties and individual differences. These differences blend into each other in an insensible series; and a series impresses the mind with the idea of an actual passage. (*Origin*, 51)

This claim could still be understood as a difficulty presented by taxonomic analysis at the species$_L$ level. Species$_H$ relationships, however, are *historical* and *ontological*. Darwin's synthesis of issues subtly transforms the *taxonomic* difficulties in one domain into evidence for *metaphysical* ambiguities in degrees of historical descent in the other domain:

Hence I look at individual differences, though of small interest to the systematist, as of high importance for us, as being the first step towards such slight varieties as are barely thought worth recording in works on natural history. And I look at varieties which are in any degree more distinct and permanent, as steps leading to more strongly marked and more permanent varieties; and at these latter, as leading to sub-species, and to species...; and I attribute the passage of a variety, from a state in which it differs very slightly from its parent to one in which it differs more, to the action of natural selection in accumulating... differences of structure in certain definite directions. Hence I believe a well-marked variety may be justly called an incipient species. (Ibid., 51–2)

By combining these two strands of species discourse into one, Darwin created an unusual conceptual synthesis with revolutionary consequences. At least two primary results can be seen to follow. First, he drew novel conclusions on the degree to which the historical relationships of lineages could be conceived to diverge from an original form, utilizing data now drawn from difficulties created by practical taxonomy. This linkage is not necessary and could be understood in ways that would not imply such radical claims, as in Hugh C. Watson's distinction between "book" and "natural" species (Watson 1843).

Second, since Buffon and Kant prior discussions of the species$_H$ concept had interpreted the historical relations of "natural-historical" species and their subdivisions to be confined within the limits imposed by an originating form – a *moule intérieur, première souche, Urtyp*, or archetype – that maintained a historical unity within a common stem, even if this unity allowed some wide divergences in response to environment and history. Bernhardi, for example, never questions this fundamental unity. Species$_H$ theorists in the German tradition in the pre-Darwinian period might be species *developmentalists*, allowing some kind of saltational succession in time of "central types" that might give rise to successive historical lineages of related forms. But they were not genuine species transformists in the Darwinian sense, understood to mean gradual divergence without natural limits through the slow adaptation to changing conditions by slight character variation.

But we see that for Darwin, the difficulties raised by practical taxonomic problems tied to species$_L$ considerations were employed to undermine the claims for the historical unity of groups within a common stem central to the species$_H$ discussions. In a curious sense, Darwin is a kind of ontological realist concerning species$_H$, in that for him there "really are" these lineages. But such species no longer have historical permanence or maintain a conservative "unity of type." Such unity is undermined by the destructive analysis flowing from arguments centered around species$_L$ issues.

In the arguments of the crucial Chapter 4 of the *Origin*, Darwin introduced his principle of divergence and used the well-known branching bush diagram (frontispiece, p. ii) to illustrate how natural selection worked over time. It is remarkable to see how this diagram functions within the framework of the issues set out here. The diagram is introduced initially as representing no more than the relations of individuals within species of a single genus forming lesser races or "varieties." It could be read initially as a claim about relationships between subgroups at the species$_L$ level: distinctions between these are often difficult because of the problems presented by character variation. One could, however, represent such intergradating horizontal relationships by geographical maps rather than by genealogical trees, as had been done by some of Darwin's predecessors.

But the diagram is offered as more than a depiction of species$_L$ relationships. Its fundamental intent is to represent a temporal, and not a spatial, relation of forms. The blurring of distinctions between taxa designated at the variety and species levels in the species$_L$ domain is summoned by Darwin as support for a more general argument about the blurring of the *historical* relations of groups. As one quickly perceives from the few pithy pages of argument that follow, the coordinates of his diagram are purely relative. The horizontal lines, representing time, can represent a hundred, or thousand, or ten thousand or millions of generations, or even geological ages.

Similarly, the degrees of divergence illustrated by the diagram can represent the relations of varieties$_H$ (or, perhaps better, "subspecies" or "races" in the German sense) within species$_H$, species$_H$ within genera$_H$, genera$_H$ within families$_H$, or any other degree of relationship designated by category names applied in a genealogical sense. And these relationships are now historical rather than horizontal.

As several recent authors have argued, although not without some controversy, the relevant analogy for depicting these relationships is that drawn from the literature of comparative philology of this period. The comparative philologists of the time used similar branching diagrams to depict the origins and filiation of natural languages, a tradition drawn upon by Darwin since his earliest notebooks (Stamos 2007, Chapter 3 and references therein). This linguistic metaphor provided him with a model for making the following points.

First, like natural languages, species have common origins in time from which they have diverged by gradual historical changes, even though all the intermediate stages of this historical development may not have survived and must be inferred by comparative study of contemporary languages.

Second, there is no strict requirement of monophyly – the claim that natural groups can have only a single historical source – although the monophyletic origins of groups is Darwin's usual presumption. The linguistic analogy explains the otherwise curious argument in the *Origin* that dogs, which can all interbreed with fertility, are still viewed by Darwin as probably polyphyletic, whereas pigeons are monophyletic (*Origin*, Chapter 1; Stamos 2007, Chapter 4).

Third, the similarity of "dialects" to "varieties" or, better, to "subspecies" in the species$_H$ sense – with many going extinct, others leading over time to new dialects and even to new language groups under the conditions of isolation – fits this model surprisingly well.

Fourth, the "reality" of different languages, like that of species, can be recognized in two ways. First, they exist horizontally in a species$_L$ sense, in that it is reasonably unproblematic to give definitions of natural languages in the present that discriminate these from one another, either through "essentialist" or "cluster" definitions. The distinction of French, Portuguese, Italian, and Spanish, for example, is a real one at the present time. Second, they have another kind of "existence," vertically in the species$_H$ sense. There is a concrete historical connection between these linguistic groups and a common historical ancestor-language (Latin), from which they all differ. We may also, in some cases, recognize "incipient" divergences in distinguishable language groups at the present time that may result in more fundamental subdivisions in the future.

The primary disanalogy in the linguistic model centers around Darwin's principle of divergence – the tendency of natural groups to diverge and fragment under the incessant Malthusian pressure of population. This means that such groups tend to subdivide, rather than coalesce, and to fragment rather than combine, in areas of overlap. It can be argued that there is nothing like this differentiating principle operative in the development of languages, where there are strong forces within the human social group that tend to create unification of several linguistic traditions into a single language in areas of overlap, or through successive stages of inclusion, such as has occurred with English. To the contrary, Darwinian divergence tends to "sympatric" speciation – that is, speciation without evident geographical isolation – as populations tend to subdivide single ecological niches and develop character divergences in order to maximize the amount of biomass possible in a given space in response to Malthusian pressures.

But with this important disanalogy recognized, the linguistic model provides several points of contact for understanding the way in which Darwin unified history and description, classification and genealogy, in his novel synthesis.

IV. SPECIES AFTER 1859

Space limitations permit only a brief pointing to some select issues that faced those trying to understand Darwin's complex species discussions during the post–1859 period. The controversies and critiques of Darwin's theory resulted in well-known modifications of Darwin's presentation of his argument in the subsequent editions of the *Origin* during the period between 1859 and the sixth edition of 1872 (final revision 1878). Revisions of the *Origin*, either minor or major, occur in nearly 75 percent of the original text (*Variorum* 1959). On the issue of species, however, there are few, if any, fundamental changes in the argument.

Although Louis Agassiz's quip – "if species do not exist at all, as the supporters of the transmutation theory maintain, how can they vary?"(Agassiz 1860 quoted in Beatty 1985, 270) – exemplifies a long tradition of those who read Darwin's arguments in the *Origin* as those of a "species nominalist" (Stamos 2007, Chapter 1), at least some influential German biological theorists clearly read him as a "species realist" within the species$_H$ tradition, with a surprising outcome. For example, the influential Swiss botanist Karl Wilhelm von Nägeli (1817–1891), who had been formed intellectually within the tradition of "philosophical" natural history inspired by Lorenz Oken's *Naturphilosophie*, originally conceptualized organic species as dynamic and holistic entities united by common reproduction in a group governed by a unifying rational Idea. In a lecture delivered on March 14, 1853, while a professor at the University of Breisgau, he remarked:

Individual plants do not occur purely as independent Beings by themselves. They are at the same time also parts of a higher Totality, elements of a general motion [*Bewegung*]. Because they generate new individuals, because these propagate themselves in turn and the procreation process is repeated continually in their progeny, there arises from this an undeterminate sum [*Summe*] of plants, which is not to be considered a loose aggregate, but forms the species, an undivided whole, which is held together by a common Idea. (Nägeli 1853 quoted in Bachmann 1906, 180)

As Darwin's *Origin* was transmitted and assimilated into the German states in a scientific context heavily influenced by Schelling's *Naturphilosophie* (Mullen 1964), Nägeli's subsequent reflections demonstrate how some German theorists of biology simply transformed their conceptualizations of Ideal natural-historical species undergoing a rational "development" in time into a thesis about concrete and ontologically real species, conceived as dynamic holistic entities analogous to organic individuals, physically transforming into new species. In a published lecture on this topic delivered in 1865 while at the University of Munich, Nägeli, now a new Darwinian convert, spoke of how species "actually die like individuals [and] new species are developed once again like an individual" (Nägeli 1865, 35). Such species could be related like the branches of a tree, with degrees of taxonomic rank related to the recency of the splitting of the stem.

> When a plant form begins to vary, it makes first notable individual Varieties, which are connected to one another through middle forms.... By constant divergence, the motion [*Bewegung*] passes from Species into Genera, and these into Orders and Classes. Between the designated categories there is no absolute difference. It is only a case of more or less.... The Genera and the higher concepts are not abstractions, but concrete things [*concrete Dinge*], complexes of interconnected forms which have a common origin. (Ibid., 31–2)

A very different reading of Darwin's arguments is found in the development of statistical and populational interpretations by the British biometricians, represented in particular by Karl Pearson (1857–1936). Originally trained in mathematics and engineering statistics, Pearson transferred his powerful mathematical skills to issues of evolutionary biology through his close association with his colleague, the zoologist W. F. R. Weldon (1860–1906). Through a series of landmark papers, Pearson and Weldon gave the analysis of the species question a statistical and populational interpretation that exploited the species$_L$ tradition in new ways. Species are defined by the clustering of character measurements around a central mean, and they constitute logical classes of individuals defined by a statistical type representing the average character values. But the statistical variation displayed by the character measurements in a population renders the boundaries of statistically defined species groups intergradating. Such groups may also be changed into new species by direct selective pressures acting on their phenotypic characters over time (Gayon 1998, Chapter 7). Such change was presumably demonstrable by the use of advanced statistical methods developed by Pearson and Weldon and applied experimentally to natural populations (Pearson 1900, Chapter 10).

Sorting out the interplay of the complex discussions of the species question in post-Darwinian discourse, and the complexities this interplay introduced into the philosophy of evolutionary theory, extends beyond the limits of this chapter. It is, for example, of some interest that Ernst Mayr traced the origins of his own species realism directly to a text in the species$_H$ tradition (Mayr 1968). Similarly, the early geneticist Hugo DeVries drew upon Nägeli in support of his own claim that species constituted real "entities in nature" (de Vries 1910, 2: 589). What can be safely stated is that the issues raised by these historical problems continue to interact in contemporary discussions among cladists, species "individualists," adherents to the "biological" species concept, and "logicists." Gaining deeper clarity about Darwin's conceptual achievement in the *Origin* itself is at least one way to sort out some of these controversies.

6 Darwin's Keystone
The Principle of Divergence

Darwin chose an apt architectural image when he wrote J. D. Hooker that 'the "principle of Divergence" ... with "Natural Selection" is the key-stone of my Book' (*Correspondence* 7: 102). In the *Origin*, the fifteen-page section on divergence is placed strategically at the end of Chapter 4 on natural selection, where it distributes the weight between the core theory and the evidence for descent. Darwin portrays adaptation and the origin of species as emerging out of the entangled plenitude of mutual relations mediated by natural selection. The principle of divergence united this ecological vision with Darwin's complementary view that evolutionary history can be read in the irregular branching of the taxonomic tree of life. However, there is an irony in the historical fate of the principle. Much of twentieth-century evolutionary biology rejected Darwin's explanation of 'speciation' as muddled (Mayr 1942, 1992; Sulloway 1979; Coyne and Orr 2004).[1] Yet the profound depth of ecological relationships and the very diversity of life that Darwin evoked through the principle can be understood as one of the *Origin's* most enduring contributions. Moreover, the standing of contemporary approaches to speciation that, like Darwin's, emphasize ecological factors – but now often supplemented by moderate isolation of various kinds – while remaining controversial, is perhaps higher than it has ever been.[2] So 'The stone that the builders rejected has become the cornerstone' (Psalm 118:22). For these reasons alone, it is important that

---

[1] See Mallet's critique (2005).

[2] This theme runs through several chapters of Coyne and Orr (2004). A case in point is the controversy over speciation in cichlid fish in African lakes, where, for example, Lake Malawi is reported to have produced over 700 species in 0.7–1.8 myr (147–56).

the modern reader of the *Origin* understand, in its own right, Darwin's principle of divergence and its dramatic role in his intellectual maturation.

The nub of the principle is ecological, and here is one of its typically concise statements:

> ...the more diversified the descendants from any one species become in structure, constitution, and habits, by so much will they be better enabled to seize on many and widely diversified places in the polity of nature, and so be enabled to increase in numbers. (*Origin*, 112)

Darwin's keystone is fashioned around the premise that more life can be supported in an area if organisms of different types occupy that area. His principle of divergence is an application to natural history of an agronomic version of the idea, originating with the economist Adam Smith, that more wealth is created where there is a 'division of labor'.[3] However, when Darwin deployed the principle of divergence, he always did so in conjunction with natural selection. The principle acts as an amplifier of selection. This coupling of divergence and selection created a special case or type of natural selection, which we may term divergence selection. This is selection where conditions favor divergent specializations among related forms sharing a common location. Furthermore, as we will see, the principle of divergence was also the centerpiece in Darwin's explanation of the origin of new species.[4] So there is much that depends on this principle.

Like natural selection itself, divergence selection is not a unitary idea, but rather a complex argument. Before I examine that argument, I will first consider the structural role of divergence in

---

[3] Darwin was familiar with Milne-Edwards (1844; see Ospovat 1981), who treated the specialized organ systems of animals as a division of physiological labor, but his own agronomic use is closer to that of Smith.

[4] As Mallet (2005, 106) shows, the persistent erroneous view (e.g., Coyne and Orr 2004, 1–2) that Darwin did not explain the origin of new species in the *Origin* traces to Mayr's disapproval of Darwin's explanation (Mayr 1942). I have avoided the term 'speciation' because Darwin did not use a one-word term, except for the infelicitous 'specification'. It is particularly important not to use the anachronisms 'phyletic', 'sympatry' (Poulton 1904; see Mallet 2004, 442), or 'allopatry' (Mayr 1942, 148-9) because they obscure what was important to Darwin. For this reason, I am introducing the term 'divergence selection', which is not part of the literature, in order to discuss Darwin's distinctive treatment of the origin of species and avoiding the increasingly common modern term 'divergent selection'.

the *Origin*. I will then return to Darwin's argument for divergence selection and compare it to his other selection types. Finally, I will discuss how the principle of divergence fits into the story of Darwin's intellectual development. There I will show that this special amplification of selection, which emerged only once Darwin had effectively set to work writing the *Origin*, constitutes the culminating move in a major repositioning of his understanding of evolution. For in the principle of divergence we see the replacement of Darwin the transformist geologist, who first puzzled over 'the stability of species' aboard HMS *Beagle* (Kohn et al. 2005), by Mr. Darwin of the *Origin*, who made ecology the keystone of evolution.

Darwin not only inserted the principle of divergence selection right *after* his core argument, he also placed it *before* the nine chapters where he presented his massed evidence for descent. In this second part of the *Origin*, he reinterpreted much of natural history – both subject areas and dominant themes – translating them into evolutionary terms; and where possible, he applied natural selection to effect the translation.[5] Darwin argued that evolutionary descent produced a divergent pattern of relationships consistent with the branching conception of systematic relations, which had already incorporated embryology and comparative anatomy into a common framework in the 1840s and 1850s (Ospovat 1981). But he insisted that evolution was more than just *consistent* with the branching tree of life; he tried to show *how* key concepts and patterns in each natural history discipline conformed to the expectations of divergent evolution. Thus, at the intersection between geographical distribution and systematics, Darwin projected onto a diagram of the evolving tree of life, numerical patterns like the high proportion of varieties per species in wide-ranging groups. The result was 'a little fan' of competing and diverging incipient varieties that evolved into species, and eventually gave rise to 'dominant' wide-ranging genera (*Origin*, 116-26; Browne 1980). Likewise, it was the pattern of 'aberrant' groups with very local distributions that Darwin translated as relics of past extinction, where divergent evolution had been terminated. In embryology, it was 'community in embryonic structure' that 'reveals community of descent' (*Origin* 1859, 449; see

---

[5] See Nyhart, this volume, who also found 'translation' an apt term.

Nyhart, this volume).[6] In systematics itself, encompassing morphology and comparative anatomy, it was through evolutionary divergence that Darwin accounted for the very heart of the natural system, the hierarchical concept of groups arranged within groups. When one considers how many potent natural history concepts Darwin tethered to evolution through divergence, one sees how profoundly these translations were woven into the binding structure of the *Origin*.

THE DOUBLE ARGUMENT OF DIVERGENCE

Selection for Darwin always amalgamated the origin of adaptation with the origin of species. Nowhere is that clearer than in the *Origin* section on the principle of divergence. Here I will discuss the structure of the argument for divergence as an adaptive mechanism, and in the next section I will consider how Darwin applied this argument to the origin of species.

Divergence as Adaptive Mechanism and Selection Type

Divergence selection combined two concepts: the principle of divergence and natural selection. Operating together, they constitute a claim for the regular existence in nature of conditions of selection and a dynamic process that leads to a divergence of forms. Darwin's argument is that the ecological situation described by the principle of divergence is itself adaptive and hence that individuals are subject to selection. As we've seen, the principle is an agronomic expression of Adam Smith's theory. Indeed, Darwin first wrote the division of 'land' and then substituted the division of 'labour' (Kohn 2006, DAR 205.5: 171). In transforming Smith's concept, he argues that 'mutual relations' – the complex of other creatures, with each group 'striving against the other' – forms an adaptive situation within which selection occurs. So the division of labour is concomitant with this struggle. It is simultaneously treated as a selective advantage and as the

---

[6] Ospovat (1981) shows the link in the literature between the theory of embryonic recapitulation and 'the branching conception of life', but he says Darwin abandoned recapitulation in the 1840s (152).

result of the selection that goes on within the context of other struggling species. Since the 'land' – the operant image is plants growing
in an agricultural field or pasture – contains resources distributed
throughout its extent, these can be most efficiently exploited by
plants of different types drawing on water and nutrients distributed
at varying densities in different areas and at different depths. Divergence selection, through the struggle that it encompasses, produces
the specialized adaptations that can divvy up the land.

It is noteworthy that the correlative of 'more' life being sustained
is that more dry weight of pasturage can be produced. And it is
tempting to think that Darwin, in applying this concept to 'nature',
was drawing on an idea that ecologists later called 'biomass'. But
no doubt in Darwin's own time, the source for these ideas is agricultural chemistry and agronomy. It is the world of Liebig and crop
rotation and the experiments that were being pursued by the Duke
of Bedford's gardener, George Sinclair, whose work Darwin studied
(*Natural Selection*, 229; *Origin*, 113).

To this premise Darwin adds the further premise, elaborated earlier in Chapters 3 and 4, that selection leads to adaptation. Here
Darwin also draws implicitly on an important idea about variation,
which distinguishes his 1844 "Essay" from the *Origin*. In 1844,
Darwin believed that there was a limited amount of variation in
nature (*Foundations*; Ospovat 1981). But by the time of the *Origin*,
he was able to draw heavily on the assumption expressed in Chapter
2 ("Variation under Nature") that species have a great deal of individual hereditary variability. With abundant variability to act on,
natural selection, and particularly divergence selection, would be a
strong force. In this situation, (1) there will be an adaptive advantage
favoring the selection of different forms to exploit different aspects
of the 'land', and (2) this situation will favor selection of the most
extreme – that is, the most divergent – forms. The resulting compounding of specialized forms, though produced by natural selection,
is achieved through the avoidance of direct struggle. This vision is
'ecological' in that the key condition of each form's existence is not
just its ability, say, to absorb water from soil, but the more specialized ability to find water at a depth to which another form does
not penetrate. So, as Darwin stressed, the key conditions of existence are the 'mutual relations' of different forms. With this insight,
Darwin transformed a geologically entrained grasp of the overall

process of evolution, which was both slow and two-dimensional, into one that stressed multidimensional biological interactions over physical causes.

Darwin concluded the *Origin* contemplating an 'entangled bank, clothed with many plants of many kinds' (489), an image of ecological richness framed in language he first used in the fiords of Tierra del Fuego, and to which he returned only in the final drafting of the *Origin* (Kohn 1996). The principle of divergence, also a late development in his conceptual vocabulary, gave the entangled bank its scientific undergirding; and in attempting to explain why there is a diversity of species, Darwin broached some of the most fundamental issues of evolutionary ecology.

Entangled Divergence and the Types of Selection

As I have suggested, the principle of divergence contributes to a distinctive category of selection. The result of that selection, a compounding of mutual relations, is summed up, as just noted, in the ecological vision of the 'entangled bank', that beautiful local image Darwin used to evoke the sublime global dynamic, which formed a significant part of the 'grandeur' that he saw in the evolutionary 'view of life' (*Origin*, 489–90). To be clear, Darwin speaks of the principle and natural selection as separate concepts (*Origin*, 116), and he never went so far as to coin the term 'divergence selection' in parallel to 'natural', 'artificial', and 'sexual' selection. Nevertheless, it is instructive to postulate a 'divergent' selection because we can then compare its powerful adaptive argument for the origins of diversity to that deployed in Darwin's named forms of selection, especially to the most fundamental of these: natural and artificial selection.

Unlike natural selection itself, which emerged in 1838, in creative opposition to the systematic selection that Darwin knew was practiced by breeders (Kohn 1980), the principle of divergence was developed in the 1850s, long after Darwin had become accustomed to presenting natural selection as analogous to artificial selection. In fact, the divergence selection argument significantly mirrors the structure of artificial selection in a way that natural selection does not. To see this, it is most important to bear in mind that one feature of natural selection differs fundamentally from artificial selection. In the *Origin*, Darwin considered natural and artificial selection to be analogous phenomena, but they were not identical. Whereas artificial selection posits the intelligence of a human selector who picks variants in order to produce results, in natural selection there is no selector – no external intelligence. Rather, adaptive 'direction' results from the intersection of two independent natural processes: hereditary variation and changing environments. Natural selection describes a natural process. The link between variation and environment in natural selection is the struggle resulting from Malthusian population pressure. This too is a natural process and utterly different from the workings of artificial selection. There is a *weak* analogy between an intelligent external selector who literally picks variants and a struggle among variants. But at their cores, the analogy crumbles: 'intelligent external selector' is irreducibly different from 'competitive struggle'.

How do these distinctions play out in the compounding interactions underlying the principle of divergence? True, there, Darwin does depend on Malthusian pressure:

And we well know that each species and each variety of grass is annually sowing almost countless seeds; and thus, as it may be said, is striving its utmost to increase its numbers. (*Origin*, 113)

Undoubtedly, natural selection's struggling individuals undergird the dynamics of this process. But Malthusian pressure is not the sole focus. Rather, it operates to enforce the particular selective force at play, which is the advantage that Darwin believes accrues to the most different, or extreme, varieties:

Hence, if any one species of grass were to go on varying, and those varieties were continually selected which differed from each other in at all the same manner as distinct species and genera of grasses differ from each other, a greater number of individual plants of this species of grass, including its modified descendants, would succeed in living on the same piece of ground. (*Origin*, 113)

What we see here is a multiplication of the forms that can occupy a shared territory because of the action of some agency.[7] What is this agency? Ultimately, it is the presence of other species competing for

---

[7] See Depew, this volume, on agent versus agency.

the same resources, which becomes clearer and eventually explicit in succeeding passages:

Most of the animals and plants which live close round any small piece of ground, could live on it . . . , and may be said to be striving to the utmost to live there; but, it is seen, that where they come into the closest competition with each other, the advantages of diversification of structure, with the accompanying differences of habit and constitution, determine that the inhabitants, which thus jostle each other most closely, shall, as a general rule, belong to what we call different genera and orders. (*Origin*, 114)

We see the same emphasis on *other* competing species when Darwin summarizes how his principle produces adaptive divergence:

After the foregoing discussion, . . . we may, I think, assume that the modified descendants of any one species will succeed by so much better as they become more diversified in structure, and are thus enabled to encroach on places occupied by other beings. (*Origin*, 116)

In these three examples – and the second one reflects an exclusion experiment conducted by Darwin at Down (*Experiment Book*, DAR 157a:5) – divergence arises from intraspecific competition nested within, or going on simultaneously with, interspecific competition. From the point of view of the individual, it may all be natural selection. But in terms of divergence selection as a type of selection, by imposing an interspecific level on top of the competition going on within a species, Darwin effectively introduced the natural equivalent of an external selector. That agency plays a role analogous to that of the intelligent selector of artificial selection. Of course, it is not the intelligence or intent, but rather the externality of the selector that counts in this creature-on-creature class of selection. In his treatment of divergence, Darwin had, in fact, found a strategy for expressing natural selection that was *strongly* analogous to artificial selection. Darwin made a similar move when he developed sexual selection in *The Descent of Man*. There, females, typically, are the external selectors, and the result is gaudy or pugnacious – but healthy – males. There too, the selection is creature-on-creature. Both these forms of selection extended the explanatory domain of natural selection by structurally mirroring artificial selection. In the case of divergence selection, the result was the 'more' in 'more life can be supported' – it was the abundance and diversity of nature.

Divergence and New Species

In the *Origin*, the principle of divergence was essential to Darwin's account of how new species are formed because that gave him the means to account for evolutionary branching. This species-forming aspect of divergence flows directly out of its function as an adaptive mechanism. As we have seen, the adaptive argument drew heavily on the assumption (1) that there is much variability in species. To make the link to the origin of species, a further assumption about variation is added: (2) varieties are incipient species. For Darwin, individual hereditary variation and varieties are different segments in a continuum of species variability. Individuals at each stage are subject to selection. The most important consideration here is that while varieties are incipient species, Darwin also thought of individual hereditary variants as if they were incipient varieties.[8] Continued selection could traverse the continuum from individual variation to divergent varieties and divergent species.

Yet a further assumption was critically important, namely: (3) incipient species could develop by divergence selection without dependence on geographic isolation. Darwin held this view despite his clear recognition that crossing in a sexually reproducing species would ordinarily be expected to swamp the new variations that were at the base of this system. Darwin's first assumption – the abundance of variation – was a sine qua non for accepting the third, counterintuitive assumption that swamping would not overpower divergence selection. Since the plenum of variation allowed Darwin to posit that selection could be an intense and constantly operating force,

---

[8] When not under selection, individual differences merely 'fluctuate'. The term 'variety' included marked varieties, geographic varieties, and races. Very rarely Darwin uses 'variety' where context shows he means 'individual hereditary difference', for which he has been severely criticized (Mayr 1992, 345–8). However, one can readily work out Darwin's intention from context. Darwin would probably not have confused his contemporaries, who even wrote of hybrids as varieties. There does not seem to have been a term like 'variant' available. So when Darwin uses 'variety' where moderns would use 'variant', it was probably to avoid a lengthier locution like 'individual difference'. A similar easily penetrated cloud of confusion surrounds the word 'variability', which today generally refers to individual differences in a population, but which for Darwin could, in addition, include the varieties in a species. Again, attention to context usually clarifies Darwin's intended, self-consistent meaning.

the more variation, the larger the scope of action for selection and also the more likely that it would be sufficient to obviate the need for isolation. Here we have to clarify an additional assumption, again concerning variation, which served to reinforce the independence of divergent species formation from isolation. As we have seen, for Darwin (4) the source of hereditary variation was always the conditions of existence. Indeed, he rejected the suggestion that variation is a spontaneous product of the reproductive system (*Origin*, 131). But, as we have also seen, with the elaboration of the principle of divergence, Darwin's emphasis shifted from physical (geological) conditions to mutual (ecological) relations as the primary source of hereditary variation. The more the principle of divergence worked to complicate the mutual relations among species living together, the more variation became available for the further refinement of adaptations. Thus the principle of divergence had the quality of a self-generating force strong enough to overpower crossing. Selection triumphed over sex to create new diverging species.

ISLANDS VERSUS CONTINENTS. We may well ask, why was Darwin so concerned to exclude isolation? Here there are more assumptions to clarify, this time coming from geographic distribution and geology. We need to consider how Darwin's mechanism worked in the context of the patterns of geographic distribution that he thought most consistent with evolutionary expectations about the origin and extinction of species and higher groups. It is here, finally, that 'mechanisms' would meet taxonomic trees.

To appreciate this, we need to recall that until he worked through divergence, Darwin had basically three models of species production, namely: (1) a linear continental model in which new forms gradually depart from an ancestral form; (2) a model of geographic replacement by representative species, when moving along a continental tract – as he had witnessed in South America; and (3) a geographic isolation model, where endemic species form, particularly on archipelagos. While island isolation could account for divergent species, by the time he wrote the *Origin*, Darwin did not consider islands to be the principal loci of new species (Sulloway 1979). In the ample discussions of geographic distribution in South America and the Galápagos, both during and after the *Beagle* voyage, Darwin was always concerned to arrive at a global interpretation of species distribution.

And once he became a transformist, that meant a globally applicable explanation for the formation of species. In the 1844 "Essay," this was resolved by the fact that Darwin thought of archipelagos as incipient continents (*Foundations*, 189). If continents were formed by the fusion of islands, then geographic isolation could indeed function as a primary condition for *all* species formation. Species formed on separate islands would be distinct from each other when the islands fused. But with time, Darwin dropped this view and was left convinced that island isolation was a special case.[9] Only continents contain enough species to be relevant, and so continuous ranges had to be the primary sites of action. So Darwin developed a divergent continental model that could explain species formation without isolation. In its intellectual lineage, this model harkens back to the representative species model of the *Beagle* and the immediate post-voyage notes. But now intense divergence selection was focused on the ecologically distinctive stations within a 'whole country'. These areas he understood to be occupied by wide-ranging species capable of bridging the relatively minor barriers to migration, but also capable of adapting to diverse situations.

THE DIVERGENT CONTINENTAL MODEL. At its core, the continental model is nothing but a restatement of the idea that a divergence of varieties – and ultimately of species and genera – arises through the action of the principle of divergence and natural selection. At the end of the divergence section, Darwin illustrated how his continental model worked by means of a branching diagram – famously, the only figure in the *Origin*. But to understand his explanation, we need to be aware that while thus far we have treated divergence selection as a pair of related arguments situated in the natural selection chapter of the *Origin*, as a matter of fact Darwin tucked a rather massive

---

[9] It is important to realize that Darwin does not deny that isolation can be important in making species, as for example in the Galápagos. But it was crucial for him to show that species *could* form without isolation, essentially by selection alone. Then it became a matter of relative importance, as he puts it in *Natural Selection* (254): 'In this way, I think, isolation must be eminently favourable for the production of new specific forms. It must not, however, be supposed that isolation is at all necessary for the production of new forms; ... I do not doubt that over the world far more species have been produced in continuous than in isolated areas. But I believe that in relation to the area far more species have been manufactured in, for instance, isolated islands than in continuous mainland.'

body of evidence in support of divergence into Chapter 2 (*Origin*, 53–8). Between 1854 and 1858, he tabulated the number of species per genus and the number of varieties per species in numerous taxonomic works. The analysis was based mostly on regional floras, such as that by Boreau (1840), which identified all the plants in the center of France to the species, subspecies, and variety level. The 'botanical arithmetic' analysis (Browne 1980; Parshall 1982) Darwin performed on such monographs supported his key assumption that varieties are incipient species, whose surviving descendants produce diverging lineages. For example, Darwin thought he found that large genera, which obviously have many species, tend to have species that are large, that is, species having 'significantly' more varieties than do small species. Browne and Parshall show how Darwin ran into serious methodological problems requiring lengthy recalculations. Indeed, Parshall has shown that the analysis doesn't stand up to modern statistical analysis. Nevertheless, Darwin, though shaken, remained convinced. Here are the principal conclusions drawn from this analysis, as Darwin employed them to set the stage for his continental model of divergence (see diagram, this volume, p. iv).

The accompanying diagram will aid us in understanding this rather perplexing subject. Let A to L represent the species of a genus large in its own country; ... I have said a large genus, because we have seen in the second chapter, that on an average more of the species of large genera vary than of small genera; and the varying species of the large genera present a greater number of varieties. We have, also, seen that the species, which are the commonest and the most widely-diffused, vary more than rare species with restricted ranges. (*Origin*, 116–17)

From here Darwin proceeds to lay out how divergent 'lines of descent' will arise from the application of the principle to Species A:

Let (A) be a common, widely-diffused, and varying species, belonging to a genus large in its own country. The little fan of diverging dotted lines of unequal lengths proceeding from (A), may represent its varying offspring.... Only those variations which are in some way profitable will be preserved or naturally selected. And here the importance of the principle of benefit being derived from divergence of character comes in; for this will generally lead to the most different or divergent variations (represented by the outer dotted lines) being preserved and accumulated by natural selection. (*Origin*, 117)

The discussion continues by describing how this first branch point will only be reinforced over time.

When a dotted line reaches one of the horizontal lines, and is there marked by a small numbered letter, a sufficient amount of variation is supposed to have been accumulated to have formed a fairly well-marked variety, such as would be thought worthy of record in a systematic work. (*Origin*, 117)

Much of the remainder of the diagram discussion is devoted to extrapolating further up the hierarchy to species and genera by reiteration of the same process. Darwin leads us to imagine the growing, branching tree over hundreds of thousands of implied generations as we envision 'creations', struggles, and extinctions at all levels of life's hierarchy.

It is important to recognize, however, that Darwin offers no additional scenario for the creation of new species than the one we have discussed. Divergence as we have depicted it here is in fact his one and only explanation of how new species are made.

However, he does invoke extinction to show how this countervailing process prunes the branching tree so that together divergence and extinction account for the characteristic shape of the tree (*Origin*, 122, 124–5). Extinction takes a number of forms, including the extinction of parental forms, extinction of competing varieties derived from the same parent, and extinction by virtue of expansion into the territory of closely related species – that is, the extinction of intermediate forms. In his discussion of extinction, Darwin supplements divergence with an important corollary of natural selection. Competition is most intense among organisms that are most alike. It is here, crucially, that the old linear continental model, which only explained a new form extinguishing its parent, is superseded in the divergent continental model – itself the successor cum descent and selection to the representative species of the *Beagle*, where incipient varieties come into multiple struggles that lead to extinction. Hence, Darwin accounts for a genuine divergent splitting of species and lineages. Thus the model has two phases: (1) divergence selection – encompassing (a) the principle of divergence and (b) natural selection – which is followed by (2) subsequent battles to extinction of parent, sibling, and intermediate forms. These battles of extinction are not inherently dependent on, or produced by, divergence selection. It is just that as a by-product of phase 1, new forms are faced with parents, siblings, and intermediates. Hence, the opportunity for phase 2 extinction arises. And this separates Darwin's 'lines of descent'.

So it was that Darwin translated the nexus of geography and systematics into evolutionary terms. As we have already seen, Darwin drew on Adam Smith's political economy for the principle of divergence, just as he had drawn on Malthus for natural selection. But as Darwin leads us through his diagram, we see the shadow of two other cultural themes. First, there is Victorian improvement:

The modified offspring from the later and more highly improved branches in the lines of descent, will, it is probable, often take the place of, and so destroy, the earlier and less improved branches: this is represented in the diagram by some of the lower branches not reaching to the upper horizontal lines. (*Origin*, 119)

'Improvement' is used dozens of times in the natural selection chapter. We have shown that Darwin's idea of divergence is linked to agronomy. So while agricultural improvement could be one form of improvement implicit here, perhaps the fundamental source is Darwin's sense of contingent progress. It can only have been the cultural presumptions of an Englishman during the heyday of empire that led Darwin to conclude the divergence section thus:

We have seen that it is the common, the widely-diffused, and widely-ranging species, belonging to the larger genera, which vary most; and these will tend to transmit to their modified offspring that *superiority which now makes them dominant in their own countries*. Natural selection, as has just been remarked, leads to divergence of character and to much *extinction of the less improved and intermediate forms of life*. On these principles, I believe, the nature of the affinities of all organic beings may be explained. (*Origin*, 119, my emphasis)

The branching diagram is the final elaboration of divergence and the climax of the natural selection chapter. And here Darwin conveys a sense of evolution's dynamism by portraying advancing and diverging lines of superior forces that conquer, dominate, and eliminate inferiors. Inevitably, he was tapping deeper and darker inclinations in himself and in his audience to transform the static image of a tree into one of struggling combatants on the march. Darwin passed over a line here. It was one thing to instruct the reader, as he does in

Chapter 3, to honestly see the struggle constantly going on beneath the face of nature, 'bright with gladness' (*Origin*, 62; see Kohn 1996). But it was another matter to sound the note of triumphal superiority we hear as the branching diagram scenario unfolds. That would have consequences.

Development of the Principle of Divergence

In his *Autobiography* (120–1), Darwin remembered divergence as something he had 'overlooked' until 'long after I had come to Down'. Indeed, the principle of divergence developed between 1854 and 1859 during the writing of *Natural Selection* and the *Origin*. Although divergence is an application of natural selection, one looks in vain for texts on divergence in the aftermath of Darwin's discovery of selection in 1838. The closest he got (Schweber 1980; Kohn 1980) is expressed in Notebook E:

The enormous number of animals in the world depends, of their varied structure & complexity. – hence as the forms became complicated, they opened fresh, means of adding to their complexity. (*Notebooks*, 422, E, 95)

Although the island and continental models of species formation in the 1844 "Essay" would be important in the 1850s, we can see why divergence itself was 'overlooked' at that stage. Since in 1844 Darwin viewed hereditary variation as dependent on slow geological change, he had to concede that there is little variability in nature. The opening paragraph of the 1844 "Essay," Chapter 2, states: 'Most organic beings in a state of nature vary exceedingly little...' (*Foundations*, 82). Ospovat (1981) saw the implications: the scope of natural selection must also be severely limited. However, if we look to the end of this paragraph, we can go a step further than Ospovat. For it seems that Darwin already glimpsed the difficult path he would follow to solve the problem of limited variability:

The amount of hereditary variation is very difficult to ascertain, because naturalists (partly from the want of knowledge, and partly from the inherent difficulty of the subject) do not all agree whether certain forms are species or races. (*Foundations*, 82)

In fact, in 1844 Darwin concluded, we don't know how much variation there is because the naturalists who describe species have not systematically recorded variation. The solution, which he pursued after completing his South American geology books in 1846, was to undertake his own taxonomic study *and* an empirical survey of variability in a major natural group.

This brings us to Darwin's monograph on barnacles, which Hooker helped initiate in 1846 after earlier challenging him to 'minutely' describe many species to earn 'a right to examine the question of species' (*Correspondence*, 3: 253). The dilemma of limited variability and consequent weak natural selection was a palpable scientific problem that forced Darwin to delay – that is, intentionally to avoid – premature publication of his theory.[10] But if we look at the attention Darwin devoted to recording levels of variability beyond those required to define good species and 'well defined' varieties, it becomes clear that the barnacles solved Darwin's problem during the 1844–54 delay. For example, in a typical Darwin species description such as that for *Lepas fascicularis*, he first recognized two taxonomic varieties, Donovani and Villosa, each distinguished by a few well-defined characters. Thus far his description is in the high-level 'lumping' tradition. Detection of the natural cleavages within and between species depended on the requisite knowledge of species that Hooker considered 'philosophical' (Endersby 2008). But something more is afoot that was not at all envisioned by Hooker's challenge. For Darwin adds a section to his description called 'General Appearance', which begins:

Capitulum highly variable in all its characters; thick and broad in proportion to its length, but the breadth is variable, – in some specimens, the capitulum being longer by one-fifth of its total length than broad; in others, one-fifth broader than long. (Darwin 1851–54, 1: 93)

Darwin recorded such *individual* variability in more than one quarter of the species he described in 1851. Three years later, in volume two of *Living Cirripedia*, he had observed so much variability that he simply summarized this class of variability for each genus. Thus did Darwin gather the evidence for a solid conviction in the plenitude

---

[10] Van Wyhe (2007) cavalierly dismiss weighty evidence that Darwin avoided publication and stressed instead that Darwin was merely too busy to begin his species book (p. 177).

of natural variability that would save natural selection and unleash the thinking that led to the principle of divergence.[11]

Upon completing the barnacles monograph in September 1854, Darwin's delay was over. He set to work sorting his notes on species (*Correspondence*, 5: 537). Drafting chapters began only in May 1856, but by November 1854 the true writing of the *Origin* had already started with an outpouring of new notes.[12] These show that the following simplifying assumptions characteristic of divergence – and familiar from our discussion in the first part of this chapter – were already then in place and may have been brewing, along with the plenitude of variation, during the barnacle period: (1) new species forming in continuous 'regions' with varying habitats is the center of his attention, and (2) the necessity of geographic isolation is repudiated:

<<If the region will support so many Compositæ, what ever genus, has sported & shown its adaptation is the most likely to yield more forms. >>

. . .

[No doubt here comes in question of how far isolation is necessary, & I shd. have thought more necessary than facts seems to show it is. –] (Kohn 2006, DAR 205.9: 303–4)

By November 1854, divergence as an 'overlooked' problem to be solved also first came into focus, in terms remarkably analogous to those recollected in the *Autobiography*.

We include all in class, as <<in>> Crustaceæ, which are connected, but yet which no definition will define, – a proceeding explicable on descent. – (Kohn 2006, DAR 205.5: 148)

---

[11] Darwin (1851–54, 2: 155). This interpretation does not sustain Mayr (1992), Sulloway (1979), and Schweber (1980), who apparently did not see the empirical survey-of-variation dimension of the barnacle work. See also Browne (1980, 72, nn. 44–6), who says these authors held that Darwin's 'interest in *varieties* attracted Darwin's attention away from individual *variations*'.

[12] The standard biographical accounts underestimate the autonomy and momentum of Darwin's scientific effort during this period as revealed in his notes. Although Darwin wrote on May 14, 1856, 'Began by Lyells advice *writing* species sketch' (Pocket Diary, *Correspondence*, 6: 522), Lyell's urgings only ignited the fire Darwin had by then carefully prepared.

Moreover, Darwin already sees that in order to solve his problem he needs to formulate a specific explanatory principle:

... For otherwise we cannot show that there is a tendency to diverge (if it may be so expressed) in offspring of every class ... (Kohn 2006, DAR 205.5: 149)

Thus 'a tendency to diverge' is the first in a series of synonyms that ended in March 1857 when he coined the phrase 'principle of divergence' in the first draft of *Natural Selection*, Chapter 6.

In the same passage, we see Darwin link his search for an explanatory principle with field observations:

It is indispensable to show that in small & uniform areas there are many Families & genera. For otherwise we cannot show that there is a tendency to diverge ... (Kohn 2006, DAR 205.5: 149)

In the following summer, Darwin and the children's nurse supplied the 'indispensable' data by counting plant species as a measure of diversity in Great Pucklands – a field adjacent to the Sandwalk. Meanwhile, from November 1854 onward Darwin initiated another form of empirical support for his exploration – botanical arithmetic calculations. These were not directly relevant to defining the 'tendency to diverge'.[13] However, they were critical as Darwin now articulated the divergent continental model of species formation. Speculating on the results of his calculations on aberrant groups '<<becoming>> extinct' brought Darwin to identify large, wide-ranging genera as the key evolving groups and to seize on local circumstances as the site of species origin (Kohn 2006, DAR 205.9: 303–5). But, to truly examine local circumstances, Darwin scales down from geographical regions and their abstract 'stations' to 'small & uniform areas' such as his neighbor's thirteen-acre field (Kohn 2006, DAR 205.9: 303–5 vs. DAR 205.5: 149). Then from 1855 onward he further delimits his focus until he has defined a classic experimental model: a patch of ground a few square feet in area (Kohn 2006, DAR 205.2: 119 begins the process). Thus botanical arithmetic and field experiments creatively intersected, as Darwin

---

[13] Perhaps because Browne's brilliant study focused tightly on the botanical arithmetic, she was led to a far-too-late date for the principle of divergence after March 1857 (1980, 53, 73).

sought sufficient magnification to think through the connection between the principle of divergence and natural selection.

Notwithstanding these two empirical projects, the main articulation of the principle occurs in speculative, synthetic notes. Thus in November 1854, three key pieces of the problem's solution were already explicit: (1) mutual relations (indirectly) cause variation (Kohn 2006, DAR 205.9: 260) – thereby loosening the link between origin and geology and thereby also accounting for the new plenitude of variation; (2) divergence in the absence of isolation brings Darwin to emphasize intense selection (Kohn 2006, DAR 205.9: 303–4); and (3) most important, Darwin invokes the division of labor by this early date, a point missed by Ospovat (1981) and Browne (1980):

There is no law of Progression, but times wd. give better chance of sports, & allow more selection; & all the organisms thus living an advantage, – *a* <<*free*>> *competition of labour*, – the result wd be more complicated & more perfect; (Kohn 2006, DAR 205.9: 250, my emphasis)

Thus, we already see the core of the principle of divergence: the ecological application of Adam Smith's division of labor.[14] In the *Origin* that is expressed as the idea that more life can be supported where divergent forms occupy a common area. Here Darwin comes close, but does not quite make that connection. It would be just three months later, in January 1855, that Darwin clearly stated the ecological equivalent of the division of labor:

On theory of Descent, a divergence is implied & I think diversity of structures supporting more life is thus implied. (Kohn 2006, DAR 205.3: 167)

This is the culmination of a first rich, and hitherto insufficiently appreciated phase of Darwin's intellectual development (November 1854–January 1855). During this highly compressed period, the principle of divergence per se was formulated (Kohn 1985).

What remained, however, was for Darwin to make explicit the link between divergence and selection that we see in the *Origin*,

---

[14] Here Darwin disavows an 'innate tendency to progress', as he also did in 1839 (*Notebooks*, 422, E, 95: 'no <<NECESSARY>> tendency in the simple animals to become complicated'). But he does believe that 'complexity' and 'perfection', which we can roughly associate with a Victorian notion of progress, are consequences of natural selection and the nascent principle of divergence.

where the principle and natural selection work conjointly to produce new, diverging forms. This happened in the middle of a new
phase that began in June 1855 and ended in March 1857, when Darwin wrote the conceptual heart of the divergence section.[15] This is a
period when field experiments overlapped with drafting and rewrit-
ing – during which the ideas of the compressed phase were consoli-
dated and tested. In June 1855, as we have noted, Darwin measured
plant species diversity in a Down field. What followed, beginning
that season, but intensifying in 1856 and continuing through May
1858, was that Darwin performed a variety of rough – 'square-yard of
turf' experiments using 2' × 3' and 3' × 4' experimental patches in
the lawn, orchard, and meadows of Down House to study diversity,
dominance, struggle, and survival. It was in parallel with this effort
that Darwin made the move that clinched his position.

We find that the tight link between divergence and selection is
forged first in a note written at the end of the 1855 growing season,
after Darwin's first year of field experiments, and then in a note
written at the end of the 1856 season. We can look at these seriatim:

Aug 19 /55/ owing to power of propagation not only as many individuals
crowded together, but "forms" for more can be supported on same area,
when diverse, than when of same species.... As where many individuals
crowded together some will die, so will forms. ... All classification follows
from more distinct forms being supported on same area. (Kohn 2006, DAR
205.5:157)

Sept 23d /1856/ The advantage in each group becoming as different as possible, may be compared to the fact that by division of <land> labour most
people can be supported in each country – Not only do the individuals of
each group strive one against the other, but each group itself with all its
members, some more numerous, some less, are struggling against all other
groups, as indeed follows from each individual struggling – (Kohn 2006, DAR
205.5: 171)

These passages have a lot in common. Divergence and struggle are
linked in both, and we can take the division of labor as a given in
August 1855. The main difference is that in September 1856 we see

---

[15] The outline of the theory that Darwin sent Asa Gray in September 1857, a few
  months after writing the first version of the divergence section, shows that all the
  essential ideas were in place and explicitly grouped as a 'principle of divergence'.

nested intraspecific and interspecific struggles, the signature structure of divergence in the *Origin*. So if there was one point where the adaptive argument of the principle of divergence clicked, this was it, just as Ospovat said nearly thirty years ago (1981). Here at last was sufficient amplification of natural selection to overcome crossing and to account for the origin of species. Here is where the principle of divergence per se – 'more life can be supported' –is transformed into the compounded inter- and intraspecific selection-driven argument for divergence that produces ecological division of labor.

Darwin had already begun writing *Natural Selection* in May 1856 and had progressed steadily through many chapters. By the time he came to the chapter on natural selection, it was March 1857 – fully ten months after he wrote the definitive 23 September 1856 note. But the ideas were firmly in place, and the development of the principle of divergence was complete. When it came time to write the natural selection chapter, and to set the keystone into the arch, Darwin was ready. But the botanical arithmetic calculations were not. Now we enter a third and final phase, between April 1858, when he supplemented the original divergence section with the famous branching diagram, and September 1858, when he produced Chapter 4 of the *Origin*, refining the explanation of the diagram to clearly stipulate the separation of lines of descent by extinction of less improved parental, sibling, and intermediate forms. Despite these revisions, the logic of the principle of divergence remained unaltered.

In devising the principle of divergence, Darwin resolved much that had long remained unsatisfactory and incomplete in the edifice of his species theory. In the end he accomplished, to his own satisfaction, a unified explanation of the origin of adapted species and a unified account of the historical generation of living creatures. Perhaps what impresses one most about Darwin's intellectual process in this endeavor is first the hard work and then the intellectual flexibility required to formulate and support just this one part of the *Origin*. From the minute anatomical study of barnacles, to the tedious calculation of taxonomic proportions in flora upon flora, to the counting of grasses in Great Pucklands, the great theorist resolutely pursued the evidence he needed to make his case. At the same time one marvels at the hard thought and change of perspective required to resuscitate the broken 1844 theory. With what effort of will did he take up this essay, which had diminished natural selection to near insignificance? And with what imagination did he replace it with a reasoned yet inspiring explanation of nature's abundance that made selection the most potent of all biological causes? And yet we have seen how this thinking scientist was both nourished and diminished by the general ideas and prejudices of his time and place. Like one of Darwin's entangled square yards of turf, studying Darwin's principle of divergence, close up, is like peering into a microcosm teeming with intellectual life.

7 Darwin's Difficulties

Chapters 6 and 7, on "Difficulties on Theory" and "Instinct," are the fulcrum of Darwin's *Origin of Species*. They come at the center of the work, after the first five chapters that explicate the theory of natural selection and outline the causes and effects of the mysterious laws of variation on which Darwin's argument for natural selection depends. These two chapters sustain the narrative drive that Darwin had been building through the first five chapters and introduce new scientific and emotional dimensions, particularly in his treatment of the theory's larger implications for the religious context in which he was writing.

"Difficulties on Theory" and "Instinct" take on two classes of what Darwin describes as "a crowd of difficulties [that] will have occurred to the reader . . . [l]ong before having arrived at this part of my work" (*Origin*, 171). The first class of problems, covered largely in Chapter Six, deals with the apparent lack of transitional forms, wide variation within taxa, organs of apparently small importance, and, most importantly, the difficulty of giving a wholly materialist explanation for the evolution of structures and organs of extreme perfection, such as the vertebrate eye. The second class, covered in Chapter Seven, deals with the evolution of behavior and instinct, phenomena for which the physical basis was quite unknown (as with the laws of variation), and which thus presented particular problems for a materialist explanation.

Moreover, these required two concurrent, qualitatively different categories of explanation. The first was technical: how could organs of extreme perfection, like the eye, or instincts of extreme perfection, like building honeycombs, have been evolved through the blundering, chancy process of natural selection? Even more challenging, how could structures and instincts that were inexplicable even on other materialist evolutionary theories – most importantly, the differentiated sterile castes of the social insects – have evolved to a high degree of perfection?

Darwin's second category of explanation was theological, responding to the arguments of the natural theologians who had formed his own early thinking about natural history and whose arguments, still in 1859 the most compelling for explaining the order and beauty of the natural world, were deeply familiar to his audience as well. The natural theologians' argument for the existence of God rested on the argument from design, the proposition that (as its most celebrated proponent, William Paley, put it in 1802), just as the existence of the human watchmaker can – and must – be inferred upon the discovery of a watch, even if found on a barren heath in the absence of any prior knowledge of the device or its purpose, so too can – and must – the existence of the Divine Creator be inferred from observation of and reasoning about the manifest order and beauty of nature. The argument from design, however, entailed difficulties of its own – in particular, its implications for theodicy, the problem of explaining how evils (like famine and death, or the appalling instincts that lead ichneumon wasps to parasitize living but paralyzed caterpillars, or cuckoos to take ruthless advantage of the parental instincts of other hapless birds, or ants to make "slaves" of other ants) can exist in the natural world at all if God the Designer is both infinitely powerful and infinitely benevolent. In the course of substituting materialist explanations for the evolution of perfect organs and instincts, and for the evolution of apparently "odious" instincts, Darwin both implicitly and explicitly substituted new ways – they might be described as natural *a*theological ways – of understanding these problems as well.

In the case of organs and instincts of great perfection, Darwin took on the two classes of explanation simultaneously. He chose as the centerpieces of his explanation two of the classic examples from the natural theological literature of the manifestation of divine ordering of the living world: Paley's great example of the structure of the vertebrate eye, and the construction of mathematically optimal honeycombs by little insects of little (though crucially not, Darwin would argue, of no) intelligence, a golden oldie of natural theology. The two instances were explicitly linked for him; in marginal notes to a discussion of honeybee instincts in a work on natural theology that he read in 1840, Darwin noted, "Very wonderful – it is as wonderful in the mind as certain adaptations in the body – the eye for instance, if my theory explains one it may explain other" (di Gregorio, 92).

In both cases, Darwin paid tribute in the *Origin* both to the perfection of the character under discussion and to the immense difficulty in explaining it otherwise than by divine fiat. "He must be a dull man," Darwin agreed with the reader, "who can examine the exquisite structure of a comb, so beautifully adapted to its end, without enthusiastic admiration.... Grant whatever instincts you please, and it seems at first quite inconceivable how [hive-bees] can make all the necessary angles and planes, or even perceive when they are correctly made" (*Origin*, 224). Likewise, to "suppose that the eye, with all its inimitable contrivances for adjusting the focus to different distances, for admitting different amounts of light, and for the correction of spherical and chromatic aberration, could have been formed by natural selection, seems, I freely confess, absurd in the highest possible degree" (*Origin*, 186).

Nevertheless, conclusions about inimitability or beautiful adaptation, just like sensations of enthusiastic admiration or absurdity, must all give way. The logical observer's "reason ought to conquer his imagination" (*Origin*, 188), difficult as Darwin admitted this subjugation to be on the basis of his own experience. If he has been able to accept the argument for natural selection and its power up until the point of confronting the problem of perfection, perfection itself should pose no greater difficulty. An entirely material explanation will suffice here as elsewhere. In the cases of both structure and instinct, the solution is the same. We may not be able to see the intermediate stages in the evolution of the human eye or the hive-bee's comb-building instinct, but we can look laterally, to simpler related forms, to build up a series of plausible intermediates that exist in the here and now. We have no fossil data on the evolution of eyes among vertebrates, and they are so highly developed among all living vertebrates that no contemporaneous series can be constructed, but among extant invertebrates a whole range of eyes still exists, from "an optic nerve merely coated with pigment, and without any other mechanism ... until we reach a moderately high state of perfection" (*Origin*, 187). The existence of this sequence of "graduated diversity" among living organisms itself makes the existence of historical gradation the more plausible, particularly "bearing in mind how small the number of living animals is in proportion to those which have become extinct" (*Origin*, 188). Moreover, so long as each eye, from the simplest light-sensitive nerve onward, is "useful to its possessor," natural selection can be constantly working on the heritable variations always present to produce greater perfection in succeeding generations (*Origin*, 186).

As with the case of invertebrate eyes, the hive-bee's honeycomb, while mathematically perfected, differs only in degree rather than in kind from combs constructed by other species of bee, which range from simple irregular round or cylindrical cells of wax among humble-bees through the honey cells of Mexican bees of the genus *Melipona*, constructed as spheres that normally adjoin with others at one, two, or three points, each of which then becomes a shared flat wall of wax, of the same thickness as a normal exterior cell wall; where three such walls come together, they form a pyramid like that forming the basis of every hive-bee cell. "Hence we may safely conclude that if we could slightly modify the instincts already possessed by the Melipona, and in themselves not very wonderful, this bee would make a structure as wonderfully perfect as that of the hive-bee" (*Origin*, 227). This would require only that hive-bees be shaped by natural selection to adhere to a few more or less absolute rules: stand a certain distance from your coworkers; always construct spheres or cylinders but construct flat walls where these intersect with the coworkers' cells; don't make the walls too thick or too thin. And natural selection could act on these simple instincts because they conferred an immediate profit on the bees at each intermediate stage, as Darwin's theory required: "bees are often hard pressed to get sufficient nectar [and] a prodigious quantity of fluid nectar must be collected and consumed by the bees in a hive for the secretion of the wax necessary for the construction of their combs.... Hence the saving of wax by largely saving honey must be a most important element of success in any family of bees" (*Origin* 233–4). Incremental improvements in the efficiency of honeycomb construction would yield continual, if incremental, advantages to the bees who made them, until that "stage of perfection in architecture" was reached, beyond which "natural selection could not lead; for the comb of the hive-bee, as far as we can see, is absolutely perfect in economising wax." Thus can natural selection take "advantage of numerous, successive, slight modifications of simpler instincts" to derive "the most wonderful of all known instincts," the honeybee's (*Origin*, 235). No Designer need be postulated.

In both these cases, of perfect structures and of perfect instincts, Darwin dodged a bullet he likewise dodged in the work as a whole. Just as "my theory" could shed no light on the ultimate origins of life, but only on the processes that led to its progressive differentiation thereafter, so Darwin explicitly disavowed theorizing either about the origins of the most primitive eyes or about the material basis and origins of instincts: "How a nerve comes to be sensitive to light, hardly concerns us more than how life itself first originated" (although Darwin went on to assert that "several facts" led him to believe that any sensory nerve could become sensitive to light) (*Origin*, 187). In parallel, for instincts, Darwin averred that "I must premise, that I have nothing to do with the origin of the primary mental powers, any more than I have with that of life itself" (*Origin*, 207). Natural selection could not speak to the material origins of absolute novelties. Subsequently, however, its power was absolute, so much so that Darwin staked his whole theory on natural selection's ability to account even for cases of apparent perfection. "If it could be demonstrated," he declared, "that any complex organ existed, which could not possibly have been formed by numerous, successive, slight modifications, my theory would absolutely break down. But I can find out no such case." In a characteristic piece of rhetorical judo, Darwin transformed the "inconceivable" and "absurd in the highest degree" with which he had begun his discussions of the idea that natural selection might shape the eye or the honeycomb into a categorical "no such case" (*Origin*, 189).

The difficulty posed by the problem of perfection for Darwin was more theological than material, since, given the initial conditions (a light-sensing nerve), natural selection clearly had the materials on which to work. The case of instincts, however, was different. There was no convincing theory about how instincts, or indeed any mental or moral powers, were rooted in the material body (and some, though not all, natural theologians denied that they were material). Their origins, in history and ontogeny, were mysterious. And finally, in certain cases, most particularly those of the social insects with sterile worker castes, these difficulties were compounded both by being combined with often flamboyant modifications of physical structure and by the inability to explain the continuance of these structures and instincts by straightfoward inheritance from parents (who did not manifest either). Darwin accordingly hived his discussion of instinct off from the "difficulties" discussed in Chapter 6, and dealt with them separately. Characteristically, as in the case of perfect organs, he structured his argument so that the problem that even to him had "at first seemed insuperable, and actually fatal to my whole theory" (*Origin*, 236) – namely, the evolution of differentiated sterile castes in the social insects – was revealed, in the capstone to the two chapters, not only to be wholly explicable on the basis of natural selection, but to be explicable on no other material basis. Indeed, it was a most gratifying vindication of his "faith in this principle ... that natural selection could have been efficient in so high a degree" (*Origin*, 242).

Darwin had further reasons for highlighting the importance of the evolution of instincts for the theory of natural selection. Not only did he have to argue that, unknown as the material basis of the powers of mind might be, the same methods of analysis should be applied to them as to corporeal structures (an argument by no means original to Darwin), he also had two further implicit ends in view. The first was, again, theological; Darwin found that his new mechanism to explain the evolution of instincts provided, if not a solution to the problem of theodicy, at least an end run around it. The second was long-range. Darwin deliberately steered clear in the *Origin* of questions of human origins and evolution, but these were always on his mind from the inception of his evolutionary thinking. Central to any explanation of human evolution, Darwin believed, would be explanations of two distinctive human characteristics: sociality, and the development of the moral sense that qualitatively distinguishes humans from other animals. A theory of instincts was vital to both.

Instinct was a particularly tricky subject because it was much harder to define than a wing or an eye. To all the enormous initial difficulties involved in discussing the origin and action of mental powers in general, had to be added some differentiation of the idea of instinct from those of habit or of reason, and this was by no means easy. Some cases, as, for example, skilful first-time cocoon building or nest building by individuals who had never seen one made and could have no conception of their ultimate purpose, were relatively clear-cut. However, often matters were not so clear. Were habits acquired during the life of an organism and practiced enough to become automatic the same as instincts, or analogous? Indeed, was the origin (by the inheritance of acquired characters) of instincts innate in subsequent generations? Could or did elements of ratiocination ever play a role in the execution of an instinct, or was the power of reasoning limited to the higher animals alone, or indeed solely to human beings?

Darwin was in a position to tackle these issues in ways that would not have been possible even fifty years before his time, thanks to a wealth of new observations, chiefly on the social insects, many of them explicitly natural theological in character, that had been made over the previous seventy-five years. The social insects (the ants, bees, wasps, and for Darwin to a much lesser extent the termites) occupied both very old and very recent positions in the annals of natural history. Since antiquity, ants and bees in particular have served as mirrors and proxies of human beings in literature and natural history.

Eighteenth-century natural historians turned their new disciplines of attention to the social insects (see Daston 2004) and found much to criticize in ancient accounts: while the sluggard was counselled to go to the ant because she, though "having no guide, overseer, or ruler, provideth her meat in the summer, and gathereth her food in the harvest," the attentive observers of northern Europe denied that she did any such thing. (Only later in the nineteenth century were observations made by naturalists of Mediterranean ant species that do indeed harvest seeds, vindicating the wisdom of Solomon, at least in myrmecology.) In building his account of the evolution of instincts, Darwin drew heavily on the work of several of these naturalists (see Drouin 2005). He had read the work of the French observer of ants Pierre-André Latreille, whose *Histoire naturelle des fourmis* (1802) he had picked up even before leaving on the *Beagle*, and his copies of the works of the Swiss father and son François and Pierre Huber, on bees and ants, respectively (*Nouvelles observations sur les abeilles* [1814] and *Recherches sur les moeurs des fourmis indigènes* [1810]), were both heavily annotated and extensively drawn upon for the *Origin* (di Gregorio, 409–13).

It was in the works of natural theologians, however, that Darwin found the strongest supports for and challenges to his emerging views on the evolution of instinct. Naturalists in general, natural theologians included, were divided on the subjects of both the nature of instincts and whether or not there was an irreducible gap between the reasoning powers of human beings and the merely mechanical manifestations of animal behavior. Paley, for example, came down on the side of an unbridgeable chasm between human and animal intelligence. Other naturalists, however, were not so sure. Henry, Lord Brougham, whose *Dissertations on Subjects of Science Connected with Natural Theology* (1839, written to accompany an illustrated edition of Paley's *Natural Theology*) Darwin read in 1840, opined on the basis of his own empirical research that even seemingly simple creatures were capable of a degree of free rational action, in which "the means are varied, adapted, and adjusted to a varying object" (quoted in R. J. Richards 1987, 138). For Brougham, writing in the British tradition of sensationalist epistemology to which Darwin also belonged, animals were not Cartesian machines without self-awareness; they had a degree of freedom of action (in Brougham's view, to be sure, granted to them by the Creator) that meant that their intelligence differed from ours in degree but not, crucially, in kind. To insist on an uncrossable gulf between human and animal mind was not a necessary component of natural theological reasoning. Darwin found a second component of Brougham's reasoning to be compelling, albeit in a way that would not have pleased Brougham himself. Brougham argued that, given that many instinctual actions were performed without an animal's having any possible experience of either their performance or their object (as in the case of solitary wasps that stocked nesting burrows with particular species of spiders or grubs for larvae they would never see, or a chick pecking out a hole from inside the eggshell to reach a world of which it had no experience), there could be no question of these instincts, or by extension any instincts, having evolved by the inheritance of repeated habit, as Lamarck and other early evolutionists believed. Darwin concurred with Brougham's challenge, but did not take this refutation of use inheritance as a disproof of his theory of descent with modification. On the contrary, while it challenged him to think more deeply about the limits of habit inheritance (which he continued to think important), it also caused him to consider how natural selection could produce these one-time marvels in an animal's existence, unaided by habit inheritance. Brougham's work, he concluded in his notebooks, was "profound" (*Notebooks*, 580).

The work that probably had the greatest influence on Darwin's thinking about instinct until about 1858, and that presented him with the starkest difficulty for his theory, was William Kirby and William Spence's *Introduction to Entomology*. The collaboration of a High Tory churchman and natural theologian (Kirby, later the author of the seventh Bridgewater Treatise) and a secular physiocratic political economist (Spence), the *Introduction* was the standard work on entomology for most of the nineteenth century. Darwin esteemed the whole highly and particularly the chapter on instinct ("the best discussion on instincts ever published," he wrote in the long manuscript "Natural Selection," which formed the basis of the *Origin* [*Species Book*, 468]). This discussion was written by Spence, in contradistinction to the avowed views of his coauthor, who explicitly disassociated himself from them in volume 3 of the work (see Clark 2006). Spence noted the difficulty of defining instinct, but gave a working definition that Darwin was to adapt for the *Origin*:

Without pretending to give a logical definition of it, which, while we are ignorant of the essence of reason, is impossible, we may call the instincts of animals those unknown faculties implanted in their constitution by the Creator, by which, independent of instruction, observation, or experience, and without a knowledge of the end in view, they are impelled to the performance of certain actions tending to the well-being of the individual and the preservation of the species. (Kirby and Spence, 2: 471)

Spence, a secularist, agreed with Brougham's view of the material basis of powers of mind; but, more to Darwin's taste than to a natural theologian's, he limited the Designer's powers of endowing these to the original creation of species, rather than agreeing with Brougham's "interposition of the Deity at each moment" in the operation of instinct (quoted in Clark, 46). Moreover, Kirby and Spence affirmed the contiguity between the operation of instinct and reason: age, for example, could bring a measure of wisdom even to insects, which gain knowledge from *experience*, which would be impossible if they were not gifted with some "portion of reason" (Kirby and Spence, 2: 415).

Kirby and Spence also brought Darwin to a most formidable obstacle to his incipient theory of natural selection, namely, the problem of the evolution of differentiated neuter castes in the social insects, where no amount of acquired variation in habit or structure among sterile individuals could possibly be passed directly to succeeding generations. Bees, and more particularly ants and termites, may have one or more worker forms, different in both habits and physical structure from their fertile parents. Accounting for their evolution was particularly sticky. The margins of Darwin's copy of Kirby and Spence were littered with memoranda of his frustration (among others – "Neuters do not breed! How instinct acquired," and "I can understand a neuter having any instinct which the female could have had, but no others cd have been acquired by <u>habit</u>" (di Gregorio, 452). Robert Richards (1987) has persuasively argued that Darwin's inability to devise a convincing account of the evolution of neuter castes, a problem potentially, as he rightly observed, "fatal to my whole theory," was a principal cause of his delay in publishing the theory of natural selection. It was only well on in the writing of the big "Natural Selection" manuscript, in 1858, that he saw his way through to a solution, shortly before Alfred Russel Wallace's letter jolted him into the frenzy of abstracting and writing that produced the *Origin*.

In redacting his thoughts on the evolution of instinct for the *Origin*, Darwin structured his argument in the way that had, by Chapter 7, become familiar to the reader. Demurring at the outset to speculate on the ultimate origins of mental powers, he then provided a loose overview of what he meant by instinct, derived, as noted, from Spence's definition in the *Introduction to Entomology*. Prefacing his remarks with the statement that he would not attempt any general definition of the term, he observed that "It would be easy to show that several distinct mental actions are commonly embraced by this term" (*Origin*, 207). The phenomenon could be gotten at only by concrete example:

every one understands what is meant, when it is said that instinct impels the cuckoo to migrate and to lay her eggs in other birds' nests. An action, which we ourselves should require experience to enable us to perform, when performed by an animal . . . without any experience, and when performed by many individuals in the same way, without their knowing for what purpose it is performed, is usually said to be instinctive. But I could show that none of these characters of instinct are universal. A little dose...of judgment or reason, often comes into play, even in animals very low in the scale of nature. (*Origin*, 207–8)

Instinct might also be compared to habit, although while this "comparison gives, I think, a remarkably accurate notion of the frame of mind under which an instinctive action is performed," it gives no indication "of its origin" (*Origin*, 208).

Having brought his reader to a point of common ground, if not perfect agreement, Darwin then turned to his most familiar, principal, method of argument, by analogy from artificial selection. The instincts of domestic animals vary, both from one individual to the next and from one generation to the next, particularly under the action of deliberate selection, as for example tumbling in pigeons or pointing in hunting dogs; they can also be bred out of a line, whether deliberately, as in the case of English dogs (in contrast to Fuegian animals) selected over generations not to attack livestock, or inadvertently, as in the loss of brooding instincts in some varieties of domestic poultry. The inheritance of acquired habit seemed to operate, at least in part, in some cases (as in pointing), while it seemed most unlikely in others (tumbling). As under domestication, so in nature, in cuckoos, aphids, honeybees, ants.

Darwin paid particular attention to the instincts of ants, because these were among the most extraordinary known. The sensational case of so-called ant "slavery" had been discovered at the turn of the nineteenth century by Pierre Huber, whom Darwin esteemed as "a better observer even than his celebrated father" (*Origin*, 219). Huber discovered that workers of certain ant species, *Formica sanguinea* and *Formica* (later *Polyerges*) *rufescens*, regularly raided nests of a related ant species, *Formica fusca*, and carried large numbers of larvae and pupae back to their own nests. These then hatched out in the alien nests and immediately set to work as though in a colony of their own, exhibiting their normal instincts for nest construction, brood tending, mutual feeding, and so on, though subverted to the advantage of the "slave-making" species. While colonies of *F. sanguinea* did not seem to be absolutely dependent on their captives, occasionally being found without any, *F. rufescens* was absolutely dependent on a regular infusion of *fusca* individuals, *rufescens* workers being entirely incapable of making their own nests, feeding themselves, or tending their own larvae. Without *F. fusca*, Darwin noted, "the species would certainly become extinct in a single year" (*Origin*, 219). Darwin, a keen and exact observer himself, devoted many hours over three successive summers to confirming and extending Huber's observations on *Formica sanguinea*, which also occurred in the south of England, along with its prey species, *F. fusca*.

The extraordinary specificity and strangeness of the slave-making instincts would have been enough, presumably, to account for Darwin's interest, but he avowed another reason in his opening discussion of them in the *Origin*, one not directly derived from the scientific interest of the behavior, but rather from its repugnant moral implications. "Although fully trusting to the statements of Huber," Darwin carried out his own observations, he claimed, because he felt the need "to approach the subject in a sceptical frame of mind, as any one may well be excused for doubting the truth of so extraordinary and odious an instinct as that of making slaves" (*Origin*, 220). Raised in a famously abolitionist family, familiar with the horrors of human slavery firsthand from his South American experiences aboard the HMS *Beagle*, Darwin was loath to believe that the institution of slavery could be in any way inscribed in nature (as some defenders of human slavery indeed claimed). The identification of ant "slavery" with human slavery, however, was evidently overdetermined for nineteenth-century observers, whether abolitionists or slavery advocates. Few naturalists, for example, argued that as the ants in question were of two different species, rather than members of the same kind, the analogy did not hold, or might be more closely made to domestication than to slavery – though the unfortunate coincidence that both of Huber's warlike "slave-making" species were large and red, while the "slave" species was small and black, probably inevitably sealed the metaphor (see Clark 1997a). Even the abolitionist Darwin succumbed to its allure; he triumphantly wrote Joseph Hooker of his first observations of the phenomenon in May 1858: "I had such a piece of luck at Moor Park: I found the rare Slavemaking Ant, & saw the little black niggers in their Master's nests" (*Correspondence*, 7: 89).

The existence of a spectrum of instincts between the still relatively independent *F. sanguinea* and the helpless *F. rufescens*

allowed Darwin to frame a conjecture as to the evolutionary origins of slave making: ants in general, he observed, would carry off
the pupae of other species back to their nests to store with other
food. If some of these chanced to hatch out, the ants "thus unintentionally reared" would naturally "follow their proper instincts,
and do what work they could. If their presence proved useful to the
species which had seized them – if it were more advantageous to
this species to capture workers than to procreate them – the habit
of collecting pupae originally for food might by natural selection be
strengthened and rendered permanent for the very different purpose
of raising slaves" (*Origin*, 223–4). Straightforward carnivory might
thus become immoral theft of both bodies and labor.

The "odious" instincts of slave-making ants troubled Darwin; so
too did other instances where animals behaved in ways viscerally
repugnant to developed morality and religion. Natural theologians
had been exceedingly troubled by such instances as cuckoo chicks
that deliberately ejected their foster siblings from their nests, to certain death, in order to usurp all the unwitting parents' resources
to themselves; or ichneumon wasps that paralyzed living caterpillars and laid their eggs in their bodies, to hatch out into larvae that
would slowly eat the caterpillar from the inside out, gruesomely
preserving their hosts' nervous systems and vital organs to the very
living last, so as not to destroy their own food supply – the caterpillar presumably suffering all the while in a horrific reenactment
of the Passion in miniature. How a benevolent, omnipotent God
could not only have allowed but ordained such suffering was a central problem of natural theology, an instance of the classic religious
problem of theodicy, the justification of God's goodness in the face
of such apparent evils. Darwin's theory could not make the problem
of natural evils vanish, but he himself found comfort in the idea of
seeing them as unfortunate manifestations of general laws leading
to general happiness, rather than as specific instances of creation
requiring particular explanation. In the conclusion to Chapter 3,
on the struggle for existence, Darwin had consoled himself and the
reader "with the full belief, that the war of nature is not incessant,
that no fear is felt, that death is generally prompt, and that the vigorous, the healthy, and the happy survive and multiply" (*Origin*, 79).
"Odious" instincts were perhaps an even bitterer pill to swallow, as
they represented instances of deliberate, perhaps even to some degree

reasoning, behavior on the part of individual organisms. In this bleak case, Darwin did not go so far as to offer his readers "consolation." Nevertheless, the theory of natural selection offered a path out of the blind alley of theodicy on this question too. "It may not be a logical deduction," Darwin wrote in the concluding words of Chapter 7, "but to my imagination it is far more satisfactory to look at such instincts . . . not as specially endowed or created instincts, but as small consequences of one general law, leading to the advancement of all organic beings, namely, multiply, vary, let the strongest live and the weakest die" (*Origin*, 244). Immoral horrors were the inevitable consequence of an amoral natural law that would nevertheless, paradoxically, lead to advancement and greater total happiness (and ultimately, Darwin would argue in *The Descent of Man*, to the appearance of the moral instincts themselves).

The last, virtuoso case study of the chapter on instinct combined all the elements, scientific and philosophical, that Darwin had brought into play in the *Origin* thus far. In his conclusions on the origins of the slave-making instincts, Darwin had already alluded to the solution he had at long last found to the problem of explaining ants' neuter castes. As already observed, this knotty problem had long retarded Darwin's final shaping of the theory of natural selection. No dependence on the eventual inheritance of long-repeated habit could explain neuters' instincts, nor could use inheritance explain their sometimes greatly modified structures. Special creation truly did seem to be the only explanation, as Brougham had triumphantly deduced. How to see a way through it on the basis of natural selection? Robert Richards (1987) has retraced Darwin's arduous journey to a solution that he found not only satisfactory, but so clear-cut an instance of the power of natural selection to elucidate what otherwise could only be credited to the Designer that Darwin made it the capstone of these two chapters, a triumphant reverse transubstantiation of the theological to the material realm of explanation.

Characteristically, the path led through artificial selection. Darwin reread William Youatt's *Cattle: Their Breeds, Management, and Diseases* (1834) in 1857 while writing the manuscript on "Natural Selection." Youatt described the methods cattle breeders employed to improve beef cattle. It was not possible to judge, naturally, the culinary quality of a beef animal without killing it. How then to improve the breed? "The breeder goes with confidence to the same family" (*Origin*, 238) and breeds from the tasty animal's relatives, on the warranted presumption that these animals will also exhibit the desired characteristics to a greater degree than the norm. Likewise, "a well-flavoured vegetable is cooked, and the individual is destroyed; but the horticulturist sows seeds of the same stock, and confidently expects to get nearly the same variety" (*Origin*, 237–8). Sterility in an individual, whether accidental or innate, is no bar to its characters being passed on through collateral lines: "selection may be applied to the family, as well as to the individual" (*Origin*, 237). This being the case, even the evolution of regular sterility among a caste of individuals in a family itself is not terribly difficult to explain, "not much greater than that of any other striking modification of structure," given that innate sterility is an occasional variant met with regularly in insects that would provide the initial material for selection to work on (*Origin* 236).

Thus far, however, the explanation covers only the inheritance of characters possessed by nonreproducing individuals and their fertile relations alike. How could modifications of instinct and structure become fixed *only* in sterile individuals? In fact, Darwin argues, the same principle applies, odd though this may seem. There are many cases, he observes, of characters exhibited in only one sex, or only at particular life stages. In cattle, there are even chance cases of characters, such as longer horns, exhibited only by artificially castrated males, so that it would presumably be possible, if any one were interested, to breed a race of long-horned oxen from their short-horned, but fertile, relations. This difficulty too, "though appearing insuperable," is thus "lessened, or, as I believe, disappears" (*Origin*, 237).

Darwin in the argument on neuter castes recapitulated in miniature the strategy of the *Origin*'s two chapters on "difficulties on theory," that is, to take the difficulties one at a time (organs of perfection, organs of small importance, peculiar instincts) and to reveal, one by one, that they not only present no difficulty, but in fact (when properly analyzed and understood) furnish superior illustrations of the power and reach of natural selection. He thus in the case of sterile ants built up one difficulty (the evolution of sterility), only to knock it down in pursuit of another greater one (the differentiation of sterile from fertile castes), only to knock this one down so he could pull the greatest trick of all out of his hat, the purest example of the power of "my theory" alone to explain the shape of the living world: explaining how the neuter castes can, as in the cases of many ants, diverge not only from the fertile castes but also from *each other* within a single species, "sometimes to an almost incredible degree," into workers and soldiers, or foragers and honey-pots, or any of the many other "such wonderful and well-established facts" that entomologists had discovered incarnated in ants over the previous century (*Origin*, 238–9).

The ontogeny of the argument recapitulates, moreover, the phylogeny of Darwin's thinking on the problem. Darwin was able to solve the problem of multiple castes only by combining his thinking about selection acting on the family or community with observations of ant species with multiple castes. He corresponded, beginning in late 1857, with the entomologist Frederick Smith of the British Museum, a specialist on the Hymenoptera (the ants, bees, and wasps), on subjects including the morphology of ant castes. Smith told Darwin, who confirmed with his own observations, that considerable variation occurred in neuter insects within a single nest, even in species that did not have differentiated castes (this, of course, accorded gratifyingly with Darwin's ideas on variation in general), and that, in cases of species with strongly differentiated castes, intermediates between them could nevertheless regularly be found, and thus "that the extreme forms can sometimes be perfectly linked together by individuals taken out of the same nest" (*Origin*, 239). Darwin's inference from these observations was that, in cases where wide variation in neuters occurred and where the variations occurring at either end of this spectrum, both proved profitable to the community:

natural selection, by acting on the fertile parents, could form a species which should regularly produce neuters, either all of large size with one form of jaw, or all of small size with jaws having a widely different structure; or lastly, and this is our climax of difficulty, one set of workers of one size and structure; – a graduated series having been first formed, . . . and then the extreme forms, from being the most useful to the community, having been produced in greater and greater numbers through the natural selection of the parents which generated them; until none with an intermediate structure were produced. (*Origin*, 241)

Only from this evolved sterility, in fact, could the ants, according to Darwin, have achieved their exemplary high degree of the division of labor and its attendant efficiencies, since, if all members were fertile, intercrossing would have caused this fine degree of specialization to be swamped out.

Thus has natural selection achieved what, through the combination of perfect structures and perfect instincts, should have appeared to be the very finest example of the argument from design, "the most serious special difficulty, which my theory has encountered" (*Origin*, 242). Moreover, no other materialist evolutionary mechanism could do so, either. Darwin indulged himself in a bit of understandable self-congratulation to close his discussion of the subject, crowing, "I am surprised that no one has advanced this demonstrative case of neuter insects, against the well-known doctrine of Lamarck" (*Origin*, 242).

Chapters 6 and 7 of the *Origin*, ostensibly dedicated to confronting the severe "difficulties on theory" that Darwin worried would bring down the theory of natural selection (and that he had himself indeed worried mightily over in the twenty years preceding its writing) actually serve as its triumphant vindication. The basis of the argument and its underpinnings having been laid out in the previous five chapters, these two chapters reveal how each of the difficulties that Darwin had confronted evanesced, or indeed – in important cases like the evolution of perfect structures, the evolution of instincts, and, in culmination, the evolution of differentiated sterile castes in social insects – was transformed into confirming evidence for his theory.

Beyond this important work, the chapter on instinct, however, laid the groundwork for the most important further extension of Darwin's ideas. From the beginning of his thinking on descent with modification, he had had two aims in view: to explain the history of life on Earth, and to explain the origins and nature of human beings. Vital to the latter project, he early saw, would be an account of the origins of sociality and of the moral instincts in man; "society could not go on except for the moral sense, any more than a hive of Bees without their instincts," he wrote in a notebook of 1838 (*Notebooks*, 609). Insects served all along for Darwin as important illustrations of both the range of possibilities in intelligence and the moral sense, and of the possibilities of illustrating these (as in the case of the evolution of the eye) in an unbroken contemporaneous series impossible for vertebrates: "[T]he mental powers of man and the lower animals do not differ in kind, although immensely in degree," he concluded in *The Descent of Man*.

A difference in degree, however great, does not justify us in placing man in a distinct kingdom, as will perhaps be best illustrated by comparing the mental powers of two insects, namely, a coccus or scale-insect and an ant, which undoubtedly belong to the same class. The difference is here greater, though of a somewhat different kind, than that between man and the highest mammal.... No doubt this interval is bridged over by the intermediate mental powers of many other insects; and this is not the case with man and the higher apes. But we have every reason to believe that breaks in the series are simply the result of many forms having become extinct. (*Descent*, 1: 186–7)

Darwin's account of the evolution of sociality in human beings was likewise analogous to the account of the evolution of insect societies he gave in Chapter 7 of the *Origin*, although he continued to rely in his thinking about the former on the eventual inheritance of long-repeated habit (in strengthening both inheritance and the faculty of sympathy), the mechanism he had been forced to abandon in the case of neuter insects. The social instincts, developed by selection through their profit to the community, operate in part by the faculty of sympathy that knits its members together. Groups, whether of insects or men, that were more cohesive through these means would ultimately displace or conquer other, less social associations, and the evolution of sociality and morality would thus become a motor leading to ever-greater perfection of societies. To be sure, the shape these would take would depend on the species' peculiarities and past history – dogs' consciences would not be the same as men's, Darwin reflected to himself already in 1838 (*Notebooks*, 564), and if, he concluded decades later in the *Descent*, "to take an extreme case, men were reared under precisely the same conditions as hive-bees, there can hardly be a doubt that our unmarried females would, like the worker-bees, think it a sacred duty to kill their brothers, and mothers would strive to kill their fertile daughters; and no one would think of interfering." These horrors as our own species would understand them would nonetheless arise from "some feeling of right and wrong, or a conscience" in bee-men with "intellectual faculties ... as active and as highly developed as in man" (*Descent*, 73).

Darwin was gratified to see his ideas on instinct taken up by other investigators during his lifetime. Darwin's protegé and intellectual son, George Romanes, took up the origin of mind and instinct, and struggled lifelong, like his mentor, with questions of natural theology and theodicy (see Richards 1987). Darwin's neighbor and frequent information source John Lubbock supported and furthered Darwin's ideas on both the social insects and the evolution of man (see Clark 1997b). A younger generation of myrmecologists (ant biologists) sought to apply Darwin's insights to new studies of ant morphology, behavior, and societies. Darwin engaged in an enjoyable correspondence beginning in 1874 with the young Swiss entomologist Auguste Forel, who had sent him a copy of his *Fourmis de la Suisse*; Darwin wrote, gratified, in the margin of his copy, "approves of what I have said of origin of slave-making" (di Gregorio, 239). Even the German Jesuit myrmecologist Father Erich Wasmann was a convinced evolutionist who developed theories of the evolution of various sorts of cooperation within ant nests (see Lustig 2004; Sleigh 2006). The tradition of using the ants as principal proxies for the study of the evolution of sociality as a general phenomenon has endured through the twentieth century, through E. O. Wilson's *The Insect Societies* (1971) and *Sociobiology* (1975), and on into the twenty-first.

Modern framings of Darwin's questions about the origins of sociality, however, have undergone a fundamental transformation that has greatly obscured their original context. William D. Hamilton, in his 1964 formulation of the idea of *inclusive fitness* or *kin selection*, argued that the evolution of altruism among members of a community (he took the hymenopteran social insects as a principal case study) resulted from the advantage accrued to each individual through the preservation of her genes through collateral lines of descent, furthered by her own self-sacrifice on their behalf. In doing so, Hamilton made a profound shift in the object of interest to the evolutionary biologist and in the language and concepts for understanding it. Where Darwin had asked how sterility would profit the community and found "no great difficulty" in answering that question (taking the evolution of the cooperative, moral instincts to be an inevitable concomitant to the evolution of sociality), Hamilton demanded to know how sterility could possibly profit a sterile individual and found great difficulty indeed in answering the question, struggling long and hard with it over a period of years, much as Darwin had done for the evolution of differentiated castes – a question that was, ironically, for Hamilton, armed with modern genetics, a question of no great difficulty once the barrier of sterility was overcome. The transformative program of selfish gene's-eye analysis that Hamilton and other biologists inaugurated in the 1960s and 1970s has been enormously successful, and though they have so altered the language and thinking that Darwin would perhaps scarcely recognize them, its proponents have never ceased to look back to their illustrious progenitor.

8 Darwin's Geology and Perspective on the Fossil Record

CHARLES DARWIN AS GEOLOGIST

In November 1859, Darwin's masterpiece, the *Origin of Species*, was published. Earlier that year, in February, he had been awarded the Wollaston Medal, the highest honour of the Geological Society of London, "for his numerous contributions to Geological Science, more especially his observations on the Geology of South America, on the Phaenomena of Volcanic Islands, on the structure and distribution of Coral-reefs, and his Monographs on recent and fossil Cirripedia" (*Correspondence* 7: 237; Darwin 1842, 1844, 1846, 1851a, 1851b, 1854a, 1854b). Also contributing to Darwin's high reputation as a geologist was his *Fossil Mammalia*, a publication from the voyage of HMS *Beagle* (Darwin, ed. 1840). This work was done jointly with the anatomist Richard Owen (1804–1892). Given Darwin's prodigious efforts as a geologist, it was natural for the subject to play an important role in the *Origin*.

Darwin learned his geology in stages. As a teenage boy, he was exposed to 'experimentation' as an enthusiastic 'assistant' to elder brother Erasmus (1804–1881) through chemistry and mineralogy in the garden tool shed at home. From 1825 to 1827, as a medical student at the University of Edinburgh, he attended lectures given by the geologist Robert Jameson (1774–1854) and the chemist Thomas Charles Hope (1794–1871) (Secord 1991). He also gained knowledge of the invertebrate animals and worked with the Lamarkian zoologist Robert Edmond Grant (1793–1874).

Such clearly innate enthusiasms, having at this time no natural channel into a future career, were partially diverted by his abandoning the Edinburgh medical course, to which he seemed temperamentally unsuited, and transferring to Christ's College, Cambridge. He took up residence at Christ's in January 1828 and began reading for an ordinary degree. At Christ's College he joined his cousin William Darwin Fox (1805–1880), who fostered Darwin's interest in many aspects of natural history, notably the collection of beetles. While at Cambridge, Darwin attended the botanical lectures of John Stevens Henslow (1796–1861), who introduced him to such leading figures in natural science at the university as the geologist Adam Sedgwick (1785–1873) and the philosopher of science and polymath William Whewell (1794–1866). Having completed and passed the examinations for his degree by January 1831, Darwin was obliged to remain in Cambridge in order to complete ten terms of residence before being allowed to graduate in April 1831 (Peile 1913).

This was a period of intense reading for Darwin – Alexander von Humboldt (1769–1859) and John Herschel (1792–1871) being authors of special importance. Reading Humboldt gave Darwin a sense of the value of highly organised, scientifically based exploration to foreign countries. Through Humboldt he was also introduced to the progress of science on the continent (Richards 2002). From Herschel he gained a useful methodology by which observational information and practical fieldwork could be used to formulate and solve larger questions in the natural sciences, with specific reference to geology and its scientific potential. Herschel's influence is apparent in the *Origin*, both in the form of its argument and in its sensitivity to geological questions (Waters 2003b; Herbert 2005). It was during this period, in 1831, that Henslow directed Darwin to the study of geology and arranged for him to accompany Sedgwick in North Wales. This training proved to be ideal preparation, not for an excursion to the Canary Islands, which Darwin had originally envisioned, but for his five-year voyage on HMS *Beagle*. This opportunity came about once again through the intervention of Henslow.

In a rapid phase of preparation for his voyage on the *Beagle*, Darwin acquired the equipment necessary for his role as an exploratory naturalist aboard the ship, and a reference library, the geological component of which was assembled on the advice of Henslow and Sedgwick. Of these books the most influential by far was the *Principles of Geology* by Charles Lyell (1797–1875). The cumulative effect of Lyell's vision of the interpretative power of

geology as a science that encompassed both the physical and biological worlds and their interwoven effects in moulding the surface of the
planet was synthetic and beguilingly powerful. Lyell was grappling
with a number of apparently either 'static' or 'dynamic' phenomena
associated with the then-known crust of the Earth. Static phenomena included: the overall inorganic complexity (extraordinary miner-
alogical and petrological diversity), the apparent ubiquity of subdivisions in the geological succession globally, the successional appearance of unique fossil types in strata, and the geometrical complexity
of particular geological formations and their occasional intercon-
nectedness (as in ribbon-like mountain ranges). Dynamic phenomena included: the evidence of apparently dramatic sea-level change,
earthquakes and volcanic eruptions, the form and distribution of
oceanic islands, the remarkable geographic distribution of organisms, the influence of climate, and ultimately, of course, the larger
implications derived from observations on the variability of species.
Rationalising, and perhaps integrating, several or perhaps all of these
phenomena (generalised Herschelian and Lyellian goals) had the
potential to lead to a more generally applicable theory of the Earth.

   During the voyage Darwin made the most of his training, his reading, and his opportunities for fieldwork. He applied the essentially
inductive programme espoused by his masters to generate synthetic
explanations for many of the pressing and problematic issues then
current in geology. Observations of weathering, gravitational separation and differential crystallisation in lava flows, and the effect
of heat on rocks provided keys to understanding the generation
of diversity in the mineral and petrological kingdoms; continental-
scale measurement of raised beaches – linked to his witnessing of
an earthquake and its associated effects – allowed him to provide a
unified explanation for sea-level change, earthquakes, vulcanism,
and continental elevation. This latter set of observations, in turn,
allowed Darwin to develop a theory that beautifully integrated biology (the life history of coral organisms) with geology (volcano formation and ocean floor subsidence equilibrated by the elevations of
the land seen on adjacent continents). Darwin's theory of coral reef
formation posited an explanation for the history of the development
of oceanic islands from volcanic island to coral atoll. With elevation
(uplift) and subsidence (fall) as the primary agencies in his understanding of the motion of the earth's crust, Darwin wrote in his Red

Notebook toward the close of the voyage, "Geology of whole world will turn out simple" (*Notebooks*, 44 [RN, 72]).

On his return from the *Beagle* voyage, Darwin presented the results of his geological researches at the Geological Society of London. The agencies of elevation and subsidence provided a unity – and a style – to much of his work (Rhodes 1991). For example, in 1838 he wrote of "the grandeur of the one motive power, which, causing the elevation of the [South American] continent, had produced, as secondary effects, mountain-chains and volcanos" (Darwin 1840). The term "grandeur" is already redolent of what would appear in the closing passage of the *Origin* many years later.

During the early 1840s, Darwin was surprised and disconcerted on one score: the advent and rapid success of 'glacial theory' as promoted by Louis Agassiz (1807–1873) (Rudwick 1974; Herbert 2005). Darwin incorporated many of these 'new' views, and glacial theory appears prominently in the *Origin*. With regard to Darwin's adoption of transmutationism in March 1837, the palaeontological component was critical. It is to that subject we turn next.

DARWIN'S PALAEONTOLOGY

While the assiduous compilation of specimens and detailed geological observation and measurement formed the core of Darwin's work each time he disembarked from the *Beagle*, an important, intellectually stimulating, component of his collection was represented by fossils, with which he was far less well acquainted. The first notice taken of fossils was his discovery of large fossil mammal bones in Patagonia toward the end of his first year aboard the *Beagle*. He wrote in his notes, "[T]he number of fragments of bones of quadrupeds is exceedingly great: – I think I could clearly trace 5 or 6 sorts" (DAR 32.1.64). Large vertebrate fossils from South America were known through the discovery of *Mastodon* (an elephant-like proboscidean) and *Megatherium* (a giant ground sloth) remains. Darwin was able to collect remains that were recognisably mastodont and megatheroid, but additional material indicated a greater diversity of life than had previously been reported. Greater resolution awaited specialist research upon Darwin's return to Britain. He was, however, aware of the general anatomical similarity between some of these fossils and the uniquely South American animals ('edentates' and its diversity

of ungulates: notoungulates, litopterns, etc). For example, Darwin
had collected patches of tessellated bony armour that, though considerably larger in size, resembled the bony plates that covered the
bodies of living armadillos. Richard Owen, the renowned British
comparative anatomist, willingly aided Darwin in the description
and assignment of these fossils. In the major report on the fossil
mammals (Owen 1840), he confirmed Darwin's suspicions by revealing the presence of various 'edentate' ground sloths (*Megatherium*,
Megalonyx, *Mylodon*, *Scelidotherium*, and *Glossotherium*); a large
armoured animal related to the armadillo (later recognised as
Glyptodon); small armadillos; the rhinoceros-sized but anatomically enigmatic form *Toxodon* (a huge, lumbering notoungulate –
sometimes referred to as a "gigantic guinea pig"); Owen's "camelid"
Macrauchenia (in reality a litoptern, a member of a group of exclusively South American ungulates that were anatomically and evo-
lutionarily convergent upon true camels); *Mastodon augustidens*
[-*Stegomastodon*], an elephant-like proboscidean; hystricomorph
rodents; and, most remarkable of all, a true horse (*Equus* sp).

Darwin (and indeed Owen) was strongly impressed by a number
of facts linked to the 'dry' taxonomic observations about these fossils: the clear biological relationship between the (generally) gigantic
but *extinct* forms and their modern counterparts; their comparative *uniqueness* inasmuch as they were representatives of a truly
South American fauna; the comparative recency of their existence –
Darwin had been able to demonstrate that the majority of invertebrate (mollusc) fossils found with these extinct mammal bones were
similar to those still living today, which argued strongly for a very
recent period in Earth's history; as well as the extraordinary insight
into the history of the horse. Unknown to human civilisations in
South America prior to the arrival of the conquistadors, it was now
clear that ancestral *Equus* had become extinct in South America
in the geologically recent past. Such observations probed the heart
of heated debates in Paris between Georges Cuvier (1769–1832) and
Étienne Geoffroy Saint-Hilaire (1772–1844). Cuvier, a catastrophist
and functionalist (seeing all similarities in anatomy as products of
common adaptation to perform specific tasks) who envisioned a creative God sweeping all aside in mass-extinction events and then
repopulating the Earth de novo, was the counterpoint to Geoffroy

(a disciple of Jean Baptiste de Lamarck [1744–1829]), who saw progression in time and continuity in nature: while Cuvier claimed that the legs of man, mice, and lizards were similar because they performed the function of walking (Desmond 1989), Geoffroy identified homology in such limbs – they all shared the same detailed component bones and muscles, indicating a common blueprint for all vertebrate animals; and this continuity could be demonstrated in time as well as space.

Owen presented his first Hunterian lectures at the Royal College of Surgeons in the summer of 1837 (Sloan 1992) and included mention of the newly acquired *Beagle* specimens. Among these was the large but enigmatic skull of *Toxodon*, which Owen had considerable difficulty placing within his general understanding of the recognised groups of Mammalia. The mystery skull had grinding teeth and a wide diastema (the condition seen in modern rodents); it also had nostrils that had migrated in position backward onto the top of its snout (as they do in the skulls of elephants and whales). Clinging to Cuvierian principles of comparative anatomy, Owen inevitably concluded that *Toxodon*

manifests an additional step in the gradation of mammiferous forms leading from the *Rodentia*, through the *Pachydermata* to the *Cetacea*.... (Owen 1840)

While this *sounds* almost evolutionary in its import, Owen was approaching the placement of this mystery animal in the context of a metaphor: a tree-like scaffolding, within which all mammals could be represented. Owen advocated the view (first propounded by Johannes Müller [1801–1858]) that extinctions were a reflection of the "life-force" within all organisms: just as an individual lives and dies, so too the species (i.e., "slothful" species use less energy and persist for longer than "vigorous" species); equally, he proposed that the past differences between species reflected their functional integration with different regional and climatic conditions.

Darwin was caught up with many of these threads of debate and discussion. He was actively meeting with Owen to talk over his South American fossils, while Owen was simultaneously preparing his Hunterian lectures (which Darwin may well have attended) and presenting short papers on Darwin's fossils to the Geological Society of London. Darwin was deeply impressed by the fact of extinction, which his South American fossil mammals revealed. While being aware of Müller's "life-force" idea, he was struck, on the one hand, by the universality of some extinctions: what could have killed off all the mastodons in the Northern and Southern as well as the Eastern and Western Hemispheres, simultaneously? And why did the horse go extinct in South America alone? On the other hand, he was also struck forcibly by the geographic distinctiveness of this fauna (both past and present). The latter became even more important as he began to perceive the South American faunal influence on the Galápagos Islands and the variability of species between the islands.

Thus, by the early 1840s, through a series of vitally important field observations and convenient fossil discoveries of his own, as well as through direct exposure to the anatomical expertise of Owen, linked to the deeply challenging philosophical struggles in which the general discipline of comparative anatomy was embroiled, Darwin was in an ideal position to rationalise and develop his own theories. While assiduously completing the reports on the findings of the voyage of the *Beagle* (which took a decade after his return from the *Beagle* voyage), he consciously developed his ideas in his notebooks on the transmutation question and wrote preliminary drafts of his species theory. Once his work on the *Beagle* material was complete, he began new projects. In order to better understand the taxonomic issues associated with the identity of species, he began his "barnacle work," which blossomed into a comprehensive review of both living and fossil species and provided an empirical foundation for his philosophical development with respect to his later species work. All of his work took place in the context of complicated intellectual and institutional developments during the 1840s and 1850s, whose history has been the subject of numerous studies (Rupke 1994; Browne 2002).

THE ARGUMENT

In arguments it is often useful to lead from a position of strength by placing one's most persuasive points first. In the *Origin*, Darwin placed variation under domestication in his first chapter. By contrast, and tellingly, his treatment of the geological record is deferred until Chapters 9 and 10. However, even in the first four chapters of the *Origin*, where Darwin lays out his basic argument, geological points surface from time to time.

The first case in which geology becomes a key point is in Chapter 4, entitled "Natural Selection." Darwin was in the process of arguing for the "continued preservation of individuals presenting mutual and slightly favourable deviations of structure." He then went on to compare the action of natural selection to the action of geological forces, as they had been understood by his mentor Lyell:

I am well aware that this doctrine of natural selection, exemplified in the above imaginary instances, is open to the same objections which were at first urged against Sir Charles Lyell's noble views on "the modern changes of the earth as illustrative of geology;" but we now very seldom hear the action, for instance, of the coast-waves, called a trifling and insignificant cause, when applied to the excavation of gigantic valleys or to the formation of the longest lines of inland cliffs. Natural selection can act only by the preservation and accumulation of infinitesimally small inherited modifications, each profitable to the preserved being; and as modern geology has almost banished such views as the excavation of a great valley by a single diluvial wave, so will natural selection, if it be a true principle, banish the belief of the continued creation of new organic beings or any great and sudden modification in their structure. (*Origin*, 95–6)

Darwin took what he believed to be the widespread acceptance of Lyell's claims for the efficacy of cumulative small changes to explain large-scale phenomena in geology to add weight to his argument that, in a similar fashion, small changes in individuals might over the course of many generations cumulatively yield great changes in morphology.

Within the core argument of the *Origin*, contained in the first four chapters of the book, Darwin also raised the subject of extinction. While deferring full treatment until his later chapters on geology, he wished to connect the issue to his core argument by suggesting that as a consequence of high geometrical powers of reproductive increase, each area is already "fully stocked" with inhabitants. Therefore, "it follows that as each selected and favoured form increases in number, so will the less favoured forms decrease and become rare. Rarity, as geology tells us, is the precursor to extinction" (*Origin*, 109).

Immediately following Darwin's brief treatment of extinction in Chapter 4 is a probing and philosophical discussion of an issue that Darwin treated under the heading of "Divergence of Character." This topic is treated separately in this volume, but it should be noted that the fanning out of separate species groups was, like extinction, an essential part of Darwin's explanation for the history of life on Earth.

From the point of view of arrangement of material, it is significant that the one diagram in the *Origin* (frontispiece, p. ii) occurs in Chapter 4, on natural selection, and that it both reflected and shaped Darwin's perspective on the fossil record. The very noticeable feature of this diagram is its high level of abstraction, for neither species nor strata are labelled. While other contemporary authors were attempting to express relationships of species across time, it was usual to label groups and temporal boundaries.[1] By contrast, Darwin's diagram is serenely abstract. (This was in contrast to several of his earlier unpublished diagrams, which were more concrete in their labelling.)[2] From bottom to top, the diagram represents the passage of time from older to younger, but there is no beginning point indicated. In the course of his book, Darwin suggested that there must have been life before the lowest Silurian beds (*Origin*, 306). (The lowest fossil-bearing strata were then known as "Silurian." Darwin did use the term "Cambrian" in the fifth edition of the *Origin* [*Variorum*, 513] [Secord 1991].) Thus there is no representation of an aboriginal group from which all else had sprung. Ultimately, Darwin chose to leave this subject aside. It was, however, taken up fairly quickly by his scientific peers in England after 1859 (Strick 2000). To some extent, of course, Darwin had raised the question himself by his use of the tantalizing term "origin" in the title of his book. He had contemplated entitling his book "On the Mutability

---

[1] For reproduction of a number of these authors' works, see Barsanti (1992).

[2] See *Correspondence*, 8: 379–80, for an example of a labelled diagram where Darwin is speculating, with remarkable perspicacity, on details of the alternative patterns of diversification (phylogenetic relationships) that might be exhibited by marsupial and placental mammals. See also Bredekamp (2006) and Voss (2007) for reproduction of a range of Darwin's diagrams, some of which are labelled as groups, some of which are more abstract.

of Species," but ultimately decided in favour of the bolder "On the Origin of Species."[3]

Another noteworthy feature of the diagram is that Darwin was not offering a transmutationist rendering of, say, Linneaus' *Systema Naturae*. In his diagram, Darwin did not represent major units of plant and animal classification. Further, he was flexible with regard to the meaning of his symbols. His letters A through L were initially designated as standing for the "species of a genus large in its own country; these species are supposed to resemble each other in unequal degrees, as is so generally the case in nature, and is represented in the diagram by the letters standing at unequal distances" (*Origin*, 116). Later in the *Origin*, he wrote that the same letters A through L on the figure "may represent eleven Silurian genera, some of which have produced large groups of modified descendants" (*Origin*, 432). Darwin's alternative lettering designations (species/genera) suggest that the uses of argument for the diagram were paramount: its function was illustrative.

Darwin's representation of the passage of time in the diagram was similarly flexible. In his chapter on "Natural Selection," he wrote that "horizontal lines may represent each a thousand generations; but it would have been better if each had represented ten thousand generations" (*Origin*, 117). A few pages later, he allowed that

In the diagram, each horizontal line has hitherto been supposed to represent a thousand generations, but each may represent a million or hundred million generations, and likewise a section of the successive strata of the earth's crust including extinct remains. (*Origin*, 124; also see 331)

Thus Darwin shifted easily, within the same sentence, between describing the horizontal lines as representing a number of generations and having them represent strata. As he wrote it to his publisher John Murray when he submitted the diagram for inclusion in his manuscript, "It is an odd looking affair, but is *indispensable* to show the nature of the very complex affinities of past & present animals" (*Correspondence*, 7: 300). If one keeps in mind the main purpose of Darwin's diagram, as he saw it, the flexibility in its use is understandable.

---

[3] See DAR 205.1.70 for the alternative title page.

Despite the labile nature of his diagram, there is important information in it regarding Darwin's conception of the pattern of life's history. There are fewer forms represented at the bottom of the diagram than at the top. This fanning out of forms was explained by the principle of divergence. There are also forms represented that have become extinct. As Darwin noted, all the forms beneath the uppermost horizontal line may be considered as extinct. These two principles – divergence and extinction – run through his presentation of the history of life. Presumed in this diagram is directionality, though this was a complicated issue for Darwin as it was for his contemporaries, particularly Lyell; we will return to it later as we discuss the chapters devoted exclusively to geology.

Before leaving discussion of the core chapters of the *Origin*, we wish to highlight Darwin's inclusion in them of two very concrete references. While his overall strategy was to remain at a high level of abstraction in the *Origin*, he did occasionally provide examples that served as guides to his thinking and indicated his involvement with the science of the day. Thus, in the final paragraph of Chapter 4, he referred to two extant, but comparatively ancient, groups: the duck-billed platypus (*Ornithorhynchus*) and the South American lungfish (*Lepidosiren*).

As we here and there see a thin straggling branch springing from a fork low down in a tree, and which by some chance has been favoured and is still alive on its summit, so we occasionally see an animal like the Ornithorhynchus or Lepidosiren, which in some small degree connects by its affinities two large branches of life, and which has apparently been saved from fatal competition by having inhabited a protected station. As buds give rise by growth to fresh buds, and these, if vigorous, branch out and overtop on all sides many a feebler branch, so by generation I believe it has been with the Great Tree of Life, which fills with its dead and broken branches the crust of the earth, and covers the surface with its ever branching and beautiful ramifications. (*Origin*, 130)

While Darwin's depiction of the "Great Tree of Life" is metaphorical, his inclusion of these two concrete examples gives his discussion topicality. *Ornithorhynchus* (the duck-billed platypus) was an enigmatic creature from Australia. Known from the very late eighteenth century and originally asserted to be a hoax, it created taxonomic confusion. The animal had mammalian fur, a duck-like beak, and was rumoured to lay shelled eggs like a typical reptile. Owen, who had studied its eggs and mammary glands, favoured the view that it was a mammal (Owen 1832, 1834). There were, however, differences of opinion concerning this conflicting set of anatomical features, and it was not until the 1880s that it was finally established that this was a mammal that laid shelled eggs and yet suckled its hatched young on milk that exuded from glands in folds on its abdomen.

Ornithorhynchus had attracted Darwin's attention from early on. In his transmutation notebooks from the 1830s he had speculated: "Perhaps the father of Mammalia as Heterodox as ornithorhyncus," and "We have not the slightest right to say there never was common progenitor to Mammalia & fish. When there now exist[s] such strange forms as ornithorhyncus." Darwin believed Owen's treatment of *Ornithorhynchus* suffered from difficulties that occurred if animals "are thought to have been created" (*Notebooks*, 192 [B, 89], 194 [B, 97], 370 [D, 115]). By citing *Ornithorhynchus* in the *Origin*, Darwin was drawing attention to a topical but unusual group whose existence, he believed, was best explained by his own theory.

Lepidosiren, the South American lungfish, was described and named by Leopold Fitzinger in 1837. This was a timely debut for the organism, given Darwin's interest in aberrant species. The lungfish is an eel-like fish with tendril (thread-like) fins and normal internal gills within a gill chamber. It has the ability to gulp air into its lungs at the water surface and to burrow down into mud during the dry season, aestivating inside a cocoon. The origin of its name, eliding as it does 'Lepido' (scaly) and 'Siren' (a mythical serpent), represents an echo of its general resemblance to similarly enigmatic New World animals known as sirens (amphibian, salamander-like creatures [*Siren* and *Pseudobranchus*]), which had been known from the mid eighteenth century and then "lately" described by James Gray in 1826. They had similar habits and appearance, being purely aquatic, and having a long eel-like body and an ability to gulp air into their lungs. Sirens, unlike the lungfish, have permanent external gills, rather than a gill chamber, and reduced limbs (rather than fins or tendrils). These animals were outwardly remarkably similar and presented the taxonomist with a challenge: what were they – fish, saurian, or something in between? Darwin read Owen's 1841 article on the lungfish and commented on them as connecting fish and reptiles (*Notebooks*, 448 [E, 168]). When Darwin wrote the *Origin*, the memory of them was still fresh, and he referred to *Lepidosiren* and *Ornithorhynchus* as "in some small degree" connecting by their affinities two large branches of life.

Darwin's full treatment of geology is contained in two chapters: Chapter 9, "On the Imperfection of the Geological Record," and Chapter 10, "On the Geological Succession of Organic Remains." There is also ancillary information on geology in Chapters 11 and 12 on geographical distribution, and in Chapter 13 on classification and embryology.

Let us first look at what Darwin set out to accomplish in Chapter 9. What he faced was a cohort of highly respected contemporary palaeontologists and geologists who, he believed, were nearly unanimous in opposition to his views. As he wrote toward the end of the chapter:

... all the most eminent palaeontologists ... and all our greatest geologists ... have unanimously, often vehemently, maintained the immutability of species. (*Origin*, 310)

Their most significant objection was that if "the number of intermediate varieties, which have formerly existed on the earth, be truly enormous," then why does geology not reveal "any such finely graduated organic chain." Darwin answered as follows: "The explanation lies, as I believe, in the extreme imperfection of the geological record" (*Origin*, 280). Darwin once wrote that he felt that his ideas came "half out of Lyell's brains," and on no point was this more true than on the subject of the imperfection of the geological record (*Correspondence*, 3: 55). For both Darwin and Lyell, the greater portion of the Earth's strata had been destroyed by the sea or worn away in the process of the rise and fall of the Earth's crust. Following Lyell, Darwin initially emphasized the role of the sea in shaping the land. In the fifth edition of the *Origin* (1859), however, he placed more weight on "subaerial degradation" (*Variorum*, 479 [39.1:e]). Whatever the agency, strata were destroyed, and the geological record was thereby rendered imperfect.

In a transmutation notebook from the 1830s Darwin embraced this view and aligned it with his theory on species:

Lyell's excellent view of geology, of each formation being merely a page torn out of history, & the geologist being to fill up the gaps. – is possibly the

same with the ... philosopher, who has trace[d] the structure of animals &
plants. – he get[s] merely a few pages. (*Notebooks*, 352–3 [D:60])

In the *Origin*, he made the metaphor of the imperfection of the
fossil record central to his argument:

Those who think the natural geological record in any degree perfect, and
who do not attach much weight to the facts and arguments of other kinds
given in this volume, will undoubtedly at once reject my theory. For my
part, following out Lyell's metaphor, I look at the natural geological record,
as a history of the world imperfectly kept, and written in a changing dialect;
of this history we possess the last volume alone, relating only to two or
three countries. Of this volume, only here and there a short chapter has been
preserved; and of each page, only here and there a few lines. Each word of the
slowly-changing language, in which the history is supposed to be written,
being more or less different in the interrupted succession of chapters, may
represent the apparently abruptly changed forms of life, entombed in our
consecutive, but widely separated, formations. On this view, the difficulties
above discussed are greatly diminished, or even disappear. (*Origin*, 310–11)

This passage is one of the most important in the *Origin*, for upon it
rests Darwin's claim that it will never be the case that all, or even
the majority of, transitional forms will be found. While Sedgwick
(1830) had also compared the strata of the fossil record to pages in
a book, Darwin and Lyell pressed the analogy hard. Interestingly, in
discussing geology, Darwin brought in the analogy of the "slowly-
changing language" in which history is written. During the nineteenth century a great deal of research was being done on the subject
of historical linguistics and the genealogy of languages. Darwin drew
on this research in making his argument (Alter 1999).

Darwin sought to strengthen his argument for imperfection in
other ways. In the section of Chapter 9 titled "On the Lapse of Time,"
he argued for the vastness of geological time. What separated him
from the majority of his contemporary geologists was a willingness
to offer a numerical estimate for the time that had elapsed since
the formation of a well-studied portion of the stratigraphical record.
Lyell had supplied a description of the thickness of the stratigraphic
succession of the Wealden Formation of southeast England, but he
had not estimated the time since its formation. Darwin calculated,
using a denudation (erosion) rate of 1 inch (2.54 cms) per century,
that at least 300 million years were required to remove the overlying

deposits it had originally contained. Darwin's purpose in supplying this estimate was not to date the geological record as a whole, much less to assign a date to the age of the Earth, but to suggest "what an infinite number of generations ... must have succeeded each other in the long roll of years!" (*Origin*, 287)

The next section in Chapter 9 is titled "On the Poorness of Our Palaeontological Collections." Darwin argued that wide intervals of time separated several geological formations, and he supplied a set of geological explanations. Areas such as sea floors, deltas, and lakes undergoing continuous (steady) subsidence would be expected to show uninterrupted accumulations of sediment; by contrast, thick deposits could not accumulate in shallow seas in which the underlying crust was stationary. In areas undergoing elevation, the erosive action of the sea would wear away sediment. To this argument regarding the vagaries of sedimentary accumulation (the necessary condition for fossil burial) he added another that was congruent with his ideas concerning the optimal conditions for the formation of new species. He suggested that during the elevation of land "new stations will often be formed," which promote the formation of new varieties and species. Yet clearly and, in a sense, paradoxically, periods of elevation are the ones during which erosion predominates: the rise in land that promotes the creation of new species is logically inimical to their preservation as fossils. He concluded that "Nature may almost be said to have guarded against the frequent discovery of her transitional or linking forms" (*Origin*, 292).

The third section of Chapter 9 is titled "On the Sudden Appearance of Whole Groups of Allied Species." From the point of view adopted by contemporary geologists, this was a specially troubling point. Darwin took two tacks in combating this objection. First, arguing from a theoretical point of view, he suggested that while it might take a long succession of ages for an organism to adapt to some "new and peculiar line of life, for instance to fly through the air, ... a comparatively short time would be necessary to produce many divergent forms, which would be able to spread rapidly and widely throughout the world" (*Origin*, 303). Second, in an argument that would have been more persuasive to geologists, he pointed out several cases where recent discoveries had shown that certain groups of animals (all of the examples he cited were animals) had been demonstrated to be present earlier in the geological record than had previously been known. Mammals were supposedly only to be found in Tertiary rocks, yet Owen (1854) had described a veritable 'fauna' of fossil mammals from the Secondary (Cretaceous) rocks of Dorset; these reinforced the observations of William Buckland (1784–1856) and others on fossil mammal jaws from the Jurassic of Stonesfield (Buckland 1836). Contrariwise, the report of whale fossils from the Upper Greensand by Lyell proved to be entirely mistaken. While Darwin had posited a Tertiary origin for his sessile barnacles, new evidence showed them to exist in Cretaceous (Mesozoic) rocks. Teleost fish supposedly originated in the Chalk period, and yet evidence was accumulating in palaeontological treatises, such as that by François Jules Pictet de la Rive (1809–1872) and others, not only that it was difficult to identify teleosts unambiguously, but that their distribution in time was arguably more ancient than claimed. Having cited these examples, Darwin concluded that given how little of the world had been explored geologically, "it seems to me to be about as rash in us to dogmatize on the succession of organic beings throughout the world, as it would be for a naturalist to land for five minutes on some one barren point in Australia, and then to discuss the number and range of its productions" (*Origin*, 306).

The last section of Chapter 9 is titled "On the Sudden Appearance of Groups of Allied Species in the Lowest Known Fossiliferous Strata." Darwin here placed himself in opposition to the views of the geologist Roderick Murchison (1792–1871), and others like him, who believed that "we see in the organic remains of lowest Silurian stratum the dawn of life on this planet" (*Origin*, 307). Darwin believed that life existed much earlier. For example, he wrote that "I cannot doubt that all the Silurian trilobites have descended from some one crustacean, which must have lived long before the Silurian age, and which probably differed greatly from any known animal" (*Origin*, 306). Why were the remains of these earlier organisms not preserved in the fossil record? Here Darwin admitted he had "no satisfactory answer," but he did allow himself to speculate. Taking note of the fact that "no one oceanic island is as yet known to afford even a remnant of any palaeozoic or secondary formation," he suggested that oceans and continents have not always existed where they do today. Should one then expect to find "palaeozoic" and "secondary" formations on the crust underlying the oceans? Not necessarily, answered Darwin, "for it might well happen that strata which had subsided some miles nearer to the centre of the earth, and which had been pressed on by an enormous weight of superincumbent water, might have undergone far more metamorphic action than strata which have always remained nearer to the surface" (*Origin*, 309). Metamorphism would likely have destroyed the fossil evidence.

Chapter 10 of the *Origin* is titled "On the Geological Succession of Organic Beings." This chapter has a strongly defined rhetorical strategy. In it Darwin sought to show how his view ("slow and gradual modification, through descent and natural selection") explained the empirical facts of what was known of geological succession better than did the "common view of the immutability of species." (*Origin*, 312). His tack is understandable, since it was the facts of geological succession that had in large part led him to espouse transmutation in the first place (Herbert 2005). However, he was aware that few contemporary geologists and palaeontologists were transmutationists, so he trod carefully. Darwin also treated cautiously a common belief that there was a law of development that governed geological succession.

Darwin began by asserting that species, once extinct, never reappear. This may seem an obvious point to a present-day reader, but it was troublesome even to his confidant Lyell. His point was that each species is modified by its history. So, even though physical conditions identical to those existing earlier may return, and even though the new species may "fill the exact place of another species in the economy of nature," "yet the two forms – the old and the new – would not be identically the same; for both would almost certainly inherit different characters from their distinct progenitors" (*Origin*, 315): life was shaped by its history. He came back to this point later in the *Origin* when he stated that the "simplicity of the view that each species was first produced within a single region captivates the mind. He who rejects it, rejects the *vera causa* of ordinary generation" (*Origin*, 352).

In his next section, titled "On Extinction," Darwin continued the subject. After suggesting that the "old notion" of catastrophic destruction of all inhabitants of the Earth at successive periods had been "very generally given up" by palaeontologists, he turned to discuss his own experience in South America (*Origin*, 317–19). Knowing full well that the modern-day horse had been introduced into South America by the Spaniards, he recalled marveling when he "found in La Plata the tooth of a horse embedded with the remains of Mastodon, Megatherium, and other extinct monsters." But, he reminds the reader, his astonishment was utterly groundless, for Owen soon informed him that the horse was not identical to living horses but was an extinct species. (For photographs of two specimens of fossil horse teeth that Darwin collected, see Gardiner 2004.) In Darwin's view, what had happened was simply that the earlier South American horse had become rare to the point of extinction: that was its history. If the horse tooth had come from a presently living species, Darwin could have accommodated that possibility as well (given the existence of modern mollusc shells), though it would have required slightly more complicated reasoning.

More complicated reasoning was indeed required in the next section, titled "On the Forms of Life Changing Almost Simultaneously throughout the World." Darwin did not challenge this notion. Rather, he asserted that "this great fact of the parallel succession of the forms of life throughout the world is explicable on the theory of natural selection." He argued that the "dominant, varying, and far-spreading species, which already have invaded . . . the territories of other species, should be those which would have the best chance of spreading still further." He allowed that diffusion would "be slower with the terrestrial inhabitants of distinct continents than with the marine inhabitants of the continuous seas." (*Origin*, 325–6). Worldwide change in marine species would occur more easily and quickly than worldwide change in terrestrial species.

The next section of Chapter 10 is titled "On the Affinities of Extinct Species to Each Other, and to Living Forms." This is one of the strongest sections in the *Origin*. Darwin opened it by commenting with elegant simplicity that extinct and living species "all fall into one grand natural system; and this fact is at once explained on the principle of descent" (*Origin*, 329). He then invoked for support the views of Buckland, Owen, Cuvier, Pictet de la Rive, and Joachim Barrande (1799–1883) on particular points. Further on in the *Origin* he returned to the theme of "one natural system" when he discussed the subject of classification. He wrote, in Chapter 13, that "the natural system is founded on descent with modification," and that "all true classification is genealogical" (*Origin*, 420).

In Chapter 10, under the heading "On the State of Development of Ancient Forms" Darwin also broached the question of whether the fossil record shows progression. He concluded that "in one particular sense the more recent forms must, on my theory, be higher than the more ancient; for each new species is formed by having had some advantage in the struggle for life over other and preceding forms" (*Origin*, 337). In this section Darwin also addressed the question of whether, as Agassiz thought, "ancient animals resemble to a certain extent the embryos of recent animals of the same classes." Darwin was conciliatory on this point, seeing its potential compatibility with his own views while at the same time suggesting that Agassiz's claim of a parallel between embryological development and geological succession had not yet been proven (*Origin*, 338).

In the last section of Chapter 10, "On the Succession of the Same Types within the Same Areas, during the Later Tertiary Periods," Darwin reiterated his view that the theory of descent with modification explained the similarity of fossil and recent species in the same geographical area, "for the inhabitants of each quarter of the world will obviously tend to leave in that quarter, during the next succeeding period of time, closely allied though in some degree modified descendants" (*Origin*, 340). In this section he also attempted to respond to critics who, in ridicule, had asked whether such present-day South American species as the sloth, armadillo, and anteater were "degenerate" descendants of the "megatherium and other allied huge monsters" (*Origin*, 341). His answer was an emphatic "no," for he was skeptical concerning laws of either development or devolution. To explain the living South American species he appealed to the existence of collateral lines of extinct animals, more similar to current species than were the "huge monsters" of the past.

Thus concludes Darwin's treatment of geology per se. However, in Chapters 11 and 12 on geographical distribution many of the same themes recur. For example, since Darwin insisted that each species had come into existence only once, the question of means of dispersal was key for him; that subject is fully treated in Chapters 11 and 12. Darwin also gave considerable attention in those chapters to discussion of "Dispersal during the Glacial Period" (*Origin*, 365–82). In summary, we can say that in the first edition of the *Origin* Darwin was on the defensive because of the evident paucity of the fossil record in the 1850s. He used generalised arguments that probed a fundamental issue, which was whether geological history better supported the view that species were immutable or the view that species exhibited "slow and gradual modification." For the most part he relied on the changing opinions (initially anti- but increasingly pro-transformation) of well-regarded authorities such as Lyell. Making the general observation that rocks of more recent age contained fossil species still recognisable as living today, and that the further back in time one went the larger the percentage of extinct species, he nevertheless strove, wherever possible, to demonstrate that there were no developmental laws governing species duration.

This view was integrated with Darwin's more general perception of extinction. Contrary to the serial universal catastrophe (mass-extinction) claims of Jean Baptiste Élie de Beaumont (1798–1874), Murchison, and Barrande (and before them of Cuvier and Buckland), it was increasingly being seen that species and groups of species disappeared gradually and intermittently across long periods of time and that their individual durations varied enormously. Although he was careful to report the "sudden extermination" of some large and widespread groups such as the trilobites at the end of the "palaeozoic period" (Paleozoic Era) and the ammonites at the close of the "secondary period" (Mesozoic Era), Darwin's personal experience of the fact of extinction related to his fossil mammals of South America: notably, the loss of the horse, which prompted him to consider the many and varied influences (lost to our knowledge today) that might have promoted its decline on that continent but not elsewhere. In this way extinction became for Darwin the norm, a necessary counterpoint in the "struggle for existence" to the continuous origination of new species.

If one were to translate his ideas into present-day terms, Darwin imagined new terrestrial species arising relatively quickly, particularly in suitable areas, such as those undergoing elevation. At the other end of the scale, he expected that species would die out comparatively more slowly. This process is similar to what would today be termed a 'clade spindle': each species' duration being depicted as a fusiform shape oriented vertically in space, indicating its origin at a point in time, with a phase of rapidly increasing abundance until it reaches a maximum before dwindling to its extinction.

After Publication

The influence of the publication of the *Origin* on the fields of geology and palaeontology was profound and immediate, and Darwin's response correspondingly nimble. Darwin's estimate of the time necessary for the denudation (erosion) of the Weald was criticized sharply. To take only one example, the Oxford palaeontologist John Phillips (1800–1874) reviewed the *Origin* in his February 17, 1860, presidential address to the Geological Society of London. He accused Darwin of "abuse of arithmetic" in his calculation of a figure for the denudation of the Weald and suggested alternative computations. Darwin responded by reducing the period for the denudation of the Weald from 300 million years to between 100 and 150 million years. He then dropped the estimate entirely from the third and subsequent editions (Herbert 2005; Morrell 2005). While Darwin's original estimate did not survive, it did promote attention to the question of absolute geological time and the duration of the stratigraphical succession.

As he revised the *Origin* Darwin also added new material reflecting progress within the field of geology. One area that had been an early difficulty for him was glacial theory. For this reason he tracked the issue closely, being particularly attentive to the theories being put forward by James Croll (1821–1890). Croll posited that changes in the Earth's orbit had been the cause of glacial periods. While Darwin had always left the door open to extraterrestrial influences, his usual focus, like that of Lyell, had been the forces operating within an Earth-bound system. By the 1860s, glacial theory required a step away from that focus, and Darwin incorporated Croll's ideas into the fifth edition of the *Origin* (*Variorum*, 593 [215.2:e]; *Correspondence*, 16, part II: 873–6).

Yet of all the geological subjects contained in the *Origin*, palaeontology was the most important. And here even Darwin's supporters had qualms. Speaking as an evolutionist, Thomas Henry Huxley (1825–1895) noted the intervals or gaps between " very distinct groups: – Insects are widely different from Fish – Fish from Reptiles – Reptiles from Mammals – and so on." He asked why the gaps exist and where might be the connecting forms:

Among the innumerable fossils of all ages which exist, we are asked to point to those which constitute such connecting forms. Our reply to this request is, in most cases, an admission that such forms are not forthcoming, and we account for this failure of the needful evidence by the known imperfection of the geological record. We say that the series of formations with which we are acquainted is but a small fraction of those which have existed, and that between those which we know there are great breaks and gaps. I believe that these excuses have very great force; but I cannot smother the uncomfortable feeling that they are excuses. (Huxley 1868, 358)

From Huxley's point of view the most satisfactory way to relieve that "uncomfortable feeling" was to fill some of those gaps, which, indeed, was the point of his article, titled "On the Animals which Are Nearly Intermediate between Birds and Reptiles." The centre-piece of Huxley's article was *Archaeopteryx*, which he presented as a candidate for a group linking birds and reptiles.

In the third (1861) edition of the *Origin* Darwin had taken up the question of bird origins. Birds had been thought to be early Tertiary (Eocene) in origin, but Owen had recently identified bird fossils in the Upper Greensand (Mesozoic Era). (Subsequently these bones were shown to be the remains of flying reptiles [pterosaurs].) Darwin was also able to refer to evidence from publications by Edward Hitchcock (1793–1864) on the supposed bird footprints from the much more ancient (New Red Sandstone) Connecticut Valley. By the fourth edition (1866) Darwin was able to add, in the paragraph on bird origins, the evidence again provided by Owen (1863) of the ancient (Jurassic Period) "strange bird" *Archaeopteryx*, noting particularly the presence of claws on its wings and a long lizard-like tail. In the fifth (1869) edition of the *Origin* Darwin called attention to Huxley's findings linking "on the one hand, the ostrich and the extinct Archeopteryx, and on the other hand, the Compsognathus, one of the Dinosaurians" (*Variorum*, 509 [194.1:*b*, 194.4:*b*, 194.4:*d*], 540 [137.2:*e*, 137.2:*f*]).

The case of the bird *Archaeopteryx* and the allied dinosaur *Compsognathus* compels interest because of the timing of the discoveries in relation to the publication of the *Origin* and the way it interweaves the views of Owen, Darwin, and Huxley. Building on work by earlier researchers, Owen had 'invented' dinosaurs in 1842 as very large, quadrupedal pachyderm-like reptiles (the zenith of vertebrate form during the "secondary period" [Mesozoic Era]) and created a lasting image of these as life-sized models in 1854 at the Crystal Palace Park in London (Delair and Sarjeant 1975; Desmond 1975, 1979). *Compsognathus* was an almost complete, chicken-sized skeleton discovered, in somewhat mysterious circumstances, in the mid-1850s in a lithographic limestone quarry near Solnhofen in

Bavaria. The dinosaur was named by Andreas Wagner (1797–1861) in 1859. The celebrated remains of *Archaeopteryx* were found in a nearby quarry in the same limestone formation; the first discovery (1860) was a feather impression, and a year later a partial skeleton, with associated feather impressions, was recovered. This was purchased by the British Museum in 1862 and studied, with remarkable speed, by Owen – the specimen arrived on October 1, 1862, and its description was received by the Royal Society on November 6, 1862 (de Beer 1954). Owen declared it to be a fossil bird in almost all its anatomical attributes (Owen 1863), but one that differed from living birds in two unusual features: a long, bony tail (more reminiscent [to Owen's eyes] of the unfused embryonic tails of living birds) and claws on its wings (explained as a relatively common variation, given that digits are found on the wings of bats, flying lemurs, and flying reptiles). During the 1860s, Huxley became interested in dinosaurs, birds, and *Archaeopteryx*. In his haste to describe *Archaeopteryx*, Owen had made a rather fundamental error in mistaking the upper for the lower side of the animal (allowing Huxley some simple point scoring). More crucially, his investigations of the anatomy of some well-known dinosaurs (*Megalosaurus* and *Iguanodon*) began to reveal some equally fundamental mistakes in the identification of the pelvic bones, which took on a decidedly avian form; these observations, linked with the relatively complete but small and superficially bird-like *Compsognathus*, provided Huxley with an argument that responded to the accusations of Darwin's detractors concerning the absence of intermediary forms in the fossil record.

Bird anatomy was highly specialised, unique, and very different from that of reptiles; this justified their separation into different vertebrate classes (Aves and Reptilia). Owen had already demonstrated that *Archaeopteryx* exhibited a range of uniquely avian features. However, in Huxley's opinion the long bony tail and separate bones and claws on its hands represented reptilian traits (rather than embryonic avian traits or functional convergences). Furthermore, when Huxley compared the anatomy of known large, land birds (ostriches) to that of dinosaurs he was able to demonstrate a surprising number of similarities (notably in the detailed structure of the sacrum, pelvis, and hind limbs); and, added to this, observations of the most bird-like (and anatomically complete) of the known dinosaurs, *Compsognathus*, seemed completely to accord with this novel view.

Alluding to the Triassic footprints of bird-like animals from Connecticut described by Hitchcock in 1858, Huxley suggested that during the "secondary period" (Mesozoic Era) there must have existed a range of bird-like dinosaurs and dinosaur-like birds that formed a range of evolutionary intermediates between two great (and apparently widely separated) classes of vertebrates. In Huxley's words:

It can hardly be doubted that a lithographic slate of Triassic age would yield birds so much more reptilian than *Archaeopteryx*, and reptiles so much more ornithic [bird-like] than *Compsognathus*, as to obliterate completely the gap which they still leave between reptiles and birds. (Huxley 1868, 365)

Discoveries over the latter decades of the nineteenth century mostly confirmed Huxley's predictions: several dinosaurs were discovered that exhibited very bird-like characteristics. There was, however, a prolonged period (the 1920s to the 1970s) following the publication of a detailed review of the subject by Gerhard Heilmann during which dinosaur-bird (ancestor-descendant) affinities were doubted on the basis of some anatomical inconsistencies. It was not until the description by John Ostrom in the 1960s of *Deinonychus*, another remarkably bird-like theropod dinosaur (Ostrom 1969) that Huxley's theory was resuscitated (Ostrom 1976). Since then, Huxley's perceptive commentary has been amply confirmed by the discoveries in China since 1997 of a remarkable diversity of filament-covered and in some instances genuinely feathered dinosaurs. These new discoveries have partially fulfilled his goal of finding "reptiles so much more ornithic."

In summary, the six editions of the *Origin* demonstrate that Darwin was deeply involved with geology and the closely allied field of palaeontology. This book lay down a gauntlet that challenged and ultimately transformed both fields of research.

9 Geographical Distribution in the *Origin of Species*

I. GEOGRAPHY AND ITS IMPLICATIONS

In 1845, Darwin wrote to Joseph Dalton Hooker predicting that he (Hooker) would soon be recognized as the first authority in Europe on "that grand subject, that almost key-stone of the laws of creation: Geographical Distribution."[1] Darwin had already included a substantial section on the topic in his "Essay" of 1844 and would interact extensively with Hooker on the topic over the next decade and more. The *Origin of Species* itself contains two chapters on distribution, occupying sixty-four pages, or just over 13 percent of the text of the first edition. These two chapters would be modified on minor points in the subsequent editions, but would remain essentially intact, serving as one of the main lines of support for his theory. Since they include his discussion of the distribution of species on the Galápagos Islands, they represent a key link in the process by which the theory was developed and then presented to the public.

It has always been recognized that the study of how species are located around the globe, on both a small and a large scale, was a key line of evidence leading Darwin toward the theory of common descent. The idea that divergence from a common ancestor was generated by a process that adapted populations to changes in their local environments (physical or organic) would eventually serve as one of his most convincing arguments in favour of natural selection. Coupled with the proposal of plausible hypotheses about how members of a species could be transported from one area to another, the theory

---

[1] Darwin to Hooker, February 10, 1845, in *Correspondence*, 3: 140.

would account for a host of facts about distribution that had become increasingly puzzling as more evidence was built up by exploration of remote regions of the globe.

But in order to understand Darwin's arguments on the topic, we need to be clear about his intentions. More specifically, we need to recognize how limited those intentions were – in a sense, to realize what he was *not* trying to do. Later in the century, evolutionists would incorporate geographical evidence into their efforts to reconstruct the history of life on Earth. In the 1870s, Alfred Russel Wallace, who had also been inspired by the topic to become a transmutationist, compiled a massive two-volume survey, *The Geographical Distribution of Animals*, backed up by a later volume on *Island Life*.[2] Over the next few decades there was a flurry of interest from scientists who used the geographical distribution of living and fossil species to indentify the points of origin, and subsequent migrations and extinctions, of all the major groups of animals and plants.[3] But, as with his analysis of the fossil record in the *Origin*, this was not Darwin's project. He doubted that the fossil evidence would ever be sufficient to allow a complete reconstruction of the course of life's development. He discussed the fossil record in order to head off claims that its discontinuity counted against evolution, and to show that when trends could occasionally be identified, they matched the predictions of his theory.

The point I want to drive home is that these are exactly parallel to the arguments he would apply to the evidence from geographical distribution. There was to be no attempt to reconstruct the origins and migrations of all the major groups, no comprehensive survey of the geography of life. What Darwin did was to choose a small number of examples of distribution about which enough evidence had become available, and to show that the patterns made perfectly good sense in terms of his own theory but no sense at all on the basis of special creation. In particular, he was concerned to refute the idea that to explain cases where the same species existed in two or more widely separate regions, it was necessary to invoke the multiple creation of identical parent forms. Darwin's case was that as we gain more knowledge of the processes by which living things

[2] Wallace (1876, 1880).
[3] Outlined in Bowler (1996), Chapter 8.

can move around the globe, we see how such cases can always be explained by dispersal rather than by multiple creation. In order to do this, he was willing to invoke limited geographical changes in the course of geological time. But even when postulating geographical and climatic changes on a global scale, he always focused on specific test cases. He did not think it would be possible, in practice, to make a complete reconstruction of the history of life on Earth. The vast project to undertake such a reconstruction, inspired by Ernst Haeckel and eventually applied to biogeography by Wallace and his successors, was not part of Darwin's research program.

II. DARWIN'S EARLY IDEAS

This is not the place for a detailed study of Darwin's response to the discoveries on the Galápagos Islands or the other biogeographical insights he gained while on the *Beagle* voyage. But once he had formulated his theory of natural selection in the late 1830s, he immediately began to use case studies from biogeography as part of the argument he would construct to defend the theory. In its initial form, as displayed for instance in the "Essay" of 1844, the theory incorporated several points that Darwin would eventually abandon before writing the *Origin*. Janet Browne has nicely described these developments, so they will only be summarized briefly here.[4] There were two key changes. First, Darwin gradually abandoned his early enthusiasm for massive elevations and depressions of the Earth's surface in the course of geological time, which had initially led him to think that whole continents could be submerged beneath the sea and then re-elevated. By the time he wrote the *Origin*, Darwin was convinced that the present distribution of the continents and oceans was permanent, at least as far back as the Mesozoic. And secondly – but only in the last years before he was precipitated into writing the *Origin* – he developed his "principle of divergence," which led him to place much less emphasis on the need for geographical dispersal on islands to initiate speciation. The Galápagos situation would become only a special case of divergence, not a model for how it normally took place. By 1859, Darwin thought that the splitting apart of species normally took place on large landmasses.

[4] See Browne (1983), Chapter 8; also Browne (1995), pp. 514–21.

Chapter 6 of the 1844 "Essay" offers several classes of phenomena for analysis. As Darwin eventually makes clear, his purpose in discussing these areas is to show that they all display unexpected features that the creationist (and he uses that term) has to accept as "just ultimate facts" for which there is no rational explanation. But surely, he argues, we should be searching for general laws, and that is what he eventually moves on to do.[5]

The first topic of interest is the distribution of the inhabitants of the different continents.[6] Here he notes the efforts that had been made to define biogeographical regions – five is his preferred number – and that these are defined by the major geographical barriers, especially deep oceans where the seabed is unlikely ever to have been raised above the surface. He notes that tropical Africa and South America have very different inhabitants, despite exhibiting the same range of physical environments. This is not what we would expect if species were designed by a supernatural Creator to be perfectly adapted to their environments. Later in the chapter he points out that these characteristic differences can be seen in the fossils unearthed in the same regions. Even more strikingly, how can the creationist explain why marsupials are found only in Australia and South America? His interest in these striking differences in the distribution of species was, of course, stimulated by his own experiences on the *Beagle* voyage.

The *Beagle* discoveries also come to the fore in his next topic: insular floras and faunas. If species are separately created, why is it that the inhabitants of oceanic islands always resemble those of the nearest mainland? Why are the Galápagos species characteristically South American, and why are there different but related species on many of the individual islands?[7] A parallel phenomenon can be found in the case of floras of mountainous regions (alpine floras), which are often identical even on widely separated peaks. Curiously, the same species are often found far to the north in Arctic regions. In other cases, similar species can be found in places widely scattered and separated by oceans. Darwin's explanation of these anomalous distributions closely follows the one that would be published by

---

[5] Darwin and Wallace (1958), pp. 192–4.
[6] Ibid., pp. 169–75.
[7] Ibid., p. 176.

Edward Forbes in 1846.[8] Inspired (like Darwin) by Charles Lyell's uniformitarian geology, Forbes appealed to the evidence of a period of intense cold in the recent past (which would become known as the Ice Age) to argue that during such a period Arctic plants would migrate south. When the climate warmed again, they would retreat northward, but would also move to higher elevations in mountainous regions of the southern localities they had temporarily occupied. These isolated populations would end up stranded on the peaks, like islands separated by a sea of lowland in between. Darwin also invoked icebergs as a possible means by which Arctic plants could be transported from the North to isolated areas in the Southern Hemisphere.

Darwin then reveals why he is so interested in these relationships. Because there seemed no prospect of these species having migrated from one place to another across the hostile lowlands in between, the phenomenon of disjunct species was widely taken as the best evidence not only of divine creation, but also of the fact that the same species could be created in more than one location.[9] Darwin's extension of the Ice Age theory has thus provided a clear alternative to the creationist position, which he reinforces by pointing out our extreme ignorance of the means by which species can be dispersed – which might conceal other possible explanations of anomalous distributions. He also notes that mammals, which can obviously migrate more readily than plants, do not show the same anomalies, although on the theory of multiple creations there is no reason why they too should not be created in more than one location. All the evidence, he argues, points to each species having a single point of origin and then spreading out to occupy whatever territory it can gain access to. Only prior occupation by another species exploiting the same resources might prevent a newcomer from successfully colonizing new territory.

Darwin has, in effect, called in the gradual modifications of climate postulated by Lyell's uniformitarian geology, coupled with

---

[8] Ibid., pp. 180–1. On the similarity between Darwin's and Forbes's views, see Browne (1983), p. 121, and Bowler (1992), pp. 275–80.

[9] Darwin and Wallace (1958), pp. 185–6. Louis Agassiz was widely recognized as the most active exponent of the idea of multiple creations, although Darwin does not name him.

the hope of extending our knowledge of dispersal mechanisms, to explain the anomalies of distribution that must remain just brute facts to the creationist. As noted earlier, however, Janet Browne has pointed out that some aspects of his explanation differ considerably from what he would later offer in the *Origin*. Although Darwin already appreciated the significance of deep oceans as more or less permanent barriers to migration (and hence as defining the main biogeographical regions), at this point he still believed it was possible for geological forces to produce gradual, but significant, changes in the elevation of any point on the Earth's crust. The theory he offered in 1839 to explain the "parallel roads of Glen Roy" in Scotland assumed that these ancient beaches were formed at a time when Europe was temporarily submerged to a large extent beneath the ocean.[10] He was also convinced, in part by the example of the Galápagos fauna, that the most favourable circumstances for speciation (the splitting up of one species into a number of related 'daughter' species) occurred when a population was divided up among isolated islands. At this point in his career, he was convinced that adaptive variations would appear only very rarely. Although there would be a better chance of such favoured individuals appearing in a widely distributed population, they would have to interbreed with a large number of unchanged individuals, and this would effectively swamp the innovation. In a small island population, on the other hand, the rare individual variant's effect would be concentrated by inbreeding and would thus more readily cause the population to diverge from the parent type.

As a consequence of these assumptions, in 1844 Darwin put forward a theory in which both the production and the dispersal of new species depended on slow but constant changes in the land surface brought about by geological forces.[11] He envisaged continental areas being alternately lowered beneath the sea and then raised again. In the course of submersion, the continent would slowly be split up into a number of isolated islands, which would gradually diminish in size and be split up further. When the once-widespread population was broken up among many islands, each subpopulation would adapt

---

[10] On Darwin's Glen Roy theory and his subsequent acceptance that the Ice Age provided a better explanation, see Martin J. S. Rudwick (1974).

[11] Darwin and Wallace (1958), pp. 195–203.

to the new conditions in its own way, and there would be extensive speciation. When the continent once again emerged from beneath the sea, the islands would gradually merge together, and the species originally confined to them would disperse and come into conflict with one another. Eventually, the most successful of them would occupy wide areas, as we find on the continents today. But once this equilibrium was reached, there would be little further change until the next episode of submergence. Evolution thus depended on submergence and fragmentation of the land surface, while dispersal depended on the subsequent opening up of land connections to allow outward migration.

III. DEVELOPMENTS BEFORE THE *ORIGIN*

Over the next decade, Darwin worked on topics that he knew would provide additional support for his theory. These included gathering information about geographical distribution and the means by which various types of organism could be distributed from one region to another. He interacted with a number of fellow naturalists who were interested in these topics, especially if they were prepared to follow the Lyellian model of dispersal coupled with gradual geographical and climatic changes. On botanical matters, he formed a close link with Joseph Dalton Hooker, who – as noted in the letter quoted earlier – he anticipated would become the leading European expert on distribution. Asa Gay would eventually fill the equivalent role in America, and both men would eventually be informed of the radical nature of Darwin's views on species. Alfred Russel Wallace would also interact with Darwin, although not in a manner that allowed Darwin to realize the full extent to which their views were moving in the same direction.

Eventually Darwin began to write his "Big Book" on species, which would eventually be cut down to form the *Origin of Species*. This too contained a section on geographical distribution, the manuscript of which was sent to Hooker for comment in 1856.[12] In this chapter Darwin was mainly concerned to outline and extend

---

[12] *Species Book*, Chapter 11. For the editor, Robert Stauffer's, analysis, see pp. 528–35. Darwin evidently intended to write another chapter on the topic, but this either was not produced or has not survived.

his account of the dispersal of Arctic forms during the Ice Age, turning it, in effect, into a global theory centered on the possibilities of dispersal during this period of anomalously cold climatic conditions. He comments on the uniformity of Arctic species around the high latitudes of the Northern Hemisphere, something made possible by the close connection between Eurasia and North America. He even seems to accept Edward Forbes's suggestion that there might once have been a land connection across the North Atlantic.[13] He also endorses Forbes's idea that much of Europe was submerged beneath the sea at some point during the period of cold conditions, plants being transported from one island to another by icebergs. The islands, of course, would become mountains on subsequent re-emergence of the land.

Darwin also accepts the possibility that – assuming both hemispheres endured cold conditions at the same time – icebergs could transport plants or their seeds across the equator to establish populations of northern plants in remote areas of the South, such as Tierra del Fuego and Kerguelen Island. His concluding remarks make it clear that his main target is still the theory of the independent creation of these isolated populations. He argues that the possibility of migration (especially given our ignorance of the modes of transportation) is strong enough to undermine the case for multiple creations in distinct areas. If one could identify two identical populations existing in regions where there was no possibility of migration, then "the whole of this volume would be useless & we should be compelled to admit the truth of the common view of <absolute> actual creation."[14]

Hooker did not think the transport of plants by icebergs very plausible (he had observed icebergs firsthand in his own Antarctic travels). He was more inclined to propose an extension of the land surface in the distant past, generating 'land bridges' across which plants could migrate by more traditional means.[15] He certainly felt that more information was needed on means of dispersal, unaware

---

[13] Ibid., pp. 536–8.
[14] Ibid., p. 566.
[15] Hooker to Darwin, August 4, 1856, and August 7, 1856, in *Correspondence*, 6: 198–200, 203–204. Hooker would continue to defend the idea of a great Antarctic continent through the 1880s; see Bowler (1996), p. 413.

that Darwin was working on such a project. In a detailed response to Darwin's 1856 manuscript, Hooker confessed that he had been impressed by the evidence for change in the natural world and had "never felt so shaky about species before."[16] He suggested many ideas on how to extend the theory of migration by bringing in factors such as the direction of ocean currents (he was now less opposed to the possibility of transport by icebergs).

As Janet Browne points out, by the time he wrote the *Origin of Species*, Darwin's ideas on speciation and dispersal would be very different from those expressed in 1844 and 1856.[17] He was becoming convinced that the geographical breaking up of a population onto separate islands was not essential for speciation. Islands certainly did generate species, of course, as in the case of the Galápagos, but he now conceived his 'principle of divergence' according to which the competition for resources within a single continuous habitat would produce diverse forms that could exploit the environment in different ways. Evolution could thus occur on a continental land mass, and indeed would take place more readily there than on islands.

At the same time, Darwin was becoming increasingly uncomfortable with his earlier assumption that geological changes had produced drastic elevations and depressions of the Earth's surface. It was unlikely that Europe had been largely sunk beneath the ocean as recently as the last Ice Age (thus undermining his Glen Roy theory) and equally unlikely that areas of what is now ocean had been elevated to form 'land bridges' between the existing continents in the recent geological past. This threw increasing emphasis onto the means of dispersal of land animals and plants across the oceans, and if icebergs were not a plausible mechanism of transport, it would be necessary to investigate what alternative means might exist. By the time he wrote the *Origin*, Darwin had become far more aware of the possibility that apparently unlikely processes could, in fact, aid dispersal and thus be used to explain how a species evolved in one area could spread to distant locations. He now had a theory that depended less on environmental change to drive speciation, but still appealed to the possibility of different conditions in the past to explain distribution. Another naturalist working along similar lines

---

[16] Hooker to Darwin, November 9, 1856, in *Correspondence*, 6: 259–64.
[17] Browne (1980, 1983).

was Alfred Russel Wallace, and in 1858 Wallace's famous paper triggered Darwin's rush to complete the single-volume account of his new theory that became the *Origin of Species*.

IV. GEOGRAPHICAL DISTRIBUTION IN THE FIRST EDITION OF THE *ORIGIN*

The two chapters on distribution in the *Origin* (Chapters 11 and 12 in the first edition) pick up on themes explored in the 1844 "Essay" but explain Darwin's new approach to the topic. The basic aim remains the same, however: to challenge the evidence for divine creation by showing that there are phenomena that cannot be explained on that basis, but that can be understood on the assumption that species evolve in a single location and then disperse to other localities. Darwin begins once again by noting the differences between the inhabitants of the various continents, especially between those of Africa and South America, even though they often exhibit exactly the same range of habitats. It is the geographical barriers that define the regions, with the oceans being the most permanent – although other barriers, including rivers and deserts, can also be effective on a shorter time scale. He also notes the affinity between the inhabitants of oceanic islands and those of the nearest continent. There is "some deep organic bond, prevailing throughout space and time, over the same areas of land and water, independent of their physical conditions."[18] This bond is simply inheritance, the only thing we positively know can produce such affinities. Darwin's point, of course, is that divine creation offers no such explanation because there is no reason why miracles should be constrained by geographical barriers. Later on in the chapter he repeats the example of oceanic islands and notes that Wallace has postulated (in his 1855 paper) that species always appear in a region where there is a preexisting allied species.[19]

In a significant modification to the explanation offered in 1844, though, Darwin stresses the role of competition in determining which species will spread out most successfully to occupy neighboring territory. Wide-ranging species that have triumphed over rivals

[18] Darwin (1859), p. 350.
[19] Ibid., p. 355.

will have the best chance of occupying any new territory they can migrate to, although even they may be checked if another species adapted to the same mode of existence is already in occupation. He emphasizes that there is no law of necessary development, so species need not change if they move to a new location with a similar environment, especially if associated species all move together so as to preserve what we now call the ecosystem.

Darwin insists that all the species within a genus must have come from a single progenitor. By the same token, scattered populations of the same species must have been derived originally from a single location – it is inconceivable that identical individuals could have been derived from parents specifically distinct. Naturalists have long discussed the question of "single centres of supposed creation," and although there are many difficulties to explain on the hypothesis of migration, the simplicity of the idea of production in a single region "captivates the mind."[20] To reject this idea is to reject the *vera causa* of ordinary generation and migration and to call in the agency of a miracle. The observation that land animals are confined to territories defined by ocean barriers while plants are not is explained by the fact that the animals cannot cross those barriers, while there are mechanisms by which plants can occasionally be transported across stretches of open sea. Darwin admits that there are anomalies that at first sight will seem difficult to explain, but he argues that they are not sufficient to cause us to give up the idea of dispersal, especially when we understand how complex are the processes that might play an occasional role in transporting living things across barriers.

There follows an extensive description of the evidence Darwin has now accumulated to show how many different mechanisms can contribute to the process by which individuals may occasionally cross what might have been thought to be impassable barriers. He stresses that what appear as barriers now may not have been insuperable under earlier, quite different conditions. But he is now very critical of Forbes and those naturalists who have "hypothetically bridged over every ocean."[21] They have no authority, he insists, for invoking such changes on this scale. There is no evidence that the continents have been united in the recent past, and indeed the

---

[20] Ibid., p. 352.
[21] Ibid., p. 357.

absence of many mainland forms on oceanic islands counts against this. Instead, he describes the many accidental (or better, occasional) means of dispersal available to species. In particular, he recounts his own and other naturalists' observations and experiments on how seeds, eggs, and even small insects can be carried across the sea. In some cases, seeds can still germinate after weeks of immersion in seawater, while floating vegetation and icebergs can transport vegetation across long distances, provided the currents are favourable. Insects can be carried for many miles out to sea on the wind, as can birds – and the latter can carry all manner of seeds and even minute land creatures in the mud attached to their feet.

Darwin now moves on to his theory of the dispersal of Arctic plants during the glacial epoch, offering this as an alternative to the theory of multiple creations to explain the disjunct populations of alpine species. Far more than in 1844, though, he develops this idea on a global scale to explain both the uniformity of Arctic plants in the Northern Hemisphere and the occasional examples of Northern Hemisphere plants that occur in isolated regions of the South. The plants of Arctic Eurasia and America may have been able to mingle during an earlier period having a warmer climate, when they lived closer to the pole, where the land surface is more continuous. They would then have migrated southward as the Earth cooled, reaching their present, more scattered locations. The northern plants found in the South may have been able to migrate across the equator during such a period of cold, using high ground where the conditions were cooler and drier, as a 'bridge'. At this point, Darwin assumes that both hemispheres would have experienced cold conditions at the same time.

Darwin argues that only the most successful northern species would be hardy enough to make the crossing, and he discusses at some length the question of why some northern forms have moved into the Southern Hemisphere, while very few species have moved in the opposite direction. The answer to this lies in the greater land surface available in the North, which has stimulated competition and diversity and led to the northern species being, in effect, more aggressive. Darwin thus joins the long tradition of naturalists (Hooker included) who invoked the metaphors of imperialism to describe the migrations of animals and plants. Any southern forms that tried to move northward would have "yielded to the more dominant forms, generated in the larger areas and more efficient workshops of the north."[22] The chapter ends with another metaphor, borrowed from Lyell, that compares the northern forms now stranded on mountains around the globe to "savage races of man, driven up and surviving in the mountain-fastnesses of almost every land, which serve as a record, full of interest to us, of the former inhabitants of the surrounding lowlands."[23]

In the next chapter, Chapter 12 of the first edition, Darwin first addresses the distribution of freshwater life, which often exhibits surprising uniformity over large areas. This is due, he believes, to the ease with which these organisms can migrate in a step-by-step fashion, each individual move covering a short distance. Local changes in the land surface are continually modifying the course of rivers, allowing them to merge and then separate. Darwin gives further details of experiments and observations on how seeds, eggs, and so forth can be transported on the feet of wading birds and ducks.

The next topic is oceanic islands, and here, as in 1844, the Galápagos Islands figure prominently. Islands remote from the mainland characteristically lack certain kinds of animal, especially those that would find it difficult to get transported accidentally across the ocean. Amphibians, for instance, are always absent even though there are conditions suitable for them – and amphibian adults and eggs are killed by even a short immersion in salt water. As Darwin says: "But why, on the theory of creation, they should not have been created there, it would be very difficult to explain."[24] One characteristic of remote islands (as opposed to those close to a mainland) is the high proportion of endemic species, that is, of species that are found there and nowhere else. Yet the species are always closely related to those found on the nearest mainland. On the Galápagos, for instance, "almost every product of the land and water bears the unmistakeable stamp of the American continent."[25] Why should this be so, on the theory of creation? The conditions on the island are unlike any found on the mainland, yet the species show a close

---

[22] Ibid., p. 380. On the use of imperialist metaphors in biogeography and related areas, see Bowler (1996), Chapter 9.

[23] Darwin (1859), p. 382.

[24] Ibid., p. 392.

[25] Ibid., p. 398.

relationship. The answer is, Darwin argues, that the island populations are established by occasional migration from the mainland, but since continued restocking by mainland forms cannot occur, the island populations have diverged and become distinct species.

The same argument can be applied to the differences between the species on individual islands. Although physical conditions on the islands are similar, new arrivals will not necessarily face the same organic environment, since they may have been preceded by other species that have already become established – and these will be different on each island because of the haphazard nature of the transportation process. There will be no regular communication between the island populations, because although they are not far apart, there are strong currents in the local waters, and gales of wind are rare. So each population evolves in isolation and becomes a distinct species, which only rarely spreads to another island.[26]

In conclusion, Darwin notes that wide-ranging genera have wide-ranging species, which is what we would expect if speciation follows dispersal to separate localities. The most wide-ranging (and hence successful) species will have the best chance of being transported elsewhere and establishing themselves in new locations. In general, it is the most lowly organized forms that are the most widely distributed. This is because they are both more ancient, and hence have had more time to spread, and less likely to change, thus preserving ancestral relationships over long periods. All of the facts cited in this and the preceding chapters "are, I think, utterly inexplicable on the ordinary view of the independent creation of each species, but are explicable on the view of colonisation from the nearest and readiest source, together with the subsequent modification and better adaptation of the colonists to their new homes."[27]

Darwin's purpose was the same in 1844, in 1856, and again in 1859. He was not trying to suggest that evolutionism should become the basis for an attempt to reconstruct the whole history of life on Earth. Indeed, he doubts that there is, or perhaps ever will be, enough evidence to allow a detailed understanding of the ancestry and original home of every major branch of the tree of life. His purpose in surveying the fossil record and the geographical distribution of life is the

[26] Ibid., pp. 401–3.
[27] Ibid., p. 406.

far more modest one of undermining the theory of special creation and suggesting that his own theory provides a better understanding of what we do know about the distribution of organisms in space and time. In the case of geographical distribution, he selects a few case studies where he can identify known patterns that make no sense at all in terms of special creation, but that fit quite naturally into his own system provided we make reasonable allowances for what may have happened in the geological past and for our ignorance of all the possible means of transportation. It is particularly important for him to undermine the case for the multiple creation of species in separate locations, and here the plausibility of the case for migration at some past time is crucial. Darwin's argument is as much negative as positive: it is an attack on the plausibility of the creation theory coupled with the suggestion that – whatever the difficulties of some individual cases – descent with modification provides a reasonable explanation of the phenomena that are so damaging for the rival theory.

The closest Darwin comes to anything like a global theory is his use of the Ice Age to explain how northern plants could reach various locations in the Southern Hemisphere. But although this argument makes use of long-range geological and climatic changes to suggest how migration routes could have opened up in the past, its purpose is still restricted to providing an alternative to the theory of multiple creations. Darwin is not trying to explain the origins of the northern plants, except in the sense that he supposes northern forms to be more vigorous than southern ones because of the more active competition they will have faced. His target is merely another case of disjunct distribution that might otherwise give comfort to the creationists. He is not trying to lay the foundations for a complete phylogeny of the plant kingdom.

V. LATER DEVELOPMENTS

The two chapters on geographical distribution underwent minor modifications in later editions of the *Origin*, but remained essentially the same in terms of their main arguments (they become Chapters 12 and 13 in the sixth edition following the insertion of the new Chapter 7 on further objections to the theory). Most of the minor changes involve the addition of new information on the locations of species and the various means of transportation postulated. Darwin continued to argue with Hooker, who published an important paper on the distribution of Arctic plants in 1861.[28]

The only major change to Darwin's arguments came in the fifth edition of 1869, where he revised his explanation of how northern plants might have crossed the equator to colonize locations in the Southern Hemisphere. This followed his acceptance of James Croll's theory of Ice Age causation, which entailed a reconsideration of the assumption that the colder conditions had affected both hemispheres at the same time. In a paper published in 1864, and later in his book *Climate and Time*, Croll developed an astronomical explanation of the Ice Ages based on variations in the amount of the sun's heat reaching the Earth's surface.[29] Because this theory depended on changes in the inclination of the Earth's axis of rotation, it required that the cooling of one hemisphere should coincide with a warmer climate in the other. If the theory were accepted – and it was certainly seen as a promising initiative – Darwin's original assumption that the cooling would have affected the whole globe would have to be abandoned. It is clear from Darwin's letters to Croll, however, that he was far from happy with his original explanation of how northern plants could have crossed the equator at a time of general cooling. He would be much happier with an explanation based on cooling in only one hemisphere – indeed, he told Croll, "it would have been an immense relief to my mind if I could have assumed that this had been the case." He subsequently wrote that Croll's results were "of more use to me than, I think, any other set of papers which I remember" and suggested that the facts of distribution were strong evidence in favour of his theory.[30]

In his fifth edition, Darwin introduced a brief description of Croll's theory and then considered its implications for his own views. He now argued that during an Ice Age in the North, some of the most vigorous temperate species would have been able to gain a foothold

---

[28] Hooker (1861); see Darwin to Hooker, February 25, 1862, in *Correspondence*, 10: 93–5.

[29] Croll (1864, 1875); for details, see David Oldroyd (1996), pp. 151–4, and Bowler (1992), pp. 227–30.

[30] Darwin to Croll, November 24, 1868, and January 31, 1869, in *Letters*, 2: 161–5. More letters between Darwin and Croll are now being edited by the Darwin correspondence project.

in equatorial regions and would have been able to cross the equator, because the southern tropical forms would have moved south owing to the warmer conditions there. When the glacial period was ended in the North, the northern forms would move back to more northerly latitudes, but might leave remnant populations isolated in mountainous regions of the southern equatorial lands they had temporarily invaded. When the Southern Hemisphere experienced a glacial period in its turn, these isolated forms would have been able to spread out across that hemisphere to reach the scattered locations in the far South where they survive to the present. The disturbances would have produced major extinctions in the original equatorial flora.[31]

The new explanation of how northern plants were able to move to isolated regions of the South was perhaps more complex than the original one. The fact that Darwin was prepared to make these additions shows how seriously he took the whole issue of geological climates and how vulnerable he felt to challenges on that score. Croll's theory allowed him to construct what he felt was a more plausible account of how plants could cross the barrier of the equator. Darwin's modified explanation retained the essential logic of his argument – that is, that there was a viable alternative to multiple creations – even though the new version was somewhat more complex than the old. In a sense, though, Darwin's position may have been strengthened by the uncertainty surrounding the question of the Ice Ages, since this reinforced his view that we simply cannot be sure about the climatic changes that may have opened up migration routes in the distant past. In these circumstances nothing could be ruled out, and the creationist position that there was no viable natural explanation of these disjunct populations was untenable.

VI. WALLACE AND THE GEOGRAPHICAL DISTRIBUTION OF ANIMALS

Darwin's very cautious approach to the evidence from paleontology and biogeography focussed on individual case studies where he felt he could either make a positive case for common descent

---

[31] In *Variorum*, the main additions are Chapter 11, sentences 215.1–9, 245.1–11, 246.1–5, and 248.1.

or undermine the credibility of the creation theory. His restraint contrasts with the increasing boldness of evolutionists in the later decades of the nineteenth century. Ernst Haeckel and others appealed to morphological evidence from comparative anatomy and embryology to fill in the gaps in the fossil record and allow the reconstruction of phylogenies for all the main groups of animals and plants. Far from restricting themselves to individual case studies, this later generation of evolutionists wanted to reconstruct the whole history of life on Earth. Biogeography would play an important role in this project, since it was hoped that the evidence from past and present distributions would allow evolutionists to identify the location in which each major group had evolved, and to trace its subsequent migrations.

Haeckel himself appealed to geographical evidence in his *History of Creation* to make the case for the monophyletic origin of every major new type. This would remain a significant issue, the alternative being not multiple creations but the anti-Darwinian theories of parallel evolution that would allow the same type to evolve separately from distinct ancestral forms.[32] But the comprehensive incorporation of geographical factors into the search for life's ancestry was inspired to a large extent by the work of Alfred Russel Wallace. His efforts in this area have been overshadowed by his role in the discovery and original publication of the selection theory, but his monumental *Geographical Distribution of Animals* of 1876 provided the first really comprehensive account of zoological biogeography and ushered in a period of intense activity among naturalists eager to identify the actual locations of the major steps in evolution and the subsequent expansions and extinctions of the major animal groups.

Wallace at first shared Darwin's fears that the available evidence was simply not enough to allow a comprehensive reconstruction of the history of life. He had been led to the idea of evolution in part by his own biogeographical studies, and he continued to work on the zoogeography of the Malay Archipelago in the early 1860s. In the Preface to his 1876 book, he records that he had then been asked by Darwin and by Alfred Newton, the Cambridge professor of zoology, to undertake a global survey of zoogeography. He had abandoned the

---

[32] Haeckel (1876), vol. 1, Chapter 14; and Bowler (1996), Chapter 8.

project in despair at the lack of information on many groups, but was revitalized by the publication of a number of detailed surveys and launched once again into the effort to produce a comprehensive overview.[33]

The result was, in effect, an extension of the Darwinian project on a global scale, far beyond anything Darwin himself had considered possible. Wallace shared Darwin's distrust of claims about 'land bridges' in the recent past and was convinced that the zoological provinces he had identified were defined by ocean barriers that had been permanent at least through the Tertiary era. He shared Darwin's view that geological changes produced their effect on life not by altering the world's geography, but by modifying the climate in a way that forced species to migrate and both opened and closed the routes available to them. From the information he reported on the distribution of the living and fossil members of each group, he was able to generate hypotheses about the original home in which the first members of each group had evolved and to suggest the time and routes of subsequent migrations. The existing state of each zoological province was, in effect, a composite built up as a result of successive migrations and extinctions. Wallace endorsed Darwin's idea that most of the successful groups had appeared first in the North, but he was able to suggest how and when they had moved to take up their present distribution.

In his later book *Island Life*, Wallace contributed to Darwin's most important case study, the population of oceanic islands. He also examined the question of changing geological climates and proposed a modification of Croll's theory of the Ice Ages.[34] His work seems to have triggered an explosion of interest in evolutionary biogeography that fed into the late nineteenth century's fascination with the effort to reconstruct the history of life on Earth. Rival theories were proposed, the supporters of the Darwin–Wallace position on the permanence of continents vying with those who preferred Hooker's hypothetical extensions of the land surface. In the early twentieth century the whole phylogenetic project came under a cloud, as a new generation of biologists became suspicious of the speculative nature

---

[33] Wallace (1876), vol. 1, pp. v–vi.
[34] Wallace (1880).

of the theories that had been proposed. Perhaps Darwin was wise to limit his attention to a small number of well-defined topics within the field, although we can hardly blame Wallace for hoping that the explosion of information that was becoming available in the 1870s might be enough to resolve the issues.

10 Classification in Darwin's *Origin*

I. A PUZZLE

Textbook histories tell us that Charles Darwin sparked a scientific revolution with the publication of his *Origin of Species* in 1859. While this revolution is obvious in many biological disciplines, it is less so in biological taxonomy or systematics – the grouping and classification of organisms. After all, the approach most of us learn in school today was first developed by Linnaeus in 1735 – 124 years before the *Origin*. That being the case, where is the revolution? Systematist Ernst Mayr expresses this doubt: "As far as the methodology of classification is concerned, the Darwinian revolution had only minor impact" (Mayr 1982, 213). Mayr is right in that the Linnaean hierarchy predated but also survived the Darwinian revolution, and has remained in use today. But it would be wrong to conclude that the Darwinian revolution was irrelevant to classification.

My intention here is to sketch out Darwin's influence on classification and the role classification played in his larger project. We shall see that despite the superficial similarity between pre- and post-Darwinian classification, Darwin's influence has nonetheless been profound. He gave systematics a theoretical foundation and an operational method. Moreover, classification was important to Darwin's project. His discussion of it in the *Origin* comes at the end, unifying and drawing together threads from important topics in earlier chapters – natural selection and divergence, comparative anatomy and morphology, and embryology and development. To

The suggestions of the editors, Robert Richards and Michael Ruse, have been very helpful. The views presented here are my own.

make all this clear, I will begin with a brief summary of the classificatory approaches predating the *Origin* (section II). Following that, I explain Darwin's views on classification in the *Origin* (section III); and finally, I will briefly describe how his views have endured in more recent approaches (section IV).

II. CLASSIFICATION BEFORE THE *ORIGIN*

Histories of classification usually begin with Aristotle. But this is somewhat misleading. While Aristotle spent great effort in observing and recording the features of animals that might have functioned in a grand classification, his methods and goals were not those of more modern systematists. He is well known for using the standard classificatory terms *genus* and *species* to divide and differentiate groups of organisms, but he used these terms in a variety of ways, and at a variety of levels, generally relative to the investigation at hand. *Genus* and *species* were not the fixed taxonomic levels for Aristotle that they would become for modern systematists (Grene and Depew, 18).

Furthermore, modern systematists observe the pattern of similarities and differences in organisms in order to group them into taxa. Feathers and beak shape, for instance, are used to group organisms into bird genera and species. But Aristotle was more interested in classifying the *character traits* of the organisms than the organisms themselves. He focused on traits because his goal was the discovery of the explanatory principles that governed their functioning. In his *History of Animals*, he gave descriptions of the various functional parts of various kinds of animals. Then, in the *Parts of Animals*, he provided causal explanations in terms of the functioning of the parts. He noted, for instance, the correlation of lungs, windpipe, and esophagus in various groups, and how this could be explained in terms of functioning. When he did bring the groupings of organisms into the discussion, it was typically subservient to his theorizing about regularities in the distribution of character traits (Lennox, 14–20).

Those who followed Aristotle turned toward the goal of a grand, overarching classification of organisms, based upon a term used by Aristotle, *scala naturae*, as the foundation for a classificatory system conceived as a "ladder of nature" or "great chain of being." With the help of the physician Galen, who saw each link of the chain as existing for the purposes of higher-level links, thinkers through to the Renaissance apparently had little difficulty in giving this idea a theological gloss, seeing the "great chain of being" as representing degrees of perfection and leading to humans and beyond – ultimately to God (Grene and Depew, 14–33).

With the Renaissance emphasis on observation, and the discovery of many new plants and animals, it became increasingly clear that no simple, linear *scala naturae* was sufficient to represent organic diversity. Credit goes to botanists such as Cesalpino (1519–1603) and to zoologists such as Pierre Belon (1517–1564), who classified birds; Guillaume Rondelet (1507–1566), who classified fish, crustaceans, mollusks, and sponges (and several sea monsters as well); and John Ray (1627–1705), who not only identified thousands of plant species, but also produced synoptic volumes on reptiles in 1693, on insects in 1705, and on birds and fishes (posthumously) in 1713 (Mayr 1982, 160–9). From the sixteenth to the eighteenth century, there was an explosion of information about organic nature, but with no established procedure for describing, naming, and classifying species. Circumstances were ripe for the most famous of systematists, the Swedish botanist Carl von Linné (1707–1778), known better by the Latinized "Carolus Linnaeus."

In 1735, Linnaeus was studying medicine in the Netherlands when he published his *Systema Naturae Sive Regna Tria Naturae Systematice Proposita per Classes, Ordines, Genera & Species*. While this first edition consisted of only a title page, eleven pages of "observations" and taxonomic tables, and two one-page leaflets, later editions expanded dramatically. The thirteenth edition of 1770 was over 3,000 pages, classifying many thousands of plant and animal species. In these thirteen editions of his *Systema Naturae*, Linnaeus worked out first, a nesting, hierarchical approach to classification that consisted of three kingdoms – animal, plant, and mineral – each subdivided into orders, classes, genera, and species; and second, a system of binomial nomenclature that identified each species on the basis of the now familiar genus-species name. Humans, for instance, were named *Homo sapiens*, and classified in the order "Anthropomorpha," along with the other primates in *Simia*.

The significance of Linnaeus comes not from his classifications, but from his streamlined descriptions, standardized terminology, and classificatory framework that is relatively unambiguous, uniform, and easy to apply (Mayr 1982, 173). The Linnaean framework was eventually adopted by Darwin, survived the Darwinian revolution, and is still widely used today, but with the addition of multiple taxonomic categories, from subspecies to subphylum, superclass, subclass, infraclass, cohort, suborder, infraorder, superfamily, family, subfamily, tribe, subtribe, and more (Ereshefsky, 215).

In the early 1800s, however, the Linnaean system was not the only system. In his 1817 *Regne Animal*, Georges Cuvier argued for a system of classification based on four "embranchements" – vertebrata, mollusca, articulata, and radiata. Each embranchement was subdivided into smaller Linnaean units – class, order, family, genus, subgenus, and species. These embranchments were based on the idea that there were four basic general body plans in nature, each associated with a particular kind of nervous system. While there was variability within each embranchment on the basis of differences in the conditions of existence to which each organism was adapted, there were unbridgeable gaps between the plans (Grene and Depew, 143–5).

By 1821, William S. MacLeay had developed a "quinarian" system whereby animals and plants were each arranged into a grand circle of five principal forms. In the animal circle there were five classes – Acrita, Radiata, Annulosa, Vertebrata, and Mollusca. Each of these five classes contained five orders, each of which contained five tribes, continuing on down in smaller nested groupings of five. Vertebrata, for instance, contained fish, amphibians, reptiles, birds, and mammals – each subdivided into five smaller groups. Quinarianism was taken seriously, especially in England. The Zoological Club of the Linnaean Society, for instance, spent much time debating and discussing it between 1825 and 1830. By the mid-1840s, however, many British naturalists (including Darwin) were coming to reject MacLeay's system (Ospovat, 101–13; Hull 93–6).

In the various debates about classification preceding the publication of the *Origin*, there were two recurring questions: first, what made a classification "natural," second, what characters should a classification be based upon? The first question was about theoretical basis: what was the classification supposed to represent or express? Most systematists from the Middle Ages up until the publication of Darwin's *Origin* gave a theological answer to the theoretical question: a "natural" classification was one that expresses

God's plan; and by studying nature, the systematist was studying God's plan. Linnaeus was of this view, seeing himself as a sort of prophet, chosen by God to reveal the deepest and most profound laws decreed in the creation (Lindroth, 13). Similarly, MacLeay was committed to a theological basis for classification. The quinarian system was "natural," according to MacLeay, because it expressed the plan of God in creation, and its mathematical beauty (based on the number five) revealed the rational design that God had imparted (Ospovat, 106). For both Linnaeus and MacLeay, because God was rational, there was a rational order to nature that could be uncovered through careful observation and reason.

It should be apparent that this theological basis for classification presents problems. First, unless we already know something about God's plan in the creation, it is unclear how we can tell we have it right. The fact that God is rational and would have expressed his design in a rational way is of little help – unless we can evaluate systems based upon their rationality. But it is hard to see how to do this. Are Linnaeus's three kingdoms more rational than Cuvier's four embranchments or MacLeay's quinarianism? The only real standards that systematists had were operational: the ability to construct a classification that was unambiguous and complete, in which every form of life has its place and only one place. But the mere ability to do this is insufficient. Suppose we could, for instance, classify unambiguously on the basis of color. It is unlikely that any systematist would regard color as a satisfactory grouping principle except as it met some merely practical need. Or perhaps, as Linnaeus argued, a natural system is one based on all characters of the organism, since each character is ultimately an expression of God's design. Believing this to be an impossible goal, however, Linnaeus used only some characters, mostly sexual, especially in his botanical groupings.

Others followed Linnaeus in using a *special similarity* approach. MacLeay based his quinarian system on a distinction already in use between "affinities" and "analogies." An *affinity*, according to MacLeay, is a shared character or similarity that is correlated with other similarities and can be seen to form an uninterrupted series. An *analogy* is a similarity that is not correlated with other similarities and does not form an uninterrupted series. For MacLeay, grouping should be based primarily on affinities, although analogies must also be represented in the groupings. In other words, an organism might be placed in Vertebrata on the basis of affinity – a series of correlated characters – but it may well have parallel similarities – analogies – with other groups (Ospovat, 104–6).

This affinity/analogy distinction led to the more familiar homology/analogy distinction. Richard Owen, who was influenced early on by MacLeay (Ospovat, 107), worked out this distinction in response to a disagreement between Georges Cuvier and Etienne Geoffroy Saint-Hilaire. Cuvier and Geoffroy were comparative anatomists at the Museum National d'Histoire Naturelle in Paris who engaged in a famous debate in 1830 about the laws of anatomy. It was standard at this time to see Newton as presenting a model for all sciences, based on the search for universal laws. Cuvier, in particular, sought a Newtonian basis, asking: "Why should not natural history also have its Newton some day?" (Grene and Depew, 134) But Cuvier and Geoffroy disagreed about the nature of these universal laws governing anatomy.

Cuvier adopted what has traditionally been described as a teleological or functional approach: the laws of anatomy are those that govern the functioning of parts in *individual* organisms so that the organism can function well relative to its conditions of existence. Cuvier explained:

Natural history, however, also has a rational principle which is peculiar to it and which it uses to advantage on many occasions; it is that of the conditions of existence, commonly called final causes. As nothing can exist unless it combines in itself the conditions which make its existence possible, the different parts of each being must be coordinated in such a way as to make possible the total being, not only in itself, but in its relations with those which surround it; and the analysis of these conditions often leads to general laws as demonstrable as those which derive from calculation or experiment. (Ospovat, 8)

For Cuvier, the laws governing anatomy relate to the functioning of the whole individual on the basis of the relation of its parts, and the relation of the organism to its environment (Ruse 1999, 12–13). Geoffroy, on the other hand, focused more on the extensive patterns of similarity *among* organisms. He thought the laws of anatomy should be understood as relating the similarities or "unities of plan" among groups, irrespective of functioning (Grene and Depew, 138).

While Cuvier and Geoffroy understood their views to be incompatible, Richard Owen did not. He focused on Geoffory's unities of plan, described them as *archetypes*, and tried to place them within Cuvier's classificatory system, arguing that they underlay the functional structures of the four embranchments (Grene and Depew, 180–1). In doing this, Owen was trying to incorporate both insights – the Cuvierian insight into function and the Geoffroyian insight into structural similarities. Owen used the terms "analogue" for the characters that had similar functions but different structures, and "homologue" for the characters that had similar structures without necessarily functioning in similar ways (Ruse 1999, 116–124). According to Owen, this implied that we could study organisms from

two distinct but by no means incompatible points of view – to ascertain first to what it may be analogous & secondly with what it may be homologous – to study the skeleton teleologically and morphologically. (Ospovat, 130)

We can, for instance, look at the forelimbs of humans and the wings of birds morphologically, and see that they exhibit a similar correspondence of parts – both relative to parts of the limb itself and relative to the rest of the organism. Or we can look at them teleologically, and see that they function in different ways – as adapted to different ends.

Despite Owen's distinction between homologies and analogies, it should be obvious that there were still very difficult problems confronting systematists in the first half of the nineteenth century: first, there was still no satisfactory account of how one could tell when a classification was "natural" – expressing God's plan; second, the recognition that some similarities were based on function and some on form was not sufficient to indicate how to construct a natural classification. Did God's plan require a classification based on homologies – Geoffroy's unity of plan – or did it require one based on analogies – Cuvier's functional adaptation to conditions of existence? Or should one employ both homologies and analogies?

The lack of satisfactory answers to these questions might encourage us to agree with Georges Buffon, Linnaeus's critic and contemporary, who believed that the only real biological taxa were species. For Buffon, species groupings were based on real processes in nature – reproduction and genealogical relationships. Higher-level groupings into the Linnaean categories, genera, classes, orders, kingdoms, and so on were based on mere abstract relations, which were not just a waste of time and effort, but antithetical to the study of nature. To truly study nature, one had to study real processes and relations (Grene and Depew, 74; Mayr 1982, 180–2).

From a modern perspective the problems in pre-Darwinian classification are evident. First, there is no way to tell when a classification has met its theoretical goal – representing God's design in the creation. Consequently, preference for one or the other system seems arbitrary. There is no obvious reason to prefer Linnaeus's system with its three kingdoms, Cuvier's system with its four embranchments, or Macleay's system with its two sets of five circles of affinity. Second, and as a consequence of the first problem, there is no way to tell which operational procedures satisfy the theoretical goal. Which similarities should we focus on in grouping organisms to represent God's plan?

III. CLASSIFICATION IN THE *ORIGIN*

Darwin began working out his views on evolution in 1837 with the first of his transmutation notebooks. At the time, he took seriously the various classificatory approaches in circulation. In his early writings, he seemed to have just assumed that MacLeay's quinarian classification was largely correct, and that it therefore required an evolutionary explanation (Ospovat, 108). But by the mid-1840s, when most naturalists in Great Britain were coming to reject McLeay's system, Darwin no longer needed to take quinarianism seriously.

Darwin had long been aware of Cuvier's views on a variety of topics, including classification, but he did not turn to a Cuvierian system after rejecting MacLeay's. This should come as no surprise. From his first transmutation notebook in 1837, Darwin had begun to think about classification in terms of genealogy. Six years later he was explicitly advocating genealogy as a theoretical principle in a series of letters to the zoologist G. R. Waterhouse (*Correspondence*, 2: 375–6). And fourteen years after that, Darwin disagreed with his friend and advocate T. H. Huxley, who was still favoring Cuvier's nongenealogical classification (*Correspondence*, 6: 461–3). For Darwin, if classification were to be based on genealogy, then only those similarities that were indicative of genealogy should be used in grouping organisms. This general framework – with its theoretical basis and operational method – demanded answers to two questions: What sort of a classificatory system best reflected genealogy? What kinds of similarities or shared characters were indicative of genealogy? These questions were answered in the *Origin of Species*.

In the first edition of the *Origin*, Darwin devoted much of the penultimate Chapter 13 to classification. He began the chapter with a statement of the basic premise of classification – that there are patterns of similarity among organisms and that these patterns can be used to group organisms. Then Darwin tells us *why* this grouping is not arbitrary: it is based on the tree-like branching of evolution. To make his case, he returned to threads from previous chapters on "Variation" and "Natural Selection," where he developed the reasoning behind his "Principle of Divergence." The basic idea is that any particular geographic area can support more life if organisms diversify and vary in their habits of life by diverging from the ancestral form. Intermediate forms will tend to be eliminated because they will be competing with both sets of divergent forms, while the *most* divergent forms will have the least competition. A form that is intermediate in size, for instance, will probably be in competition with both its larger and smaller cousins in terms of food and shelter, while the largest and smallest forms will be in competition only with the intermediate form. This divergence, Darwin claims, leads to a branching process in evolution. He refers to a tree figure (frontispiece, p. ii) that appeared in the chapter on "Natural Selection" (the only illustration in the first edition of the *Origin*).

I request the reader turn to the diagram illustrating the action, as formerly explained, of these several principles; and he will see that the inevitable result is that the modified descendants proceeding from one progenitor become broken up into groups subordinate to groups. (*Origin*, 412)

Darwin summarizes how this branching produces a group-in-group classification:

So that we here have many species descended from a single progenitor grouped into genera; and the genera are included in, or subordinate to, subfamilies, families and orders, all united into one class. (*Origin*, 413)

What is important in this tree figure, and the accompanying text, is that genealogy and the grouping of species into higher categories – genera, families, orders, and classes – are to be understood in terms of this branching of the tree.

But this does not, by itself, give us a classification. Does a particular branch represent a genus, a family, an order, or a class? This is the "ranking" problem: at what level in the classificatory hierarchy do we place each group? According to Darwin, we should rank on the basis of degree of modification:

I believe that the *arrangement* of the groups within each class, in due subordination and relation to the other groups, must be strictly genealogical in order to be natural; but that the *amount* of difference in the several branches or groups, though allied in the same degree in blood to their common progenitor, may differ greatly, being due to the different degrees of modification which they have undergone; and this is expressed by the forms being ranked under different genera, families, sections, or orders. (*Origin*, 420)

Implicit in this idea that classification should be based on the branching of the evolutionary tree is Darwin's theoretical basis – the purpose and goal of classification is to represent genealogy. And because genealogy is ultimately the product of heredity, branching, and divergence, the genealogical system will be hierarchical, with organisms placed in groups representing the branches of the evolutionary tree. For Darwin, this is what makes it a "natural" system. It is worth noting here, however, that Darwin did not believe that he was proposing an *entirely* new approach to classification. First, because we already group together males and females of a species, as well as organisms at different stages of development – no matter how different they may be in form – we are using a genealogical classification.

With species in a state of nature, every naturalist has in fact brought descent into his classification; for he includes in his lowest grade, or that of a species, the two sexes; and how enormously these sometimes differ in the most important characters, is known to every naturalist: scarcely a single fact can be predicated in common of the males and hermaphrodites of certain cirripedes, when adult, and yet no one dreams of separating them. The naturalist includes as one species the several larval stages of the same individual, however much they may differ from each other and from the adult.... (*Origin*, 424)

Second, Darwin believed that we already classify genealogically those organisms whose genealogies we know – domestic varieties.

In confirmation of this view, let us glance at the classification of varieties, which are believed or know to have descended from one species. These are grouped under species, with sub-varieties under varieties; and with our domestic productions, several other grades of difference are requisite, as we have seen with pigeons. The origin of the existence of groups subordinate to groups, is the same with varieties as with species, namely, closeness of descent with various degrees of modification. Nearly the same rules are followed in classifying varieties, as with species. (*Origin*, 423)

Third, Darwin believed that some classifications of species had been genealogical all along, because some systematists had inadvertently used characters that indicated ancestry. When Linnaeus grouped organisms on the basis of similarities in their sexual organs, and when MacLeay used "affinities," both were unwittingly using characters that Darwin thought were good indicators of genealogy. This last point can be seen in Darwin's operational method.

One of the chief advantages of Darwin's approach to classification over those of Linnaeus, Cuvier, and MacLeay is that its theoretical basis provided the foundation for an operational method. Because classification is based on natural processes – the branching and divergence of evolution – we can use what we know about these processes to group organisms. Darwin does this through a reinterpretation of distinctions used by MacLeay and Owen. Macleay had argued that classifications should be generated on the basis of "affinities" – the series of correlated similarities – rather than analogies – isolated, noncorrelated similarities. And Owen had distinguished structural similarities – "homologies" – from functional similarities – "analogies." Darwin reinterpreted MacLeay's "affinities" and Owen's "homologies" to be the similarities due to common ancestry. The idea is that some characters common to different species are due to common ancestry – inherited from a common ancestor. Other characters are due to an adaptive response to the environment. If we can tell which characters indicate genealogy or common ancestry, we can group on the basis of them, rather than on those that are the products of an adaptive response.

Based on this idea, Darwin gave us one general rule and a set of corollaries for distinguishing homologies and analogies. He introduced the general rule by rejecting the view that characters important to habits of life are relevant to classification:

It might have been thought (and was in ancient times thought) that those parts of the structure which determined the habits of life, and the general place of each being in the economy of nature, would be of very high importance in classification. Nothing can be more false. No one regards the external similarity of a mouse to a shrew, of a dugong to a whale, of a whale to a fish, as of any importance. These resemblances, though so intimately connected with the whole life of the being, are ranked as merely "adaptive or analogical characters." ... (*Origin*, 414)

The idea lurking here is that if a character has functional significance for specific habits of life, it is likely to have been the more immediate product of natural selection, and is thus irrelevant to classification. We might express this rule as follows:

General Adaptation Rule: A shared character trait that is likely to be an adaptation to a particular form of life is not likely to be homologous and is therefore irrelevant to classification.

Darwin followed this with a corollary:

It may even be given as a general rule, that the less any part of the organization is concerned with special habits, the more important it becomes for classification. (*Origin*, 414)

According to the *"special habits" corollary*, the more traits or characters are associated with special habits of life such as flying, swimming, digging, and so on, the more likely they are to be recent products of adaptation rather than of common ancestry. A second, related corollary was that reproductive organs are more useful in classification. Darwin quoted Owen on this matter:

As an instance: Owen, in speaking of the Dugong, says "The generative organs being those which are most remotely related to the habits and food of an animal, I have always regarded as affording very clear indication of its true affinities." (*Origin*, 414)

The basic idea of this *reproductive organs corollary* is that since organs of generation are usually not associated with special habits of life and the associated requirements of survival, they are unlikely to be special adaptations, and are therefore relevant to classification.

Another corollary is related to the *constancy of characters*. Darwin suggested that systematists have often been using this principle correctly:

If they find a character nearly uniform, and common to a great number of forms, and not common to others, they use it as one of high value; if common to some lesser number, they use it as of a subordinate value. (*Origin*, 418)

The idea behind the constancy corollary is that characters that are constant across species may be of importance to the species in which they are found, but not as adaptations to local circumstances. The similar skeletal structures of birds, dugongs, and humans, for instance, are not likely to be adaptive responses to special conditions, because birds, dugongs, and humans occupy very different conditions of life.

Two more corollaries of great importance to Darwin were related to *rudimentary organs* and *embryological characters*. Rudimentary organs are those that appear to be atrophied instances of functional characters. They are significant because in the rudimentary form they are of little functional importance.

No one will say that rudimentary or atrophied organs are of high physiological or vital importance; yet, undoubtedly, organs in this condition are often of high value in classification. No one will dispute that the rudimentary teeth in the upper jaws of young ruminants, and certain rudimentary bones of the leg, are highly serviceable in exhibiting the close affinity between Ruminants and Pachyderms.... (*Origin*, 416)

Darwin thought rudimentary organs were important enough that he devoted an entire section to them at the end of his chapter on classification. He concluded this section:

As the presence of rudimentary organs is thus due to the tendency in every part of the organization, which has long existed, to be inherited – we can understand, on the genealogical view of classification, how it is that systematists have found rudimentary parts as useful as, or even sometimes more useful than, parts of high physiological importance. Rudimentary organs may be compared with the letters in a word, still retained in the spelling, but become useless in pronunciation, but which serve as a clue in seeking for its derivation.... (*Origin*, 455)

Darwin similarly devoted an entire section to embryological characters. Like rudimentary organs, they are significant because it is unlikely that they are adaptations to local conditions. And the earlier the characters appear in development, the less likely they are to be adaptive responses. An embryo in a womb or an egg is, after all, typically not subjected to the selection pressures of a particular environment.

The points of structure, in which the embryos of widely different animals of the same class resemble each other, often have no direct relation to their conditions of existence. We cannot for instance, suppose that in the embryos of the vertebrata the peculiar loop-like course of the arteries near the branchial slits are related to similar conditions, – in the young mammal which is nourished in the womb of its mother, in the egg of the bird which is hatched in a nest, and in the spawn of a frog under water. (*Origin*, 440)

A grouping based on embryological characters, then, is more likely to reflect genealogy, because these kinds of similarities are more likely to be homologous – due to common ancestry – and less likely to be analogous – due to the proximate operation of natural selection.

It is likely that Darwin worked out many of these principles during his eight-year-long study of *Cirripedes* or barnacles – sea creatures that cement themselves to surfaces, grow hard shells, and catch food with feathery appendages. Darwin began studying barnacles in 1846, at least in part because he was worried that he lacked sufficient scientific authority for his theory of evolution to be taken seriously. One way to get this authority was by the careful and detailed study of a single group of organisms. In this he was successful, receiving a Royal Society medal for his barnacle work in 1854 (Stott, xx–xxv).

The study of barnacles was significant for Darwin first, because it revealed the great variability in nature. Darwin had earlier assumed that there was relatively little variation within a species in nature – except as a response to changes in conditions of existence (Ospovat 78–9). But his careful study of barnacles revealed great variability, even under similar conditions of existence. This was important because it provided support for the principle of divergence and its implied branching that Darwin used to introduce his chapter on classification. Constant variability throughout nature implies a constant tendency for the divergence and branching of the evolutionary tree that was the basis for Darwin's genealogical classification.

But his study of barnacles was also important for the development of his operational method. Barnacles had been classified with mollusks by Linnaeus, Cuvier, and Owen on the basis of the hard shells in adults. But they are more similar to crustaceans such as shrimp at the earliest stages of development. In confronting his hundreds of specimens, Darwin had to decide which similarities were most important to classification, both of the group as a whole and of groups within the group. He concluded that the embryological similarities to crustaceans implied that barnacles were really related to crustaceans. But in dividing barnacles into smaller groups he had great difficulty in settling on a satisfactory classification (Ruse 1999, 186; Stott, 53). The particulars of his eight-year project are less important here than the fact that it was probably his detailed observation and recording of all the glorious nuances of barnacle anatomy that led Darwin into thinking about and developing his operational principles as outlined earlier (Ghiselin, 103–30).

In the introductory section to this chapter, I suggested that classification came at the end of the *Origin* in part because of its importance in terms of unification. I cannot explain this idea adequately here, but we can see what it involves. First, in order to group organisms Darwin relied on assumptions about evolutionary processes. He did this through the general adaptation rule and all its corollaries, embryology, rudimentary organs, constancy, and so on, that were intended to distinguish homologies from analogies. In effect, Darwin was not grouping just on the basis of observable similarities, but also on the basis of a series of auxiliary theories he had about the processes that cause and explain change, natural selection in particular. A natural classification for Darwin was to be constructed at least in part on the basis of an explanatory framework.

But once a genealogical classification was constructed, it could in turn serve an explanatory role:

All the great facts of morphology become intelligible, – whether we look to the same pattern displayed in the homologous organs, to whatever purpose applied, of the different species of a class; or to the homologous parts constructed on the same pattern in each individual animal and plant. (*Origin*, 456–7)

The idea here is that once we have a genealogical classification based on the branching of the evolutionary tree, we can explain all the morphological facts that had seemed to be unconnected and inexplicable otherwise: facts about how mollusks and barnacles are similar in some ways and different in others; facts about how the forelimbs of dugongs, moles, and bats have similar structures yet serve different functions, and more. Moreover, once we have this understanding, we can then reconstruct evolutionary genealogy – with even greater confidence. In this "reciprocal illumination," causal processes reveal insights into historical patterns, which then reveal insights into these processes, which then reveal further insights into the patterns, and so on. Moreover, facts about embryological development, rudimentary organs, biogeography, and more can also be explained by evolutionary processes and history. In short, a genealogical classification can provide the basis for an understanding of facts from a broad range of organic phenomena. Darwin thought this explanatory power to be most important, serving as evidence for his theory of common descent. He concludes the chapter on classification:

The several classes of facts which have been considered in this chapter, seem to me to proclaim so plainly, that the innumerable species, genera, and families of organic beings, with which this world is peopled, have all descended, each within its own class or group, from common parents, and have all been modified in the course of descent, that I should without hesitation adopt this view, even if it were unsupported by other facts or arguments. (*Origin*, 458)

IV. AFTER DARWIN

In the fifty or so years after the *Origin*, interest in systematics gradually declined, first because a genealogical classification required a knowledge of phylogenies that was in short supply, and second because interest turned to other biological topics related to the cell, biochemistry, and the mechanisms of heredity (Mayr, 217–20). But by the early middle of the twentieth century, "evolutionary systematists" such as Ernst Mayr and G. G. Simpson were following Darwin by adopting a genealogical basis, ranking taxa according to degree of divergence, and using assumptions about evolutionary processes to identify homologies and analogies. Like Darwin, they assumed that similarities that are probably due to adaptive change by natural selection are less likely to be homologies (Mayr, 212–13). By the 1970s, however, this use of process assumptions had come under attack from two sources.

First came a challenge from "phenetics," an approach that rejected both Darwin's theoretical assumption that classification should be genealogical and his special similarity grouping method. Pheneticists argued that classification should be an "all-purpose" way of "storing" information about species and should therefore be based on all characters. They classified organisms into *operational taxonomic units* derived from a single quantity – *phenetic distance* – intended to indicate overall similarity (Hull 1988, 117–30; Panchen, 132–51).

Phenetics never became widely adopted for three reasons. First, a purely phenetic classification would typically group individuals of different sexes and developmental stages into different operational taxonomic units even if they were of the same species; second, different similarity algorithms produced different operational taxonomic units, and it was not obvious how pure observation could identify the best algorithm; third, by combining all characters to produce a single quantity – phenetic distance – much information was simply being suppressed. If the goal of classification is information storage and retrieval, as claimed, the use of phenetic distance is counterproductive.

The second challenge to the neo-Darwinian evolutionary systematics came from "cladistics," named for its emphasis on cladogenesis, the branching associated with evolution. It was based on the approach developed by Willi Hennig, an East German entomologist, first in a 1950 textbook in German and then in a 1966 English revision, titled *Phylogenetic Systematics*. Cladists (or "phylogeneticists," as many prefer) returned to the Darwinian genealogical (or *phylogenetic*) theoretical basis, classifying species on the basis of ancestry, and adopting a group-in-group system based on the branching process in evolution. Cladistics places species in *monophyletic* groupings – groups that contains an ancestral species, all of its descendent species and only its descendent species. This is to be contrasted with *paraphyletic* groupings, which include only descendent species, but not all descendent species; and *polyphyletic* groupings, which include species not descendent from the ancestral species (Hull 1988, 130–55; Panchen, 151–8). Monophyly as illustrated by Hennig is shown in Figure 10.1.

Here each set of circled taxa, and the corresponding branch, represents a monophyletic group. This emphasis on monophyly has

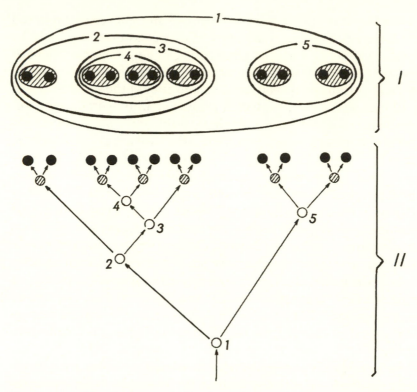

Figure 10.1. The phylogenetic kinship relations between the species of a monophyletic group, represented in two different ways (Hennig, 71). From *Phylogenetic Systematics*. Copyright 1966, 1977 by the Board of Trustees of the University of Illinois. Used with permission of the author and the University of Illinois Press.

resulted in modifications to traditional classifications that recognize paraphyletic taxa. For instance, *Reptilia* is not a monophyletic grouping, as it does not contain birds or *Aves*, which share common ancestry. Consequently, *Reptilia* is no longer recognized by most systematists.

Cladistics is Darwinian in that it employs a genealogical theoretical basis and an operational grouping method that uses only those shared characters thought to be homologies. Cladists, though, have rejected the Darwinian use of assumptions about evolutionary processes to identify adaptive similarities and homologies. Instead, they have adopted a "parsimony" approach, based on the idea that we can take a group of taxa and identify their similarities and differences and then use an algorithm to determine which way of grouping the taxa requires the fewest evolutionary changes. The most parsimonious grouping is then used to generate a branching "cladogram," that can then be interpreted to represent the genealogy in a phylogenetic tree[1] (R. A. Richards 2007).

More recently, systematists have started to turn from pure parsimony-based methods to statistical and molecular approaches based on "distance" and "maximum likelihood" (Panchen, 226–37). In part this is because parsimony seems to have its own problems. First, there are different parsimony algorithms that generate different groupings with no obvious way to adjudicate among them. Second, parsimony seems to rely on problematic assumptions about the parsimoniousness of nature and the individuation of characters (R. A. Richards 2003). Understandably, a Darwinian-style analysis of characters based on function still persists, albeit as arguably a minority approach (Ridley 364–73).

Cladistic classification, however, is still clearly non-Darwinian in that it rejects Darwin's approach to ranking. Darwin ranked taxa on the basis of degree of divergence: those groups – branches – that had apparently diverged the most from a common ancestor got a higher Linnaean ranking – class, family, and so on. Cladists reject this, arguing that the best classification is one based solely on genealogy or phylogeny. But if we rank solely on the basis of the branching in evolution, then wouldn't there will be an increase in rank for each new branch? But given the staggering complexity of the great evolutionary tree of life, the traditional Linnaean system is fantastically inadequate – even though there are now at least twenty-one categories from kingdom to subspecies, including subphylum, superclass, subclass, infraclass, cohort, suborder, infraorder, superfamily, family, subfamily, tribe, and subtribe, and a proposal to add nine more categories between the rank of family and superfamily (Ereshefsky, 215).[2]

---

[1] Some "transformed" cladists have since taken a phenetic step backward, denying that cladograms represent evolutionary branching – only character distributions (Panchen, 170–81).

[2] For proof of this complexity, see the online "tree of life project" at <http://www.tolweb.org/tree/.>

While there are some efforts to revise the Linnaean hierarchy, usually these revisions make it less Linnaean, often by introducing new categories without giving them Linnaean ranks or names. One proposal, based on phyletic sequencing, simply inserts new taxa within a Linnaean rank, but without giving them Linnaean names. Other proposals include one to replace the Linnaean hierarchy with a procedure based on indentation, and another with positional numbers indicating phylogenetic relationships (Ereshefsky, 241–6). At this point, it is still not clear how to represent the innumerable branches of the tree of life.

V. CONCLUSION

The words of Ernst Mayr at the beginning of this chapter seemed to imply that Darwin's revolution did not extend to systematics. We can now see in what ways that may be right and where it goes wrong. It is right in that Darwin did not need to create a new framework to represent genealogy. The Linnaean system was well suited to his theoretical basis in genealogy. The group-in-group classification in Linnaeus's hierarchical framework of species, genera, orders, classes, and so on, could be used to represent the branching of an evolutionary tree, where the branches on branches represent the group-in-group structure. It is also true that the operational approach of Linnaeus, as well as those of Cuvier and MacLeay, sometimes suited Darwin's theoretical goals as well. The sexual organs that Linnaeus used to group organisms seemed to indicate genealogy, as did the affinities of Macleay and Cuvier. In effect, Darwin could adopt the Linnaean hierarchy and use many of the similarities used by pre-Darwinian systematists in the service of both his theoretical goals and his operational methods.

But history may be misleading here. The mere fact that the Linnaean system was predominant both before 1800 and now does not mean that its success has been independent of the Darwinian revolution. At various times the systems of MacLeay and Cuvier might well have seemed more promising. In the absence of the Darwinian revolution, it may be that one of those approaches might well have become dominant. Or yet another may have come along. The Linnaean system might well have endured primarily because it suited the theoretical goals of an evolutionary system best. If so, the persistence of Linnaeanism is partly a consequence of the Darwinian revolution. But Darwin's use of the tree heuristic and its branching, along with the cladistic rejection of his ranking procedure, may ultimately still lead to the demise of the Linnaean hierarchy.

Perhaps most important here is the fact that Darwin gave systematics a theoretical basis, which in turn provided operational guidance. In pre-Darwinian classification, it was not at all clear which similarities should be used for grouping organisms and which should be ignored. The placement of barnacles with mollusks on the basis of one similarity makes as much sense as grouping them with crustaceans on the basis of another. Appealing to God's plan won't give guidance unless we can know something about that plan on independent grounds. In effect, we would need to know what God thought about barnacles. While, in the end, we cannot know what would have happened if Darwin had not published his *Origin*, it may well be that systematists would have gone on debating indefinitely whether nature should be grouped in three kingdoms, four embranchments, or five circles – and why.

11   Embryology and Morphology

INTRODUCTION

On December 14, 1859, less than a month after the publication of the *Origin of Species*, Darwin wrote to his confidante Joseph Hooker, "Embryology is my pet bit in my book, & confound my friends not one has noticed this to me" (*Correspondence*, 7: 431–2). Given the overwhelming mass of material presented in the *Origin* and its range across geology, geographical distribution, artificial and natural selection, hybridism, instinct, and classification, perhaps his friends could have been forgiven for having failed to recognize Darwin's pet bit. Indeed, the study of individual development in the *Origin* presents something of a paradox. As a special aspect of morphology, the study of the laws of organic form, embryology offered key evidence for community of descent. Darwin wrote that morphology was "the most interesting department of natural history, and may be said to be its very soul" (*Origin*, 434). The comparative study of the embryo receives similarly heavy rhetorical weight: "community in embryonic structure reveals community of descent" (*Origin*, 449), and "Embryology rises greatly in interest, when we thus look at the embryo as a picture, more or less obscured, of the common parent-form of each great class of animals" (*Origin*, 450). Yet the sections expressly on morphology and embryology together take up less than half of a single chapter (Chapter 13), comprising only 17 of the *Origin*'s 490 pages, or 25 if we add the section on "Rudimentary, Atrophied, or Aborted Organs" and the chapter summary. These topics show none of the weight of detail and example displayed in, for example, the chapters on geographical distribution and hybridism. Although a few references to comparative anatomy and embryology

194

may be found scattered elsewhere in the book, the space devoted to these topics seems rather meager compared to the rhetoric attached to their significance.

Yet Darwin insisted on such rhetoric, clinging to it even in old age. In his *Autobiography* he wrote (125), "Hardly any point gave me so much satisfaction when I was at work on the *Origin*, as the explanation of the wide difference in many classes between the embryo and the adult animal, and of the close resemblance of the embryos within the same class." Not only did Darwin's friends fail to recognize it, but historians have not known what to make of this. Just what was it about his embryology that gave Darwin such satisfaction?

As we will see, embryology did indeed hold a key place, both in the *Origin* and in his larger program, for it served as a kind of doorway between existing ways of relating embryology to morphology and classification and Darwin's own picture of the natural world and the problems it posed. Morphologists studied homologies, that is, organs they considered "the same" across different species, with the goal of uncovering the fundamental laws of organic form, which, they hoped, would provide for the organic realm the same kind of foundation that Newton's laws provided for mechanics. They hoped that such laws, in turn, would provide secure grounds for classifying organisms and thus producing (or uncovering) the true "system of nature." Morphology was thus understood by its leading practitioners across Europe and America to be, as Darwin put it, the "very soul" of natural history. For many morphologists, the developing individual presented a particularly compelling problem: what was the connection to be drawn between the course of individual development and the "affinities" or similarities seen among living groups of organisms?

Since the early nineteenth century, naturalists had struggled to discover these connections and wrangled with one another over their differing answers. At one level, Darwin's theory of descent solved this problem straightforwardly: since organisms were related to one another, their embryonic forms, like their adult forms, could be read as a record of their relatedness. The more similar two organisms were in their embryological development, the more closely they were related. Problem solved.

However, embryology was also a focal point of the new problems that Darwin's theory raised. If evolution proceeded by natural selection acting upon variations among organisms, then when in development did variation occur? How did variation affect the subsequent course of development? How did natural selection act upon the developing organism? What was the effect on the offspring of the modified parents? And what effects might all these events have upon scientists' reading of the embryo as a record of common ancestry (and thus upon classification)? In the *Origin*, Darwin knitted together the relations of embryology to classification and morphology, on the one hand, and to variation, inheritance, and selection, on the other, as he sought to recast the traditional problems of morphology in his own terms.

CHARLES DARWIN, EMBRYOLOGIST?

Why did Darwin devote so little space in the *Origin* to embryology? Shouldn't we take this absence seriously? One might argue that Darwin was a geologist and breeder and so had much more experience to draw on in those areas than he did in morphology and embryology. This argument falters on the evidence, however: Darwin was interested throughout his career in these topics (Richards 1992). Questions and notes concerning individual development appear in his earliest notebooks; embryology and morphology held a significant place in his essays of 1842 and 1844 and in his researches of the 1850s. Already in the late 1830s and more intensively in the mid-1840s, he began reading closely naturalists who linked classification to the study of form, including English-language writers such as Martin Barry and Richard Owen, French scholars such as the father-son duo Etienne and Isidore Geoffroy St.-Hilaire and Henri Milne-Edwards, and German scholars such as Johannes Müller. As he read, he commenced hands-on morphological work as well: from 1846 to 1854, he dissected countless barnacles to establish their morphology and classification. He directed a good deal of his attention to the development of this creature, and he made active use of embryological development to establish the classification of this group (Richmond 1985). Moreover, he was no tyro in vertebrate development, either. He devoted considerable effort in the later 1850s to measuring bird and dog neonates to determine their degree of variation as compared with adults. So although Darwin did not base his authority as a man of science on his publications in embryology and morphology, he was certainly deeply acquainted with these subjects, both from his reading and from his own hands-on research.

But he was pressed for time. He had not yet gotten to writing up these topics, in his deliberate way, when the threat of being scooped by Alfred Russel Wallace hurried him into publication, causing him to squeeze these important subjects into Chapter 13. Embryology and morphology were undoubtedly more significant for him than the space devoted to them in the *Origin*. His claims for embryology, as for evolution in general, constituted "one long argument" that extended well past 1859 through subsequent editions of the *Origin*, *Variation of Animals and Plants under Domestication* (1868), and *The Descent of Man* (1871).

EMBRYOLOGY IN THE *ORIGIN*

Darwin's main discussion of embryology appears in Chapter 13 of the *Origin*, under the title "Mutual Affinities of Organic Beings: Morphology; Embryology; Rudimentary Organs." The title is unfortunate, for it suggests a certain grab-bag quality to the chapter, as if he were cramming in all the subjects he hadn't gotten to yet in order to finish the book. As true as this may be, in fact the position of the chapter – the very last one before the book's recapitulatory conclusion – offers a clue to its unifying and generalizing quality. This is the chapter in which Darwin connects his theory to the natural system as a whole, in which he argues that indeed the natural system is none other than the genealogy of nature, explained and structured by natural selection. It is the book's punch line.

Morphology, embryology, and rudimentary organs all offer evidence that the natural system is genealogical. Conversely, Darwin's theory explains the known facts in these areas in a new and coherent way. Here Darwin embraces within his system the most intensely pursued questions of philosophical natural history of the previous half-century: What is the order of nature? How are we to understand the similarities and differences in form among different organisms (especially animals)? How is individual development related to the great patterns evinced by the animal world as a whole? His answer, as with everything else in the book, is that these topics are united and made coherent through the conception of descent driven by natural selection but are incomprehensible on the theory of separate creation of species.

As he did so often, Darwin worked by first enumerating the various classes of facts that he viewed as requiring an account in his theory. First, there was what he and his contemporaries referred to as "unity of type" and morphological "affinities" – the fact that organisms of the same class resembled one another in structural features that could not be accounted for on strictly functional grounds. These resemblances were strong, consistent, and complete enough that morphologists, following the French museum naturalist Étienne Geoffroy St.-Hilaire, could identify "the same" bones across the different vertebrates and give them the same name. Darwin's compatriot, the leading comparative anatomist Richard Owen, attributed such similarities to a common "archetype," an underlying ideal form. (In 1818, Geoffroy called his theory covering these similarities his "theory of analogues," but in 1843 Owen renamed these similarities "homologues." "Analogous" features, in Owen's recasting, were those that served the same function but used different structures, such as the wings of insects and birds. The distinction stuck.) Geoffroy also provided Darwin with a related idea that the latter would run with: such homologous parts might look entirely different and even serve different functions for different organisms, but they shared a common underlying form. For Darwin, these similarities were inexplicable if one assumed that different species were independently created, but were readily assimilated into his theory: homologies between organisms indicated common ancestry, and their variants demonstrated modifications that were adaptive to particular circumstances, culled by natural selection. His theory of descent thus accommodated both similarities and differences in form at one stroke.

In the case of embryology, comparison also yielded evidence favoring common descent. Animals of different groups within the same class, Darwin argued, often resembled one another more closely in their early embryological stages than in their adult forms. For example, embryos of mammals, birds, and frogs all shared a "peculiar loop-like course of the arteries near the branchial slits" (*Origin*, 440) despite the remarkably different conditions in which they develop – evidence to Darwin, like adult homologies, of common ancestry. A

close study of embryological stages could reveal surprising commonalities. Darwin's earlier examination of barnacle (cirripede) embryos
not only showed that the embryos were much more similar than the
adult forms, but also demonstrated to his satisfaction that they were
crustaceans, although their widely varying adult forms, most often
encased in a calcareous fortress attached to a rock, did not reveal this
fact so clearly (Origin, 440). (Indeed, Darwin's longtime foe Richard
Owen, who favored comparison of adult forms over that of developing forms in establishing homologies, considered cirripedes as a
distinct class, which he placed between the Crustacea and Annel-
ida [Richmond 1985, 394].) However, embryos did not always show
ancestral resemblances, and sometimes earlier developmental stages
could even appear higher in organization than mature ones.

After summarizing these diverse facts, Darwin came to the point:

How, then, can we explain these several facts in embryology, – namely the
very general, but not universal difference in structure between the embryo
and the adult; – of parts in the same individual embryo, which ultimately
become very unlike and serve for diverse purposes, being at this early period
of growth alike; – of embryos of different species within the same class,
generally, but not universally, resembling each other; – of the structure of
the embryo not being closely related to its conditions of existence, except
when the embryo becomes at any period of life active and has to provide for
itself; – of the embryo apparently having sometimes a higher organization
than the mature animal, into which it is developed. (Origin, 442–3)

Not surprisingly, his answer was that "all these facts can be explained, as follows, on the view of descent with modification" (Origin, 443). And several pages later, he was still more blunt: "the
embryo is the animal in its less modified state; and in so far it reveals
the structure of its progenitor.... Thus, community in embryonic
structure reveals community of descent" (Origin, 449).

At a general level, Darwin's argumentative strategy here was in
line with his handling of other classes of facts throughout the Origin,
in which common descent and natural selection accounted for a
wide variety of phenomena in nature for which the theory of independent species creation had a less satisfactory explanation (or none
at all). But there was a practical point, too, that tied these subjects
together. "We have no written pedigrees," Darwin noted (Origin,
425); "we have to make out community of descent by resemblances

of any kind. Therefore we choose those characters which, as far as we can judge, are the least likely to have been modified in relation to the conditions of life to which each species has been recently exposed." To establish true evolutionary relationships – the now-transformed task of classification – the naturalist must seek clues to ancestry in the commonalities of form, by definition those that had changed least over time. Some of these commonalities would be found in structures so vital to the organism that they could not change much at all, but other characters could also reveal those commonalities, provided they had not been modified through natural selection. Embryonic forms were often protected from selection, in Darwin's view, if they were in eggs or wombs or otherwise not actively exposed to the struggle for existence. Developing forms not only revealed the fact of common descent, then, but could also provide the clues to specific questions of classification.

But there was still more to Darwin's discussion. Right after he listed the main embryological facts he sought to explain (given in the long quotation cited earlier), he introduced some new issues. A long paragraph addressed the question of when variations appear in individual development, and concluded that

> it is quite possible, that each of the many successive modifications, by which each species has acquired its present structure, may have supervened at a not very early period of life; and some direct evidence from our domestic animals supports this view. But in other cases it is quite possible that each successive modification, or most of them, may have appeared at an extremely early period. (*Origin*, 444)

In other words, modifications could appear early on or not. "[A]t whatever age any variation first appears in the parent," Darwin continued, it seemed likely to him that "it tends to reappear at a corresponding age in the offspring." Elevating these two statements – the first quite vague, the second more specific – to principles, Darwin wrote that they could account for "all the above specified leading facts in embryology" (*Origin*, 444).

What was going on here? Why did Darwin insist that the moment at which variation appears is an important issue, and that the corresponding age of appearance in the offspring was also something that needed to be confronted? The answer, I believe, is that introducing these considerations allowed him to do three things that would assist in translating embryology into his new framework. First, he sought to account for what had formerly been seen most often as the product of a transcendental Law of Development in historicist and materialist terms, by explaining just how embryos might come to reveal their ancestral history. Second, to do this, he needed to connect his understanding of the embryo-ancestor relationship to other elements of his theory, especially variation, selection, and inheritance. And third, he sought to account in these same terms not only for cases in which embryos revealed the organism's ancestry but also for cases in which they did not, for embryos' developmental stages could bear complex relationships to variation, selection, and the representation of ancestry. Darwin's two principles of embryology and inheritance provided a crucial hinge-point between solving an old problem – the nature of the relationship between embryological development and classificatory affinities – and resituating that problem itself within a framework that decentered its importance.

EMBRYOS AND THE ORDER OF NATURE

Darwin was well aware of the different kinds of relationships his contemporaries and predecessors had drawn between individual development and the order of nature. The choices were far more diverse than the stark opposition between creation and descent that he posed in the *Origin*, and the means of choosing among them not at all clear-cut. Many naturalists, especially in the German-speaking lands, believed that there was a general "law of development" in nature that governed both individuals and the overall history of life. In such formulations, typically, individual development reflected a macrocosmic trend toward increasing progress and complexity. Conversely, naturalists thought (though not without contestation), they could be confident of the overall increase in progress and complexity of the broader organic world that was gradually being revealed in the fossil record in part because they saw a parallel in the individual embryo (Richards 1992; Nyhart 1995; Gliboff 2008).

Many naturalists shared this broad conviction, but not all were so certain that simply positing a "law" of development was a satisfying way to account for the parallels. What did it mean to say that such a law "governed" both the development of the organic world as a whole and individual development? Was this law simply an empirical generalization, or did it have causal efficacy? If the latter, how did that causal connection work? Many naturalists sought a more specific connection between the pattern of development of embryos and the larger patterns of organic nature as a whole, even as they were working out the details of both. Two ways of connecting up these two patterns had become prominent by the mid-1840s, when Darwin started to engage the topic with some intensity. One approach, which we will call the recapitulationist approach, took a primarily linear perspective on both the order of nature and the understanding of individual development. This view held that as individuals developed, they worked their way up the chain of being from less to more complex. This was the perspective held by Friedrich Tiedemann, who believed that the brains of mammalian fetuses passed through the adult stages of the lower vertebrate classes as they advanced in development. Darwin would have been familiar with this approach as far back as his reading of Charles Lyell's *Principles of Geology*, for in volume 2, Lyell (1832; reprint 1991) noted Tiedemann's finding, "most fully confirmed and elucidated by M. Serres, that the brain of the foetus, in the highest class of vertebrated animals, assumes, in succession, the various forms which belong to fishes, reptiles, and birds, before it acquires those additions and modifications which are peculiar to the mammiferous tribe." Lyell specifically characterized Tiedemann's views as transformist: "So that in the passage from the embryo to the perfect mammifer, there is a typical representation, as it were, of all those transformations which the primitive species are supposed to have undergone, during a long series of generations, between the present period and the remotest geological era" (Lyell 1991, 63). Tiedemann's argument was reinforced by Serres and several other important French students of development, who interpreted monstrosities as "arrests of development." That is, many monsters resulted from a failure to develop beyond a certain lower stage of development that paralleled the hierarchy of being. The logical connection between monstrosity and transformism was this: monstrosities subtracted levels of complexity from the end of their development, stopping earlier and representing a lower form. Perhaps all a creature needed to do in order to create a newer, higher form was to extend the end of its development. However, Lyell objected, animals in fact "never pass the limits of their own classes to put on the forms of the class above them. Never does a fish elevate itself so as to assume the form of the brain of a reptile; nor does the latter ever attain that of birds; nor the bird that of the mammifer" (Lyell 1991, 63). The logic of inversion was false.

This linear view, which had earlier incarnations, had previously been objected to, perhaps most famously by the Estonian "father of embryology" Karl Ernst von Baer. Von Baer's alternative, just becoming available to British naturalists in the early 1840s, did two things. It rejected a single scale from monad to man, supporting instead the view of Georges Cuvier, France's leading zoologist, that there were four basic and distinct kinds of organization in the animal kingdom (called "Types" by von Baer and "Embranchements" by Cuvier). Von Baer's view further interpreted development as a successive process of differentiation that paralleled the successively smaller classificatory groups to which an individual belonged. The embryo first exhibited the characteristics of the vertebrate Type, then its class (e.g., bird), then the order, and so forth down to the individual.

Although Darwin had encountered this view by 1838, he was most struck by the gloss on it presented by Henri Milne-Edwards in an 1844 article, which Darwin read in 1846, just as he was beginning his barnacle work. To the general idea that the embryo first exhibited the broadest characteristics of the Type and then successively the more particular characteristics of the class, order, genus, and species, Milne-Edwards added the corollary that the more characteristics two organisms shared in development, the more closely were they related. Embryology could thus be used to ascertain how closely or far apart two organisms should be classified.

Echoing the divisions of the nineteenth century, historians have long viewed the two systems of linear recapitulation and developmental differentiation as sharply distinct and opposed to one another. Recapitulation went with the linear view and a strong notion of absolute progress; differentiation was associated with branching and specialization. Yet Darwin, and at least one naturalist before him, thought that they could be reconciled. To see how, we must note one key issue to which historians have devoted intense scrutiny. To what, exactly, did naturalists compare the stages passed through by present-day embryos of higher forms?

There are four possibilities. The stages of present-day higher embryos might be comparable to present-day adults (especially of lower forms), to present-day embryos, to adults of past forms, or to embryos of past forms. In the pre-Darwinian linear recapitulation-ist perspective, early stages of present-day embryos were compared to present-day adults of lower forms; in Tiedemann's presentation, these also "represented" or were presumed to be analogous to (or even "the same as") adults of past forms. The differentiationist point of view compared embryos only to other embryos, not to adults. Von Baer and Milne-Edwards were mainly interested in comparisons among living organisms, not in interpreting past organisms. No one was talking about comparing present-day embryos to embryos of past forms – until *Vestiges*, it would appear.

In 1844, the anonymous author of the *Vestiges of the Natural History of Creation* sought to draw explicit parallels (and connec-tions) across fetal brain development, the present-day hierarchy of organic complexity, and the fossil record. The parallels, represented in his accompanying chart, appear clear: the human fetus's resem-blance in the fifth month to that of a rodent lines up exactly with the appearance of Rodentia in the lower Eocene. It would seem as though the author of the *Vestiges* wanted the embryological stage to resemble a past adult. Yet the Vestigiarian's language on the ana-logical target of embryological resemblance was mixed. On the one hand, he declared straightforwardly, "It is only in recent times that physiologists have observed that each animal passes, in the course of its germinal history, through a series of changes resembling the *permanent forms* of the various orders of animals inferior to it in the scale" (Chambers 1994, 198). On the other hand, he later stated (212), "But the resemblance is not to the adult fish or the adult reptile, but to the fish and reptile at a certain point in the foetal progress." Fur-thermore, as represented in an accompanying diagram (Figure 11.1), his concept of development was not strictly linear: each class within the vertebrates followed a common path to a certain point and then branched off into a path unique to its group. Struggling to accom-modate both a linear perspective and a branching one, he imagined a main line of ascent leading to humans, which lower types partially followed before branching during their development.

[I]t is apparent that the only thing required for an advance from one type to another in the generative process is that, for example, the fish embryo should not diverge at A, but go on to C before it diverges, in which case

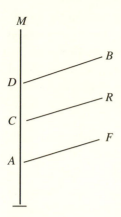

Figure 11.1. Representation of development, in which an embryo (at A) that might develop into a fish (at F), continues to advance (through C, D, and M), giving rise over time to reptiles, birds, and mammals (R, B, and M); from Chambers's anonymously published *Vestiges of the Natural History of Creation* (1844).

the progeny will be, not a fish, but a reptile. To protract the *straightforward part of the gestation of a small space* – and from species to species the space would be small indeed – is all that is necessary. (Chambers 1994, 213)

This was a critical early instance of a scientific writer seeking to combine a linear view of progress with a branching conception of development. As much as Darwin may have detested the *Vestiges*, it constituted part of the picture he was gleaning as he developed his own ideas during the key period of the mid-1840s, and it may have given him something to think about.

So, what did Darwin think? To what did he believe present-day embryological development should be compared?

DARWIN AND EMBRYOLOGICAL RESEMBLANCE

Historians differ in their interpretations of Darwin's views on the recapitulationist versus differentiationist understandings of the embryo. Most follow the argument first set out by E. S. Russell in 1916 and reinforced sharply by Stephen Jay Gould in 1977, that Darwin rejected the former in favor of the latter. In 1981, Dov Ospovat refined this view, arguing that Darwin followed the linear recapitulation model in his early work but that in the mid-1840s he was persuaded by more recent work, especially that of Milne-Edwards, arguing that successive differentiation more accurately expressed the parallel between embryos and the larger order of nature. Broadly interpreted within Darwin's developing framework, branching and differentiation in embryonic development mirrored the branching and differentiation in varieties, species, and the larger classificatory hierarchy. In this interpretation, embryos resembled neither the adults of lower present-day types nor historical adults, but only other embryos, present-day and in the past (Ospovat 1981, 166–7).

By contrast, Robert J. Richards has argued (1992) that Darwin believed that present-day embryos tended to resemble ancestral *adults*. In this view, embryonic stages revealed a sequence of adult ancestors – more primitive stages of development in the historical development of life. Critical to this interpretation is a strong reading of Darwin's two principles of embryology and inheritance. If one understands the first principle, that new variations "supervene at a not very early period of life," to mean that such variations appear as end stages of development, and then, following the second principle, they reappear at a "corresponding stage in the offspring," it follows that new *evolutionary* variations will be tacked on to the end of *individual* development in subsequent generations. Stephen Jay Gould dubbed this the doctrine of "terminal addition" (though he excluded Darwin from his list of recapitulationists who took this view). By this logic, a present-day individual will run through adult ancestral stages as it goes through development because that is how the developmental sequence itself came into being, by adding new stages to the end of what was a mature (if primitive) organism. In Richards's reading, Darwin was a recapitulationist, who viewed present-day embryos as primarily comparable to ancestral adults.

Few historians have agreed with this interpretation of Darwin as a recapitulationist, for that would seem to tie his ideas to linear thinking rather than branching, and to a progressive hierarchy rather than a view of change as differentiation (see, e.g., Bowler 2003). After all, Darwin made repeated statements such as "the embryo is the animal in its less modified state. . . . In two groups of animal, however much they may at present differ from each other in structure and habits, if they pass through the same or similar embryonic stages, we may feel assured that they have both descended from the same or nearly similar parents.... "(*Origin*, 449). Surely this is evidence for differentiation, and for embryos resembling embryos rather than ancestral adults.

But Darwin also offered the example of the forelimbs becoming modified in different directions over evolutionary time from a common ancestral pair of legs, becoming in different descendants hands, paddles, and wings:

and on the above two principles – namely of each successive modification supervening at a rather late age, and being inherited at a corresponding late age – the fore-limbs in the embryos of the several descendants of the parent-species will still resemble each other closely, for they will not have been modified. But in each individual new species, the embryonic fore-limbs will differ greatly from the fore-limbs in the mature animal; the limbs in the latter having undergone much modification at a rather late period of life, and having thus been converted into hands, or paddles, or wings. (*Origin*, 447)

Clearly, differentiation is occurring here, but the stress on the "rather late age" and on the modifications "being inherited at a corresponding late age" (the two principles) can also be read as Darwin insisting on terminal addition as the means by which this differentiation took place, and as the reason why embryos resemble one another.

In fact, the contradiction between linearity and branching is only apparent, and may be resolved. Darwin did so, and it is reasonable to believe that he did so via adult ancestors. If two organisms have a common ancestor, then that adult ancestor may be *both* the end product of a particular course of development (up to that point) *and* an earlier stage of the existing course of development of a present-day organism. The doctrine of terminal addition is not necessarily tied to a strictly linear view of nature, but is compatible with a branching one.[1]

To see this most clearly, consider that both development and evolution may be viewed from two ends: from the base of a branching Darwinian tree, or from a twig at the living end. From the base, moving upward, we see branching. But looking back from that twig,

---

[1] I have made a similar argument in explicating Ernst Haeckel's views on recapitulation (Nyhart 1995, 134–5).

the organism views a linear history back into the past – the history that led to itself. From another endpoint/twig, a different organism also sees a linear history leading backward. Where these two lines meet in a common ancestor, their two backward-leading histories become one. Branching is what we see as we move forward; linearity and joining are what we see as we move backward in time.

From Darwin's standpoint of evolutionary recapitulation, the same held true in individual development. Individual development, read forward from the deep past, would take place as follows: as evolutionary history proceeded, new variations would tend to be added on to the end of embryonic development. Natural selection would then tend to cause divergent variations to be selected, following Darwin's principle of divergence. Thus as new stages were added on to individual development, incipient evolutionary divergence would simultaneously take place. Suppose that two different end-stage variants eventually resulted in two new, modified species. Each would be able to trace back through its individual development a record of its own evolutionary history, and these two courses of individual development would join up at the point at which the evolutionary histories joined up. So embryos would tend to share a longer common developmental history the more recently they had branched off from one another, and even embryos of the same class (but of different families and orders) would tend to share common features at the very beginning of their development, reflecting their distant and early common ancestry.

Although Darwin said all of these things separately, it is difficult to find him putting the whole picture together – which may be one reason why his friends (and most later historians) did not fully appreciate his achievement. Yet if this reasoning is correct – and I am convinced that this does indeed reflect an important part of Darwin's reasoning – then his achievement with respect to embryology was substantial indeed. Darwin resolved the opposition between the linear and differentiationist approaches to the problem of embryonic resemblance, and he did so within his own framework of evolution by natural selection working on variations. All it took was a couple of ancillary principles to make the shift.

And yet I would suggest that this was not what satisfied Darwin himself most about his interpretation of embryology. His two principles of embryology and inheritance not only explained embryos' resemblance to other embryos and to ancestral adults but also offered an account of the very nature of development itself.

Consider what Darwin remembered in his *Autobiography*: he was especially proud of his "explanation of the wide difference in many classes between the embryo and the adult animal, and of the close resemblance of the embryos within the same class" (*Autobiography*, 125). What constitutes "the wide difference in many classes between the embryo and the adult animal"? Development. Terminal addition plus inheritance at a corresponding age explained development itself: "This process, whilst it leaves the embryo almost unaltered, continually adds, in the course of successive generations, more and more difference to the adult" (*Origin*, 338). This was why embryos resembled ancestors. But it was also why organisms developed at all: development was a consequence of the process of evolution. By using his two principles, Darwin could derive from evolution the very process of development.

DEVELOPMENT, VARIATION, SELECTION, AND INHERITANCE

Darwin's two principles did not just allow him to establish a material connection between individual development and evolution; they worked to weave this connection deeply into the vocabulary of his overall theory. The idea that modifications tended to supervene at a not-very-early stage of life was a claim about variation, a central component of his theory. That such modifications would tend to be inherited at a corresponding age in the offspring was a claim about heredity. Both were entwined in the physiological problem-complex of development, variation, and inheritance, known at the time as "generation," one of Darwin's most abiding interests, which would culminate in his hypothesis of pangenesis, published in 1868 (see Hodge 1985; Sloan 1985). These concerns plunged him into the nitty-gritty of the material processes by which organisms might become transformed over time.

The mass of facts he had gathered before him did not present a simple picture. Discerning regularities among the varied facts of heredity and variation was one of Darwin's most intractable problems. His need to wrestle with this problem was part of what made Darwin's interest in and claims about development different from those of his morphologist predecessors, and ultimately what reduced its place in his overall system from the central topic to one among several important areas of consideration.

Despite the stronger claims he sometimes made at moments of summary, Darwin's discussion of embryology and variation was filled with equivocation. Virtually every single time he discussed the tendency of variations to supervene at a late stage of development, he followed with a counterexample: they could also appear at an early stage. Or development might not proceed very far at all. Everything was qualified; Darwin's language was littered with expectation-lowering phrases. "It is quite possible" that new characteristics "may" have appeared late in life. But then again, it was "quite possible" that they may not have.

This equivocation has two important aspects: one has to do with Darwin's confidence in his claims, the other with their generality. First, at the time of writing the *Origin*, Darwin appears not to have been fully confident about the temporal relationship of variation to development. He hinted at the difficulty he faced in coming to a resolution at the beginning of the passage introducing his two principles of embryology and inheritance, when he treated the assumption that "variations necessarily appear at an . . . early period" – a view he opposed, but without complete conviction. It turns out that his equivocation was significant, and that Darwin had changed his mind from an earlier, opposing position. The story of Darwin's shifting stance shows his lack of certainty about the relation of modification to development, his continued efforts to link this problem up with classificatory relationships, and his desire to connect both issues to selection.

As early as his "Sketch" of 1842, Darwin had suggested that variations could enter in at different times during development, but that it did not matter just when they did so if they were protected from selection (*Foundations*, 42). In his 1844 "Essay," he elaborated upon the point, noting that all that counted was that an adaptive structure be in place at a time when selection could act to preserve it, which typically occurred only in the mature state. Thus if it could be shown that variation did not always take place early on in development but tended to take place later on, and that selection too tended to take place at later stages, this would account for differences in the adult organism; at the same time, the lack of selection at early stages would leave those early stages resembling one another more than their mature forms (*Foundations*, 220-7). Darwin's measurements of newborn greyhounds and bulldogs showed their legs and noses to be the same length, confirming that even greatly varying breeds resembled one another more closely at birth. (In the 1850s, he would gain further confirmation of his views with measurements of neonate pigeons.) Using his characteristic analogy between artificial and natural selection, he argued that if nature's selective hand worked on mature individuals as human selection did, then it would make sense that in nature, too, younger individuals would tend to resemble one another more than adults.

However, in the manuscript of his "Big Book" – the one for which the 490-page *Origin* served as a mere abstract – Darwin changed his position, based on reading he had done just after his 1844 essay. There he wrote that "modifications in the mature state will almost necessarily have been preceded by modification at an earlier age" (*Species Book*, 302). He then cited the French entomologist Gaspard Auguste Brullé, who in 1844 had argued that the more complex an organ would become in its adult stage, the earlier it must appear in development. The botanist François Marius Barneoud had found something similar in plants. Darwin then immediately turned to discussing Milne-Edwards's 1844 paper, which used the criterion of embryological differentiation to establish classificatory affinities. As Darwin interpreted Milne-Edwards, "the more widely two animals differ from each other, the earlier does their embryonic resemblance cease" (*Species Book*, 303). Darwin concluded the section with the following:

If the foregoing principle be really true & of wide application, it is of importance for us; for then we might conclude that when any part or organ is greatly altered through natural selection it will tend either actually first to appear at an earlier embryonic age or to grow at a quicker rate relatively to the other organs than it did before it had undergone modification.: consequently, . . . this early formation will tend to act on the other & subsequently developed parts of the system. (*Species Book*, 304)

Darwin copied out this section of his manuscript and sent it to Huxley for review on July 5, 1857.

Especially I want your opinion how far you think I am right in bringing in Milne Edwards['] view of classification. I was long ago much struck with the principle referred to: but I could then see no rational explanation why affinities should go with the *more or less early* branching off from a common embryonic form. But if MM Brullé and Barneoud are right, it seems to me we get some light on Milne Edwards['] views of classification; and this particularly interests me. (*Correspondence*, 6: 420–1)

What exactly Darwin thought that light would be remains unclear. In his reading notes on Milne-Edwards's article in the mid-1840s, Darwin had written that "to mature an organ, a certain time is required, & that the earlier changes can alone be hurried. This at once nearly explains the gradual loss of embryonic characters.... " (*Correspondence*, 4: 393). By 1857, his worrying over the problem may have made its implications more expansive: perhaps he inferred that complex forms – those most modified from an ancestral one – would share an embryonic resemblance only early in development. This would associate early branching from a common ancestor (Darwin's interpretation of Milne-Edwards) with modification at an early stage of development. In any case, it is clear that Darwin was working to combine and reconcile the results of Brullé and Milne-Edwards while also translating them into his own terms. At this moment, in 1857, he was thinking that new variations normally supervened early in development, not at the end. (Rachootin [1984] treats at length how Darwin developed Brullé's ideas.)

Huxley's reply was scathing – as he put it, he "brûler'd Brullé," arguing that every bit of the latter's evidence was wrong and that his logic was, if anything, worse. Moreover, he corrected Darwin's apparent interpretation of Milne-Edwards: "he seems to me to say that, not the *most highly complex*, but the most *characteristic* organs are the first developed" (*Correspondence* 6: 424–7). Thus Brullé's argument did not reinforce Milne-Edwards's, as Darwin would have it.

Darwin not only omitted the passage in the *Origin* but, as we have seen, tilted in the opposite direction, returning to his views of the early 1840s. Significant new variations tended to appear not early in development but late. This he put to work to explain embryos' resemblance to ancestors (and, significantly, this discussion appeared in the chapter on affinities, morphology, and classification, not in the chapter on laws of variation, as had the earlier version in the big species book). Yet the uncertainty remained – Brullé had found cases where modifications seemed to supervene early, and Darwin continued to take those instances seriously. The general rule that he came up with in the end was not one in which embryos *must* mirror ancestors or *must* retrace the successive features of the class, the order, the genus, and the species. Instead, it was one that emphasized contingency, variation, and the intervening hand of natural selection.

And this brings us to the other aspect of Darwin's equivocation over the timing and nature of the appearance of heritable modifications in development: the question of their generality. Darwin wanted to account not only for those cases in which embryos of related forms resembled one another, but also for cases in which they did not, as well as for different amounts and moments of resemblance. To succeed, his framework had to accommodate all the different cases. So he proceeded to show how they all might be understood as the product of evolution by natural selection acting on variations, whenever they might appear.

Sometimes embryological development exhibited a succession of modified ancestral adults. This would be the case in many vertebrates, especially in those organisms, such as birds and mammals, that were protected from the pressures of selection in the egg or womb, where there had been much modification from an original primitive ancestor, and where successive modifications were inherited at a corresponding age in the offspring. But in some organisms, early developmental stages did not benefit from the protections of egg or womb and had to fend for themselves. The pressure of selection on these free-living larval forms meant that adaptations would appear earlier on in development, tending to efface ancestral connections (*Origin*, 440) and also in some cases producing distinct metamorphic stages (*Origin*, 448). In yet other cases, embryos could display variations early on in their development that made them resemble more closely the adult forms. Drawing on his own research measuring pigeon neonates, Darwin noted that this was the case in the short-faced tumbler pigeon, which revealed its adult facial characteristic at the moment of hatching, whereas other pigeon varieties all closely resembled one another when newly hatched. He used this case to move away from the question of ancestral resemblance to focus more closely on a larger group of cases in which there was little developmental difference between young and adults; and again, he explained these cases via the pressure of selection. If the young, instead of being protected from selection, were exposed to the same environmental pressures as their parents, they would tend to display early on in development the same adaptive characteristics.

In this discussion, the resemblance of some embryos to ancestors represented one pattern among many – the only one that required modifications to supervene at "a not very early period of life." It may have been Darwin's default pattern, but the others were also significant and demanding of explanation. One way or another, evolution by natural selection could account for all the various patterns of development seen in the organic world.[2] Embryo-ancestor resemblance was one important consequence of evolution, and Darwin's theory explained it in a nonidealistic way. But in Darwin's scheme, all those other developmental patterns required – and received – explanation in evolutionary terms.

In the fourth edition of the *Origin* (1866), Darwin clarified and strengthened his claims about recapitulation, while also broadening his discussion of nonrecapitulatory cases. Thus he explicitly mentioned that embryonic resemblance corresponded to an adult ancestor:

[I]t is probable, from what we know of the embryos of mammals, birds, fishes, and reptiles, that [these animals] are the modified descendants of some one ancient progenitor, which was furnished in its adult state with branchiae, had a swim-bladder, four simple limbs, and a long tail fitted for an aquatic life. (*Variorum*, 702, 306.10.d)

Crustaceans showed a similar phenomenon. Here Darwin leaned heavily on a small book, *Für Darwin*, published in 1864 by the German émigré zoologist Fritz Müller, who lived in Brazil and had closely studied the crustacea there. In his book, Müller demonstrated a common larval stage for a range of crustaceans with widely differing adult states and argued that this larval stage represented an adult common ancestor. His position, based on impeccable embryological research, so pleased Darwin that he paid for the translation and publication in English of Müller's little book (Müller 1869; West 2003, 120–1). The fourth and later editions would be peppered with new references to Müller.

Even as he clarified his claim that embryos resembled adult ancestors, however, he also expanded his discussion of cases in which they did not. Again, the fourth edition elaborated further on cases

---

[2] Darwin included plants in his discussion but devoted much less space and attention to them.

of early adaptation, especially in insects – cases, for example, in which the need to survive in unprotected environments produced distinct metamorphic stages, which might sometimes be "higher" than later ones. Thus, to his first-edition declaration that "community in embryonic structure reveals community of descent" (*Origin* 449), he added, "but dissimilarity in embryonic development does not prove discommunity of descent, for in one of two groups all the developmental stages may have been suppressed, or may have been so greatly modified as no longer to be recognized, through adaptations, during the earlier periods of growth, to new habits of life" (*Variorum* 703, 312:d) Adaptation and selection would account for cases in which embryos did not reveal their ancestral heritage.

From first to last, Darwin believed that when embryos resembled others of related classificatory groups, they did so because they shared a common ancestry. But his understanding of variation, selection, and the general contingency of nature led him away from an understanding of morphology as the study of the *laws* of form in the strict sense of earlier (and later) continental morphologists. Nature did not make laws of form. She might have some rules governing form – frequent regularities – but these could be broken, and the organic world was littered with such breakage. The only law was evolution; all else was contingent. Even as Darwin solved the problem of embryology and ancestral affinity, he dissolved it into the larger complex of evolution by natural selection.

12 Darwin's Botany in the *Origin of Species*

He moons about in the garden, and I have seen him standing doing nothing before a flower for ten minutes at a time.
If he only had something to do, I really believe he would
be better.

– Charles Darwin's gardener[1]

DARWIN'S BOTANY: INTRODUCTION

Taxon-based studies defined much of natural history in the nineteenth century, with botany and zoology serving as the two major
realms of such inquiry. Darwin himself was taxonomically promis-
cuous, flitting from organism to organism much as his curiosity
dictated but also out of a utilitarian need for particular examples to
support a generalizable theory that explained the diversity of living
organisms. Thus, in the course of his scientific career Darwin studied
a range of organisms and familiarized himself with related sciences
like geography and geology. But increasingly after the *Origin*, his lifelong interest in botany not only revealed itself but came to dominate
his research.

That interest had started early. In fact, one could say that he inherited it; his grandfather Erasmus was a translator of Linnaeus, while
another relative, John Wedgewood, was one of the founders of the
Royal Horticultural Society. Almost serving as a prophetic image of
the role that botany would play in his life, an early portrait of the
young Charles shows him seated next to his sister Catherine holding a pot of plants. At the age of twelve or so, he was given the task

---

[1] As cited in Morris et al. (1987, 48).

of counting peony blossoms in his father's garden; and later, while engaged in what became a failed study of medicine, he was exposed to the study of *materia medica* (or plants known to be useful for medical or healing purposes) (Kohn 2008b). As a young Cambridge student, Darwin formally studied with John Stevens Henslow and spent so much time with him that he became known as "the man who walks with Henslow." Under Henslow's tutelage, he performed anatomical dissections on plants like *Geranium*, and familiarized himself with taxonomic studies (Kohn 2008b). His exposure to Henslow's herbarium and to its special emphasis on "collated" specimens, which stressed multiple collections, was especially important. It helped lay the groundwork for his understanding of variation and indeed for his understanding of speciation (Kohn et al. 2005).

Later, while exploring unfamiliar terrain during the *Beagle*'s various layovers in South America, Darwin collected numerous specimens of local floras and sent them back home for permanent storage as well as identification. Later still, while enjoying his life as the celebrated "squarson-naturalist" of Downe, he began to undertake detailed observational and experimental studies on select plants. Recognizing their importance to his larger theoretical project, Darwin also formed strong professional ties with leading systematic botanists of his day such as Joseph Hooker, the director of Kew Gardens, and Asa Gray, a botanist at Harvard University. He also cultivated an extensive correspondence network with breeders, horticulturalists, collectors, and compilers of floras the world over. These professional networks helped fuel his growing interest in plants, especially by providing some of the best examples in support of his theory of descent as set forth in the *Origin of Species*.[2]

Plants lent themselves readily to Darwin's investigations as examples in support of his theory, but they were especially valuable for the kinds of observational and experimental studies that characterized his research in the mature phase of his career. They were tractable "model organisms" (to use a presentist term), easy to grow (depending on the plant chosen) and stationary (and therefore easy to observe); knowledge of them from horticultural, agricultural, and

[2] The critical year for Darwin's shift to botany is 1860. See David Kohn, "Darwin's Botanical Research," in Morris et al. (1987, 50–9). For an overview of Darwin's life and work, see Janet Browne (1995 and 2002).

breeding practices was well developed by the middle decades of the nineteenth century, as was knowledge of their morphology, anatomy, and even cytology thanks to new microscopic, sectioning, and staining techniques. Plants additionally bore a staggering assortment of vegetative and reproductive structures, with complex mating patterns that included self-fertilization, cross-fertilization, and elaborate pollination mechanisms. These last often required other organisms, like insects and birds, thus making them ideal for studies of co-adaptation. And with what we now recognize as "open" or indeterminate growth patterns that reflected readily the direct effects of the environment, plants also drew attention to the general process of adaptation, making them ideal systems for exploring the direct and indirect effects of the environment (Briggs and Walters 1997). For all these reasons, Darwin increasingly devoted his efforts to drawing on the study of plants not only to fortify, but also to extend his theoretical insights as first developed in the *Origin*.

Though it would be hard to consider Darwin a botanist in the strict sense of the term, he employed examples of plants in at least three interrelated ways: (a) in his thinking about evolution, (b) in his own researches, and (c) in his professional life as a whole. By the end of his long and productive career, he had completed no less than six books exclusively devoted to botanical subjects, published between the years 1862 and 1880, in addition to botanical articles published in the weekly *Gardener's Chronicle* and journals like the *Agricultural Gazette* (Ornduff 1984). Some drew extensively on the work of colleagues, while others drew exclusively on his own observations and experiments performed in the hothouses and experimental gardens of his home in Downe (Morris et al. 1987).

PLANTS IN DARWIN'S *ORIGIN OF SPECIES*

The importance of plants for Darwin's theory is manifested by their prominent appearance in the first four chapters of the *Origin*, the chapters that lay the groundwork for his theory of descent. Plants do not figure significantly in the chapters on geology (Chapters 9 and 10) or in the chapter dealing with instinct (Chapter 7); but in all other chapters of the *Origin*, plant examples appear frequently, customarily following mention of a phenomenon demonstrated in animals.

As early as the very introduction to the *Origin*, Darwin set this comparative rhythm in motion by making reference to how "preposterous" it would be to attribute to "mere external conditions, the structure, for instance of the woodpecker, with its feet, tail, beak, and tongue, so admirably adapted to catch insects under the bark of trees" (*Origin*, 3). He immediately followed the animal example with a comparable plant example, the "misseltoe" (or mistletoe), which "draws its nourishment from certain trees, which has seeds that must be transported by certain birds, and which has flowers with separate sexes absolutely requiring the agency of certain insects to bring pollen from one flower to the other." It was "equally preposterous," Darwin concluded, "to account for the structure of this parasite, with its relations to several distinct organic beings, by the effects of external conditions, or of habit, or of the volition of the plant itself" (*Origin*, 3).

Chapter 1: Cultivated Plants

Chapter 1, titled "Variation under Domestication," relied heavily on examples from cultivated plants. Darwin focused at great length on the interplay of two well-known aspects of the biology of plants: their vegetative reproductive habits and their variability in different environments. Variability was especially problematic. Darwin pointed out that "seedlings from the same fruit and young of the same litter, sometimes differ considerably from each other," though they have had exactly the same conditions of life; but determining how much of this variability to attribute to the direct action of the environment was not easy. Darwin nonetheless maintained that "some slight amount of change, may I think, be attributed to the direct action of the conditions of life – as, in some cases, increased size from food, colour from particular kinds of food and from light, and perhaps the thickness of the fur from climate" (*Origin*, 10).

When seeking "laws of variation" or attempting to explain the causes of such variation, Darwin acknowledged that they were "quite unknown, or dimly seen," though he thought it worthwhile to explore historical treatises on older cultivated plants like the potato, the hyacinth, and the dahlia for what they revealed. He was surprised especially by the "endless points in structure and constitution in which the varieties and subvarieties differ slightly from each other" in these genera. This was especially striking to Darwin: "The whole organization seems to have become plastic, and tends to depart in some degree from that of the parental type" (*Origin*, 12). Darwin's use of the word "plastic" to describe plant variation deserves special notice, for it presaged the notion of plasticity in general and of pheno-typic plasticity in particular, a phenomenon observed especially in the plant world and understood only well after the 1920s and 1930s (Smocovitis 1997). But Darwin, of course, knew little about the laws of inheritance, let alone the distinction between genotype and phe-notype; he expressed his frustration with the well-known remark that the "laws governing inheritance are quite unknown" (*Origin*, 13). He simply stipulated that "any variation that is not inherited is unimportant for us" (*Origin*, 12).

The chapter then turned to "man's selection," where Darwin employed examples of human modification of animals and plants. He considered "man's" ability to accumulate in organisms traits useful to himself:

We cannot suppose that all the breeds were suddenly produced as perfect and as useful as we now see them; indeed, in several cases, we know that this has not been their history. The key is man's power of accumulative selection: nature gives successive variations; man adds them up in certain directions useful to him. In this sense he may be said to make for himself useful breeds. (*Origin*, 30).

In this way Darwin stated that new forms have come into being, summoning up Youatt's famous metaphor of "the magician's wand, by means of which he may summon into life whatever form and mould he pleases" (*Origin*, 31). As in the case of animals, new forms of plants have so arisen; but in plants, the "variations are...more often abrupt" (*Origin*, 32).

Plants also evinced a staggering diversity of parts within the same species. While Darwin made allowances for "the laws of the corre-lation of growth," he stated that "as a general rule, I cannot doubt that the continued selection of slight variations either in the leaves, the flowers, or the fruit, will produce races differing from each other chiefly in these characters" (*Origin*, 33). He further added that the horticulturalist, however, must have patience, since changes, even

in plants, can occur only slowly over many generations (*Origin*, 36).
Darwin concluded the chapter with the following closing thought:
"Over all these causes of Change, I am convinced that the accumulative action of Selection, whether applied methodically and more
quickly, or unconsciously and more slowly, but more efficiently, is
by far the predominant Power" (*Origin*, 43).

Chapter 2: Importance of Data from Floras to the Argument That Varieties Are Incipient Species

Plants figured most prominently, and were most important in laying out the argument for Darwin's theory, in Chapter 2, "Variation
under Nature." In this chapter, Darwin explored the nature and character of variation in the natural world. This is where he explored the
notion of individual differences and argued for continuous gradations going from individual differences to varieties, incipient species,
and "good" species. Early on in the chapter, Darwin recognized the
puzzling phenomenon of "protean" or "polymorphic genera." These
were widely varying species with multiple forms of "inordinate variation" that were especially prevalent in plant genera like *Rubus*,
Rosa, and *Hieracium* (this last plant, known as the hawkweed, was
later to give Mendel a headache and possibly led him to drop his
experimental studies of inheritance in plants), as well as in some
insects and Brachiopod shells. For Darwin, these were "perplexing
cases" because they appeared to be polymorphic in all "countries"
and therefore seemed to vary independently of the conditions of life.
He wrote, "I am inclined to suspect that we see in these polymorphic genera variations in points of structure which are of no service
or disservice to the species, and which consequently have not been
seized on and rendered definite by natural selection, as hereafter
will be explained" (*Origin*, 46). The preponderance of polymorphic
genera in some large plant genera aside, Darwin drew heavily on
botanical data, notably from the huge number of flora that had been
compiled or were actively being compiled by systematic botanists.
That data, in Darwin's mind, was crucial in supporting his argument
that varieties are incipient species.

As Karen Parshall (1982) has shown, Darwin relied on mathematical calculations based on a series of floras compiled by botanists

like Joseph Hooker (and his flora of New Zealand), Asa Gray (and his celebrated flora of temperate North America), Hewett C. Watson, and others to lay the groundwork for his belief that varieties were indeed incipient species (see also Browne 1980). From the summer of 1857 to the spring of 1858, Darwin worked studiously on an enormous compilation of botanical data that he then analyzed mathematically. He knew that such use of numerical data to determine relationships of genera, species, and varieties was actually fairly common at the time: Alphonse de Candolle had compiled extensive data in his *Géographie biologique raisonnée*, and Asa Gray had written "Statistics of the Flora of the Northern United States."[3] Though he was not mathematically inclined, Darwin was aware of these studies and looked to them to provide numerical evidence of a correlation between extensiveness of plant distribution and variability. The larger the genus, the larger the species that it contained, or so his theory suggested.[4]

Darwin was of course famously vague about the definition of 'species,' writing on page 44: "No one definition has as yet satisfied all naturalists: yet every naturalist knows vaguely what he means when he speaks of a species. Generally the term includes the unknown element of a distinct act of creation. The term 'variety' is almost equally difficult to define; but here community of descent is almost universally implied, though it can rarely be proved." He held that "the term species, as one arbitrarily given for the sake of convenience to a set of individuals closely resembling each other, and that it does not essentially differ from the term variety, which is given to less distinct and more fluctuating forms" (*Origin*, 52). For Darwin, therefore, species did not differ in kind from varieties; and while the naturalist's definition of species appeared "vague," it served his purposes well: it supported his argument that that there were no clear lines separating species from varieties or varieties from individual differences. Taken as a whole, furthermore, such a view

---

[3] Both had been recently published. See Alphonse de Candolle (1855) and Asa Gray (1856–57).

[4] There were three additional patterns possible: large genera having small species, small genera having large species, and small genera having small species. Darwin argued that the pattern of large genera having large species would be the likely outcome if his theory were true.

also argued against the notion that species were the products of special acts of creation.

To illustrate these points, Darwin wrote: "Guided by theoretical considerations, I thought that some interesting results might be obtained in regard to the nature and relations of species which vary most, by tabulating all the varieties in several well-worked floras" (*Origin*, 53). He considered first what he termed "dominant species," or those that were at the same time the most widely diffused and the most common in a particular geographical region. It was generally known that the most widely ranging species exhibited the most varieties, but Darwin's use of available data in the context of his theory took this point even further:

in any limited country, the species which are most common, that is abound most in individuals, and the species which are most widely diffused within their own country (and this is a different consideration from wide range, and to a certain extent from commonness), often give rise to varieties sufficiently well-marked to have been recorded in botanical works. Hence it is the most flourishing, or, as they may be called, the dominant species, – those which range widely over the world, are the most diffused in their own country, and are the most numerous in individuals, – which oftenest produce well-marked varieties, or, as I consider them, incipient species. (*Origin*, 53–4)

The data also indicated to Darwin that these "dominant species" tended to belong to genera that were of proportionately larger size. Based on his belief that species differed from varieties in degree rather than in kind, Darwin finally conjectured that species in larger genera presented more varieties than species in smaller genera, just as he had postulated if his theory held true. And since his theory held true, it could account for these phenomena better than any creationist interpretation of species.

Chapter 3: Plants and the Struggle for Existence

Strangely enough, Darwin began this chapter, titled "Struggle for Existence," stressing the relative unimportance of being able to distinguish the 300 or so species of British plants as species, subspecies, or varieties (*Origin*, 60). Far more important to him was understanding how species arise in nature and the process by which adaptation to the environment and to other interactions with species takes place. Critical to this process was the inevitable competition that took place in the "struggle for existence," a phrase that Darwin stressed was used in a "large and metaphorical sense, including dependence on another, and including (which is more important) not only the life of the individual, but success in leaving progeny" (*Origin*, 62). Once again, Darwin referred to the "beautiful co-adaptations" seen in the woodpecker and the mistletoe, but took special care to use plants to make the point that such competition is subtle as well as inordinately complex:

Two canine animals in a time of dearth, may be truly said to struggle with each other which shall get food and live. But a plant on the edge of a desert is said to struggle for life against the drought, though more properly it should be dependent on the moisture. A plant which annually produces a thousand seeds, of which on an average only one comes to maturity, may be more truly said to struggle with the plants of the same and other kinds which already clothe the ground. The missletoe is dependent on the apple and a few other trees, but can only in a far-fetched sense be said to struggle with these trees, for if too many of these parasites grow on the same tree, it will languish and die. But several seedling missletoes, growing close together on the same branch, may more truly be said to struggle with each other. As the missletoe is disseminated by birds, its existence depends on birds; and it may metaphorically be said to struggle with other fruit-bearing plants, in order to tempt birds to devour and thus disseminate its seeds rather than those of other plants. In these several senses, which pass into each other, I use for convenience sake the general term struggle for existence. (*Origin*, 62)

Darwin thus painted an ornate picture of interactive relations among organisms in nature; in the process of doing so, he also drew the distinction between interspecific and intraspecific competition, only later formally recognized by twentieth-century ecologists. But the plant examples he noted also served to give a more nuanced meaning to his use of the metaphor "struggle for existence." As he noted, plants could not rightly be said to struggle against drought, but only against other similar plants; nor could they be properly said to "struggle" at all, since that employed a kind of anthropomorphism difficult to uphold in the case of nonsentient organisms like plants. The "convenient" metaphor of "struggle" was thus subject to critical interpretation; Darwin used it loosely to depict the frequently unseen interactive forces at work in the natural world. Indeed, this chapter as a whole, which was devoted to the complex relations of living organisms, made especially notable use of such metaphors, which later scholars from ecologists to literary critics have scrutinized (Beer 1983, 2000 revised edition).[5]

The chapter also revealed Darwin's own clever experiments that eventually inspired the field of plant ecology (Harper 1967). In one critical experiment, Darwin followed the fate of native seedlings on a cleared piece of ground three feet long by two feet wide. Out of 357 seedlings that germinated, 295 were destroyed by organisms like slugs and insects. In yet another experiment on a little plot of turf (three feet by four feet) that had been mowed for some time, he followed the fate of plants allowed to grow freely without mowing and grazing. Darwin found that under these conditions, the more vigorous plants gradually killed the less vigorous plants, even though the latter were fully grown. He counted some nine species that perished as a result of competition from other species that were allowed to grow freely. Such complex interactions were also seen as a result of the enclosure of the heathlands, where Darwin had observed a takeover by the Scotch fir tree. This was due to the protection such enclosures afforded from grazing cattle. In yet other examples, Darwin showed how insects were required for the existence of various plants such as orchids, which required visits from moths to remove pollen masses for fertilization. He also noted yet another set of experiments and observations he conducted on common red clover pollination from humble-bees.[6] One of the most ecological chapters of the *Origin*, it culminated with the famous description on page 74 of the "entangled bank." A metaphor for the complex relations among diverse organisms, the entangled bank described by Darwin was the result of a natural process that obeyed well-defined laws. Thus, an image that appeared disordered or chaotic on the surface

---

[5] The most famous of these evokes a disturbing image: "The face of Nature may be compared to a yielding surface, with ten thousand sharp wedges packed close together and driven inwards by incessant blows, some one wedge being struck, and then another with greater force" (*Origin*, 67).

[6] Darwin performed a series of experiments on the nectary structure and on various flowers and insect pollinators that he published as brief articles or reports for the *Gardener's Chronicle* in the 1850s.

was in fact deeply structured, orderly, and emerged from well-defined natural laws. Its importance not only to this chapter but also to Darwin's thinking in the *Origin* overall is made apparent by its appearance (though somewhat altered in form) in the final dramatic paragraph closing the book.

Darwin concluded this important chapter by once again drawing attention to the subtleness of competition in the plant world, in contrast to that of animals. Just as the structure of the teeth and talons of the tiger served adaptive functions that enabled the animal to compete, and just as the legs and claws of the parasite that clung to the tiger's body enabled it to survive in a competitive world, so too did the "beautifully plumed seed of the dandelion" have competitive value (*Origin*, 77). With reference to such seeds, Darwin then explained that in many plants the store of nutriment therein may at "first sight" bear "no sort of relation to other plants." But then he added: "from the strong growth of young plants produced from such seeds (as peas and beans), when sown in the midst of long grasses, I suspect the chief use of the nutriment is to favour the growth of the young seedling whilst struggling with other plants growing vigorously all around" (*Origin*, 77). Darwin thus also understood the nutritive value of seeds and their role in enabling the plants to survive in a competitive environment.

Chapter 4: Plants and Support for Natural Selection

Chapter 4 finally explored Darwin's "principle of natural selection," following the development of the argument laid out for it in Chapters 1 to 3. Darwin used plant examples in three important sections of this chapter: (a) as illustrations of the action of natural selection; (b) as examples of the intercrossing of individuals; and (c) as examples demonstrating divergence of character.

IMAGINARY ILLUSTRATIONS. Darwin's best evidence in support of his theory was indirect, based on an analogy with "man's selection" and on the geographical distribution of plants and animals on continents and oceanic islands. He had no real direct evidence for natural selection and instead relied in this section on two famously "imaginary illustrations." The first was the case of the predating wolf, while the second, intended to be a more "complex" case, involved the excretion by a plant of a sweet juice that would serve to draw insects for pollination. In the latter example, Darwin devoted over two pages of text to what was mostly a hypothetical discussion on the origin of the complex parts of flowers coadapted as pollinating mechanisms that would serve to enhance cross-fertilization of the plants.

INTERCROSSING OF INDIVIDUALS. The elaborate mechanisms proposed for cross-fertilization then gave way to a discussion of its advantages. Here Darwin built on earlier insights on the subjects of generation, reproduction, and sexuality, broadly construed (*Notebooks*, 170–71). For Darwin, sexual reproduction or intercrossing between unlike individuals was potentially advantageous because it was a means of increasing the variability of organisms. Such variability, alongside generation (which included not just birth but also death), enabled the constant process of adjustment to the changing conditions of life. The alternative mechanisms of vegetative (asexual) reproduction and self-fertilization (inbreeding), though potentially offering temporary advantages (such as enabling organisms to colonize new habitats), limited variability and were in the long term detrimental to organisms. Plants again offered stunning examples of diverse reproductive or mating systems that could serve as examples for Darwin's theory; but what counted as "like" individuals, or what counted as proper, good, or true species, was not easy to ascertain in the plant world, a problem long recognized, and at times also capitalized on, by plant breeders. Indeed, hybridization between "unlike forms," while serving to increase variability, could also disrupt the integrity of species. Darwin was aware of the complex issues raised by hybridization and reserved that discussion for a later chapter devoted exclusively to the subject; in the section on intercrossing, what he needed to show was the advantages of intercrossing, meaning here mostly sexual reproduction. He therefore scoured the works of well-known plant breeders like F. C. Gaertner and J. G. Koelreuter for their knowledge and for relevant examples. He also drew extensively from C. C. Sprengel's work on pollination mechanisms and floral anatomy to engage in a substantive discussion of plant sexuality and the advantages of intercrossing. He was also aware of the phenomenon of hybrid vigour, which was later recognized by twentieth-century geneticists as "heterosis," and used it to explain the advantages of intercrossing:

In the first place, I have collected so large a body of facts, showing, in accordance with almost universal belief of breeders, that with animals and plants a cross between different varieties, or between individuals of the same variety but of another strain, gives vigour and fertility to the offspring; and on the other hand, that close interbreeding diminishes vigour and fertility; that these facts alone incline me to believe that it is a general law of nature (utterly ignorant though we be of the meaning of the law) that no organic being self-fertilizes itself for an eternity of generations; but that a cross with another individual is occasionally – perhaps at very long intervals – indispensable. (*Origin*, 96)

He also added to the store of knowledge from Sprengel his own observations from experiments on *Lobelia fulgens* (on what twentieth-century biologists call self-incompatibility) and his observations on a contrivance that prevents the stigma from receiving its own pollen. In *Lobelia*, he wrote,

there is a really beautiful and elaborate contrivance by which every one of the infinitely numerous pollen-granules are swept out of the co-joined anthers of each flower, before the stigma of that individual flower is ready to receive them; and as this flower is never visited, at least in my garden, by insects, it never sets a seed, though by placing pollen from one flower on the stigma of another, I raised plenty of seedlings; and whilst another species of Lobelia growing close by, which is visited by bees, seeds freely. (*Origin*, 98).

Darwin then noted how varieties of various common vegetables like cabbage, radish, onion, and other plants, if allowed to seed near each other, will see "a large majority . . . of the seedlings thus raised . . . turn out mongrels" (*Origin*, 99). This he confirmed with his own data. Darwin attempted to understand the "mongrelization" of forms by conjecturing that pollen from another variety had a prepotent effect over a flower's own pollen, but he then added that "when distinct *species* are crossed the case is directly the reverse, for a plant's own pollen is always prepotent over foreign pollen" (*Origin*, 99). The flowers of many species, in Darwin's view, therefore had some kind of physiological or structural mechanism(s) in place to ensure that they mated preferentially with their own kind (for the most part, that is). While crosses served to increase variability, they ideally also took place in way that ensured the integrity of species. Darwin thus set the stage for his discussion of hybridization or "hybridism," a phenomenon common in plants but that needed explanation in the context of his theory.

DIVERGENCE OF CHARACTER. By divergence of character, Darwin had in mind a principle of "high importance" to his theory, the process by which individual differences among members of the same species gradually increase so as to form varieties, subspecies, and finally distinct species. Simply stated, the principle was described thus: "the more diversified the descendants from any one species become in structure, constitution and habits, by so much will they be better enabled to seize on many and widely diversified places in the polity of nature, and so be enabled to increase in numbers" (*Origin*, 112). Outlining the process was crucial to Darwin's theory as it explained how species originated; he devoted some fifteen pages of the *Origin* to the complex subject (see Kohn 2008 for extensive discussion of this topic).

Plants once again figured prominently in Darwin's examples demonstrating this important principle. He described a novel natural experiment he had performed himself on turf plots three feet by four feet in size that followed the diversification of the plants therein. He wrote of his results: "it has been experimentally proved, that if a plot of ground be sown with one species of grass, and a similar plot be sown with several distinct genera of grasses, a greater number of plants and a greater weight of herbage can thus be raised" (*Origin*, 113). So too under "natural circumstances" the "greatest amount of life can be supported by great diversification of structure" (*Origin*, 114). Such natural circumstances included small ponds or islets or limited environments where plant and animal inhabitants "jostled" or competed with each other closely. Darwin further noted that introduced or "naturalized" plants that were able to compete and colonize successfully were often of a highly diversified nature, thus presaging insights from the science of the biology of invasive species, the area that is now known as "invasion biology" (*Origin*, 115).

Chapters 5, 6, and 7: Acclimatisation in Plants

In Chapter 5, titled "Laws of Variation," Darwin played on themes referred to especially in Chapter 1, "Variation under Domestication." Only heritable variation was important for his theory, but without detailed knowledge of the laws of inheritance, little could be done to discern variation that was due to direct or indirect actions of the environment (though Darwin did of course also think that direct actions of the environment were often heritable). Darwin nevertheless thought it possible to "dimly catch a faint ray of light" (*Origin*, 132). He repeated his belief that the direct effect of the environment – a Lamarckian process – is extremely small in the case of animals but perhaps rather more pronounced in the case of plants. He yet believed that often the direct effects of the environment would be seized on by natural selection, and so preserved by his principal device.

Chapter 8: Hybridism or Hybridization

Hybridization (and mechanical grafting) in plants was a time-honored agricultural and horticultural practice. By the eighteenth century, plant hybridization had drawn the attention of horticulturalists and practical breeders like Koelreuter and Gaertner, and Darwin relied on their insights extensively in the *Origin*. Darwin once again drew heavily on both for their understanding of hybridization to support his view of the relations among varieties, incipient species, and species. He especially needed to explain why species, which, according to his theory, had descended from each other, usually could not interbreed. Darwin offered a quick survey of the diverse and complex patterns observed in the phenomenon of sterility and its correlate, fertility in hybrids and mongrels (mongrels being the progeny of varieties). He noted especially striking peculiarities in hybridization patterns; species crossing with facility could give rise to sterile hybrid progeny, while species that crossed with difficulty could give rise to fertile hybrid progeny; and reciprocal crosses between two species sometimes gave very different results. The view that barriers to hybridization existed so as to "prevent confusion of all organic forms," the standard explanation of why species did not interbreed, was clearly not a viable explanation (*Origin*, 245). For Darwin, the patterns of sterility and fertility observed instead provided further evidence in support of his theory and the view that there were no clear lines between varieties, incipient species, and species. It was also a way of arguing against special creation. He concluded: "Finally, then, the facts briefly given in this chapter do not seem to me opposed to, but even rather to support the view, that there is no fundamental distinction between species and varieties" (*Origin*, 278). This was to prove crucial in the penultimate chapter on classification.

Chapters 11 and 12: The Geographical Distribution of Plants

In keeping with his observations while aboard HMS *Beagle*, Darwin's best evidence in support of his theory was drawn primarily from the distribution patterns of the plants and animals he had observed while he traveled on the celebrated voyage. The patterns – and the implied relationships thereof – on continents and oceanic islands in particular inspired crucial reflection and provided evidence not so much directly in support of his theory, but against the view that special creation could account for the observable distribution and variation patterns. Plants once again figured prominently in these chapters, especially since the problem of dispersal and dissemination in plants (using seeds or vegetative structures in reproduction) could be readily tested even by gentleman naturalists. Here again, Darwin's experimental efforts to test his theories of dispersal among continental and oceanic forms of plant species figured prominently. He explored what he termed "occasional means of distribution" through experiments on seed survival during ocean dispersal. He tested 87 kinds of plant seeds and found that 64 were able to germinate after an immersion of 28 days in salt water, while a few survived 137 days. He drew distinctions between small seeds (without capsules and fruit, which could increase buoyancy as well as provide reserves) and vegetative structures like branches and leaves, performing experiments that also took into account the rates of Atlantic currents, which he found listed in Johnson's *Physical Atlas*.[7] He even tested the ability of seeds to survive and be transported above flotsam and jetsam

---

[7] Darwin here was referring to a popular atlas by Alexander Keith Johnson that was available at the time. "Johnson's *Physical Atlas*," as it was popularly known, had been inspired by Alexander von Humboldt. Darwin may have used either the first edition, which appeared in 1848, or the second, extended version, which appeared in 1856; he may have also used both over a period of time. See Alexander Keith Johnson, *Physical Atlas of Natural Phenomena* (London and Edinburgh: W. Blackwood and Sons, 1848 edition and 1856 edition).

in the crops of pigeons and on the feet and in the beaks of birds. He also took into account potential transport on icebergs and then extended his discussion to include historic means of dispersal during the glacial period. The discussion of the diverse means of transport and the survivability of plants stretched to no less than twenty or so pages of the *Origin* and revealed much about Darwin's experimental prowess (see Bowler 2008).

Chapter 13: Classification, Morphology, and Embryology in Plants

Plants make brief – but notable – appearances in Chapter 13 on classification, morphology, and embryology. This was not so much because plants were not important to these concerns, but because Darwin had already given abundant evidence to support the arguments of the chapter. By Chapter 13, the penultimate chapter, Darwin had carefully built his argument for a rethinking of classification based on a more natural plan that took into account community of descent.

Darwin agreed with other systematists that vegetative structures that varied greatly were not ideal characters for classification and that reliance should instead be placed on the reproductive parts. He wrote with some emphasis: "organs of vegetation, on which their whole life depends, are of little signification, excepting in the first main divisions; whereas the organs of reproduction, with their products the seed are of paramount importance!" (*Origin*, 414).[8] In this statement Darwin also revealed an insight from what is now known as "plant developmental biology," which he later developed on pages 418–19. By speaking of "main divisions" Darwin recognized the process by which the fundamental character used in the major division of plants into the Dicots and the Monocots originated embryonically. He wrote:

The same fact holds good with flowering plants, of which the two main divisions have been founded on characters derived from the embryo, – on

---

[8] Darwin had earlier resisted this idea but changed his mind because his theory of descent upheld the view that essential parts (or characters) were common in related groups. These were therefore the least likely to vary.

the number and position of the embryonic leaves, or cotyledons, and on the mode of development of the plumule and radicle. In our discussion on embryology, we shall see why such characters are so valuable, on the view of classification tacitly including the idea of descent. (*Origin*, 436)

He also recognized yet another significant fact of plant development that was most closely associated with German morphologists like Johann Wolfgang von Goethe:

It is familiar to almost every one, that in a flower the relative position of the sepals, petals, stamens and pistils, as well as their intimate structure, are intelligible on the view that they consist of metamorphosed leaves, arranged in a spire. In monstrous plants, we often get direct evidence of the possibility of one organ being transformed into another; and we can actually see in embryonic crustaceans and in many other animals, and in flowers, that organs, which when mature become extremely different, are at an early stage of growth exactly alike.

This latter point was especially important because it suggested recapitulation, which was a demonstration of descent. Darwin then resumed his reverse argument in support of his theory, and against special creation, to account for this; he also demonstrated the taxonomic promiscuity and the comparative rhythm that characterized his style of argumentation in the *Origin* overall:

How inexplicable are these facts on the ordinary view of creation! Why should the brain be enclosed in a box composed of such numerous and such extraordinarily shaped pieces of bone? As Owen has remarked, the benefit derived from the yielding of the separated pieces in the act of parturition of mammals, will by no means explain the same construction in the skulls of birds. Why should similar bones have been created in the formation of the wing and leg of a bat, used as they are for such totally different purposes? Why should one crustacean, which has an extremely complex mouth formed of many parts, consequently always have fewer legs; or conversely, those with many legs have simpler mouths? Why should the sepals, petals, stamens and pistils in any individual flower, though fitted for such widely different purposes, be all constructed on the same pattern? (*Origin*, 437)

The answer to all of these questions is that the mentioned structures have been derived from common descent.

CONCLUSIONS

What if any general conclusions can we draw from this brief study of Darwin's botany in the *Origin*? For one thing, plants were absolutely foundational to the development of his argument. Plant examples appear in greatest abundance throughout Chapters 1 to 4, the critical chapters laying out the argument for his theory, and especially in Chapter 2, where Darwin used botanical data in support of his argument that varieties may be seen as incipient species. Darwin's use of hybridization in Chapter 8 played a similar foundational role in establishing varieties as incipient species; it also supported his natural or genealogical (or what we now term "evolutionary") classification.

Yet another conclusion we may draw is that Darwin's own researches into botany deserve more credit than has been given him. Historians have always known that botany figured prominently in his work after the *Origin*, but have not fully appreciated the extent to which the study of plants served a foundational role in formulating the argument for his theory. As one recent study has revealed, historians still have much to learn about the early influences that teachers like Henslow had upon Darwin (Kohn et al. 2005).

Along these lines, historians have also not fully appreciated Darwin's adroitness in the use of experimentation (though his powers of observation have been duly noted and appreciated). A closer reading of Darwin's botanical research in the *Origin* reveals the extent to which he was engaged in clever and crucial experiments to test or buttress key points of his theory well before 1860. Study of Darwin's botany also reveals the more practical, "down-to-earth" aspects of his work, which complements the traditional historical portrait of Darwin as a great theorist and synthesizer (he was, of course, but this was possible because he was also a keen observer and experimentalist).

Taking this point a bit further, historians might wonder why Darwin did not consider himself a botanist, and why botanists did not readily claim him as one of their own, given that it was the one area of natural history where he arguably made the most notable contributions. One reason for this is that his proximity to full-time systematists like Hooker and even Gray (who were entirely devoted to the systematic study of plants) may have thwarted his identification as a botanist. Compared to them, and other like-minded specialists, Darwin lacked the kind of single-minded devotion to one particular taxonomic group. Yet another reason may be that his experiments lacked the kind of laboratory setting that was increasingly taking center stage in botanical practice as it embraced the "new botany" associated with microscopy and table-top instrumentation. In at least one celebrated exchange with the noted German plant physiologist and experimentalist Julius Von Sachs, Darwin was criticized, if not belittled, for his lack of experimental rigor (Heslop-Harrison 1979; De Chadarevian 1996). Caught between the full-time systematists, whose botanical knowledge of individual groups was incomparably greater than Darwin's, and the proponents of the "new botany," whose experimental methods appeared to be more rigorous, Darwin's contributions to botany resided in a peculiar place that did not gain real status or legitimacy until the second half of the twentieth century, with the emergence of the new field known as "plant evolutionary biology." Even his contributions to the area of "evolutionary ecology" or "plant ecology," areas that were clearly articulated in the *Origin*, had to wait until those fields came to fruition in the latter half of the twentieth century. His equally keen insights into the biology of invasive species and his contributions to the new field of "invasion biology" have been recognized only in recent years (Hayden and White 2003).

Plant evolutionary biology itself did not properly come of age until the 1930s and the 1940s, during the period of the "evolutionary synthesis." It was during this period that Darwinian evolutionary understanding was integrated with Mendelian genetics in a manner that could account for the origin of biological diversity. A number of questions and problems that Darwin had encountered were resolved by this integration of heredity and evolution, broadly construed. For plant evolution more specifically, this integration took place in 1950 with the publication of G. Ledyard Stebbins's *Variation and Evolution in Plants*; it was only after this synthesis that many of the phenomena that had piqued Darwin's interest but that had bedeviled him and his contemporaries interested in formulating a general theory of plant evolution were resolved, or at least re-problematized (see Arnold 1997 for new views on hybridization and evolution). Thus, Darwin's botanical efforts did not really come to fruition until enough was understood about the evolutionary biology of plants; just the same, in the context of their day and in the context of the *Origin*, Darwin's insights into plant evolution proved to be critical to formulating the most coherent theory then available for general evolution.

13 The Rhetoric of the *Origin of Species*

I. THE *ORIGIN* AS RHETORICAL ARGUMENTATION

In 1828, the Oxford don Richard Whately, having in his *Elements of Logic* (1826) defined argument as the connection between a claim and its support, proposed in its newly published counterpart, *Elements of Rhetoric*, to

treat argumentative composition as a species of rhetoric. . . . The office of the logician is to infer, but the office of the rhetorician is to advocate by adducing proofs. . . . Thus even the philosopher who undertakes by writing or speaking to convey his notion to others assumes for the time being the character of advocate of the doctrines he maintains. (Whately 1963, 5)

The idea that rhetoric is situated argumentation – argumentation giving reasons why a specific audience should come to the speaker's side of a disputed issue or controversial topic – is old. Aristotle took this line when he defined rhetoric as the "ability to discover the available means of persuasion in each particular case" (*Rhetoric* I.2.1355b27–8). The idea still has champions (Toulmin 2003). This argument-centered conception of rhetoric contrasts sharply with the equally ancient view that the specific task of a rhetor is not to find or set out an argument but skillfully to deck one out with whatever ornaments of speech might induce a presumably less-than-rational audience to be affected, even infected, by it. Scientists and philosophers assume just such a conception when they hold that logically valid, context-free, undecorated arguments are the only sort that count. Whately's view also contrasts, however, albeit less severely, with a third conception of rhetoric. Emerging in the eighteenth century, I will call it the expressive view and will explicate it by citing Charles Darwin's *Journal of Researches*.

The title page and first line of the *Origin of Species* identify its author as the Charles Darwin who in 1839 began to publish his observations as a naturalist onboard HMS *Beagle* in an engaging travel narrative that at times displayed the visual vivacity first demanded of British writers by the influential eighteenth-century critic and essayist Joseph Addison. More particularly, the style of Darwin's *Journal of Researches* reflects the Romantic zeitgeist that he had imbibed from the youth culture of his day, with its Wordsworthian conviction that words can be as vivid as pictures only if they arise from the overflow of powerful emotions into an expressive medium. Writers who embraced this ideal seldom acknowledged models; following models was generally taken to be slavishly imitative. In point of fact, however, Darwin's travel narrative did have a model for its use of emotional experience and language as a way of penetrating nature's secrets and enlisting the sympathy of its readers. Alexander von Humboldt's *Personal Narrative* affected both the substance and the style of Darwin's Romantic naturalism (Manier 1978; Kohn 1996; Richards 2002; Sloan 2005).

A case can be made that the *Origin* is just as expressive as the *Journal of Researches*. Readers of Darwin's *Origin* can hardly miss its author's personal warmth, ethical sensibility, and capacity for wonder (Levine 2006). Nonetheless, the "I" who in the first line establishes his credentials by referring to his voyage on the *Beagle* is not quite the same as the "I" of the earlier book. To be sure, the attractive character of the speaker (what rhetorical scholars call *ethos*) is still designed to elicit in his readers a well-disposed affective state (*pathos*) that will open them to what he has to say. But this time the speaker is painfully anxious to secure the good will of readers whom he assumes are as skeptical as he once had been about claims that made Victorian Britons pretty nervous. The *Origin*'s readers are asked to abandon strong prejudices that had been built up by the British press and clergy ever since the decidedly French idea of species mutability had begun to circulate in the 1820s. They are to do so by constructing, through the act of reading, a counterpart of their real selves who will be responsive to the speaker's appeal to their ability to reason, to assess evidence fairly, to enter into the spirit of imaginative thought experiments, and who exhibits other flattering traits. Expressive as it sometimes is, the *Origin*'s rhetoric is unmistakably situated argumentation in Whately's sense.

Consider that in its recapitulative last chapter the speaker refers reflexively to the text itself as "one long argument." This means that it is not, or not primarily, a narrative (*Origin*, 459; Secord 2000, 510). Initially, this disconcerted even so discriminating a connoisseur of fine arguments as George Eliot, who in the course of informing a correspondent in December 1859 that she and her partner, George Henry Lewes, had begun reading Darwin's book wrote: "It expresses thorough adhesion, after long years of study, to the Doctrine of Development. The book is ill written and sadly wanting in illustrative facts.... This will prevent it from becoming popular, as the *Vestiges* did" (quoted in Secord 2000, 512–13; Beer 2000, 146).

George Eliot apparently expected her evolutionary reading to spin out grand narratives stretching from the nebular formation of the universe to the imminent reduction of psychology to brain science, as popular-science audiences often still do. She expected the *Origin* to be as satisfying to the imagination as the sensational *Vestiges of Creation*, which had appeared anonymously in 1844 and was later shown to have been written by the skilled journalist Robert Chambers (Secord 2000). The phrase "Doctrine of Development" suggests that Eliot also expected the *Origin* to repeat much the same evolutionary story that Lamarck, Chambers, and her friend Herbert Spencer were telling, a tale of inevitable complexification and progress (Beer 2000, 146). She did not know that the trashing of *Vestiges* by people whose opinions he valued had led Darwin to search for arguments that would not be vulnerable to the criticisms leveled against Chambers and, when at last he made these arguments public, to couch them in a decidedly unsensational way. Accordingly, although I make no case for Whately's influence on Darwin, I will focus in this chapter on the argumentative rhetoric of the *Origin*. Darwin's use of metaphor will come into the question, to be sure, but as serving argumentative purposes.

In order to highlight these purposes, I will make use of the distinction between real and textual speakers and audiences that I have already been drawing. In spite of the fact that it will at times seem clumsy and artificial, this distinction offers a useful analytic tool. In the last several decades, Darwin scholars have brought a range of influences to bear on interpreting the *Origin* and Darwin's other works. But not everything that went into the making of Charles Darwin necessarily went directly into the making of the "I" of the *Origin*. This voice serves to filter, transform, muffle, or even disguise whatever might be found unpersuasive to the implied audience.

What about that audience? Darwin's actual readers did not simply disappear into the textual construct that had been designed to attract them. They wrote back, creating a feedback loop between Darwin and his audience. This was possible because of the rapid expansion of the Victorian post and press at the time (Young 1985, 153). In his empirical study of the *Origin*'s reception, Alvar Ellegård counted reviews in no less than 155 newspapers, magazines, and quarterlies (Ellegård 1958, 369–84). Darwin's letters show him taking anxious note of most of these, as well as of numerous foreign reviews, registering in each case how far the reviewer "goes with me" and sometimes writing to thank this one or that one for going as far as he did. Replies, refutations, rebuttals, and concessions are obsessed about, debated with correspondents, and handed over to the speaker for inclusion in subsequent editions. In the process, the speaker's voice becomes overlaid with Darwin's bobbing and weaving defense. Perhaps these complications are partly responsible for recent tendencies to think of editions other than the first as a fall from scientific, philosophical, or literary grace. In some ways they are. But the intense feedback between the real Darwin and his real audience means that the *Origin* is not reducible either to its author's intentions or to the history of its reception. It is a temporally extended process of discursive interaction located in and responding to a constantly shifting rhetorical situation largely of its own making.

II. ONE LONG (RHETORICAL) ARGUMENT

Darwin's preferred scientific methodology demarcated inductive science from the mere hypothesizing that he and many other critics saw in *Vestiges* by requiring causes to have real existence in nature prior to exhibitions of their explanatory elegance, parsimony, or fecundity. This so-called *vera causa* (real cause) ideal dictated that Darwin's positive case for common descent driven by natural selection would have a three-part exposition. It would move from arguing for the existence of natural selection, to its competence to account for common descent, to showing, finally, how beautifully common descent through natural selection can integrate up-to-date results in biology and how elegantly it can resolve its most disputed issues (Hodge 1977). At every point in this process, however, the solicitous "I" who does the arguing pits his evidence for descent with modification through natural selection against what he repeatedly calls "the ordinary doctrine of creation." The *Origin's* "I" denies that new species arise by way of instantaneous acts of creation through non-natural means. He denies, too, that instances of such acts may be found here and there or now and again. Rather, species arise only once at a single geographical point of origin. They gradually depart by way of normal reproduction and natural selection from varieties of existing species and are dispersed across time and space by causes no less natural. All higher taxa emerge by protraction of the same process. Accordingly, the speaker directly contravenes the Cambridge polymath William Whewell's claim that "Nothing has been pointed out in the existing order of things which has any analogy or resemblance of any valid kind to that creative energy which must be exerted in the production of new species" (Whewell 1840). The *Origin's* "I" contends that a real cause observable in the existing order does indeed bear such a valid analogy – natural selection. Commentators from Darwin's American friend Asa Gray to the late Stephen Jay Gould have pointed out, usually with relief, that this claim is logically compatible with a range of religious beliefs (Brooke, this volume). But by the lights of Darwin's day, special creation was a scientific, not a theological or Biblical, claim. And by these lights it was flatly incompatible with common descent by natural selection. The *Origin's* "I" acknowledges this contradiction by arguing from start to finish against "the ordinary doctrine of creation."

Darwin probably had particular people in mind in framing his anti-creationist argument: his mentors in natural history, especially those who had slammed *Vestiges*; the professional colleagues to whom he was promising something more than the "abstract" they would soon be reading (*Origin*, 1); his wife, Emma, whose sensitivity on the subject of religion had long served as an internal check on his intimations of disbelief and who in consequence "became Darwin's model of the conventional Victorian reader" (Kohn 1989, 226). But these and other individuals, as well as the anonymous audience that Alfred Russel Wallace dubbed "the general naturalist public" (*Correspondence*, 14: 227), were all addressed by means of the internally conscripted audience with which they were asked to identify. As early as the first sentence, this audience is called "we" (Beer 2000, 60). "We" are close observers of the living world. "We" see, for example, that domesticated plants and animals vary more than those "in a state of nature" (*Origin*, 7). Increasingly, "we" become reflective about how to explain this and other phenomena that the speaker reports. Being reflective, "we" diligently follow his exposition. "We" are asked to bear with him as he proceeds further (*Origin*, 59). When he summarizes points already covered, the "I" becomes one of the "we" himself: "We have also seen that the most flourishing and dominant species of the larger genera on an average vary most" (*Origin*, 59).

The *Origin*'s "we" are not assumed to be members of an elite community of professional naturalists. On the contrary, although the speaker is clearly one of them, he sometimes refers in the third person to "naturalists" as holding this or that view (*Origin*, 44–7, 413, 469, for instance). This slight objectification of the scientific community explains several of the text's most subtle rhetorical qualities. His peers, once conscripted as readers, are asked to judge the issue as if they were overhearing an argument addressed to an open-minded but nonexpert third party, thereby prying these experts away from claims they might otherwise dismiss out of hand by virtue of their professional identities. By the same token, the general public becomes Wallace's "general *naturalist* public" by being invited to judge a scientific issue. Early in 1860, a correspondent of Emma Darwin reported that she had overheard someone asking for the *Origin* at a bookstall in a railway station, only to be told that it was sold out (*Correspondence*, 8: 35). The notion that the *Origin* was only accidentally taken up by a diverse reading public is, I think, exaggerated. The book opens itself to the public. This fact also helps explain the feeling of some professional naturalists that the *Origin* displayed an antiquated character (Ruse, this volume). In the decades since Darwin had begun secretly writing about the mutability of species, biologists had taken a turn toward continental structural morphology and away from Paley's functionalist, utilitarian view of biological traits. But British naturalists had so deeply implanted in the public mind the idea that traits are intentionally designed for specific purposes that they were compelled to remain decorously faithful to it, even while the best of them, such as Richard Owen, were furtively searching for ways to split the difference. By taking seriously an out-dated but revered view, the *Origin* had the effect of removing an obstruction to the flow of scientific discourse.

Why, we may ask, was the book addressed to "the general natu-ralist public" at all? It was, I think, the fact of having to intervene in the delicate rhetorical situation created by receipt of Wallace's paper that made the *Origin* a rhetorical performance on a public stage. This situation also turned the impressive body of inductive evidence that Darwin had collected in his "big species book" into an inventory of examples. Aristotle says that the rhetorical counterpart of induc-tion is the telling example (*Rhetoric* I.2.1356b2–5). The exemplary status of Darwin's factual evidence implies an incompleteness that the speaker acknowledges, even exaggerates, at the outset (*Origin*, 2). Incompleteness in turn places weight on the inferential aspects of the "one long argument." These include, in addition to exhibitions of hypothetical-deductive reasoning and thought experimenting, spon-taneously generated forms of rhetorical argumentation well known to rhetoricians since antiquity, all of which are aimed at loosening and then reversing the initial presumption in favor of special creation imputed to the text's implied audience.

One of these figures is the use of reductio ad absurdum as ridicule. We can see it at work in the three passages in which the phrase *vera causa* actually occurs. In all three, the phrase is set in polemical opposition to "the ordinary doctrine of creation," which the speaker takes to be no true cause at all. "The ordinary doctrine" appeals to miracles when perfectly good natural means of generation and dis-persal are right in front of our eyes (*Origin*, 352). It inconsistently appeals to natural or "secondary causes" in the case of varieties but "arbitrarily rejects them" when it comes to species (*Origin*, 482). It trivializes the Deity by attributing the slightly different stems of three species of the lowly turnip to "three separate acts of cre-ation" that are too closely related for any supreme being other than a perverse micromanager to bother with (*Origin*, 159).

By their nature, appeals to the absurdity of an opposing claim exude contempt. This is not the best way to court the good will of an audience. Thus it is unsurprising that the solicitous "I" uses it sparingly. By contrast, advancing one's argument by anticipating and allaying an audience's objections, the emotional counterpart of which is sympathetic identification with their difficulties, pervades the *Origin*.[1] From start to finish the speaker interrupts his exposition to put himself in the reader's place. He not only raises important objections but puts the case for each so strongly that he seems to take his readers' worries on himself. He does this in the expectation that the reader, having been painted as intelligent and dispassionate enough to raise such intelligent objections, will reciprocate by treating his answers to these objections as no less intelligent and, by this route, will eventually come around to his own painfully acquired view that the burden of proof should be shifted onto the "ordinary doctrine." This dimension of the *Origin*'s "one long argument" was noticed by its first anonymous reader, a friend of the publisher who had been asked to assess the manuscript. "Mr. Darwin," he reported, "has brilliantly surmounted the formidable obstacles which he was honest enough to put in his own path" (see Browne 2002, 75).

So deeply is the argument of the *Origin* structured by the strategy of anticipated objection and response that the three stages by which it seeks to meet the *vera causa* ideal – the existence, competence, and responsibility of natural selection – are marked off by episodes in which objections are raised and allayed. In the Introduction, the speaker concedes that he is "well aware that scarcely a single point is discussed in this volume on which facts cannot be adduced often apparently leading to conclusions directly opposite to those at which I have arrived" (*Origin*, 2). The case for competence proceeds throughout by refuting a "crowd of difficulties" first raised in Chapter 6. The most pressing difficulties are the absence of intermediates in the fossil record; the perceived implausibility of natural selection's power to evolve "organs of extreme perfection" like the eye; similar doubts about instincts such as the wonderful behavior of the bee and the slave-making behavior of certain ant species; and the sterility of hybrids, which seems to firm up species boundaries and to be of no use to individuals (*Origin*, 171–2). Chapter 6 itself rebuts the first two objections, although the speaker returns to the first in Chapter 9. Separate chapters are then devoted to the third and fourth objections. All this clears way for three chapters in which the theory's explanatory fecundity is set out positively. The recapitulative Chapter 14 promises a summary, but first launches into yet one

---

[1] The Greek rhetorical term for anticipating objections is *prolepsis* or more specifically *prokatalepsis*.

more run through the objections (*Origin*, 469). In contrast to most addressed arguments, then, which typically leave begrudging room for replies to a few objections at the end, it is fair to say that the *Origin* is pervasively structured by anticipation of and response to objections.

This anticipatory structure contributes not only to the division of the "one long argument" but to its unity as well. By the end, the speaker's good will has brought into existence a reader to match. Moving toward his peroration, he at last explicitly asks this reader to concur that special creation seems highly improbable (*Origin*, 471). Only at this point does he overtly place responsibility for failure to look at the massive evidence for species mutability on "a load of prejudice" and "the blindness of preconceived opinion" (*Origin*, 482–3). Disciples of Bacon could easily be shamed for having worshipped such idols. But to have raised this charge earlier would have undercut the speaker's claim that the objections are indeed serious and that he is taking them seriously. The audience is now construed, however, as having come far enough with the speaker to cast off its blinkers and let the full weight of his case sink in. The rhetorical argument is thus the matrix within which the reader is invited to embrace the scientific argument as a grounded conclusion. Questions have been raised about just how unified the "one long argument" actually is (Waters 2003). They will persist, I think, as long as the rhetorical arc that textually constructs that unity is discounted.

III. PRESUMPTION, BURDEN OF PROOF, AND THE *ORIGIN*'S NATURAL THEODICY

In opposing special creation, Darwin shrewdly avoided Lyell's mistake in *Principles of Geology* of so fully instructing his readers in Lamarck's evolutionary theory that it intrigued rather than, as Lyell had intended, horrified them.[2] Even without the speaker explicating its strong points, however, "the ordinary doctrine" had a good deal going for it, especially when matched against the *Origin*'s alternative. There were those missing fossil intermediates. Moreover, the same species were found in geographically discontinuous places. So Darwin's claim of their original contiguity had to rely on hypotheses

---

[2] I owe this point to Robert Richards.

about warming climates leaving species stranded on mountaintops that became islands and the dispersal of seeds by wind, waves, or in the beaks of birds, bolstered by some backyard experiments (*Origin*, 358). When direct evidence is missing, suspicions, not just difficulties and objections, easily arise. The speaker's way of dealing with them is to use the goodwill accumulated by his anticipatory rhetoric to shift the burden of proof onto special creation.

The *Origin*'s readers could be assumed to be familiar with and responsive to the strategy of securing consent to one side of a contradiction by pushing the burden of proof onto the other. This figure of argument had long been common in British public discourse, and not just in the legal sphere in which it still operates. British discourse generally was, and still is, *a posteriori* in spirit and, in matters not open to demonstration, appealed to standards no more (or less) strenuous than probability, moral certainty, common sense, and, not least, presumption. A presumption, wrote Whately, is not exactly a probability. "It is a pre-occupation of the ground... that must stand good till sufficient reason is adduced against it.... The burden of proof lies on the side of him who would dispute it" (Whately 1963, 112). Shift the burden, and the presumption moves to your side.

Since the end of the seventeenth century, this way of arguing had played an important role in portraying the modern British state to itself as a moderate, reasonable regime coordinating a moderate, reasonable people. Its moderation was exhibited, first, by the importance its intellectuals accorded to natural theology, which established the existence, power, and providence of God by way of evidence about the functional order of the universe that all could see. On the foundation of this "argument from design" was erected a moderate interpretation of revealed truth that pushed both Puritan "enthusiasm" (too much revelation) and tepid deism (too much natural theology) to the margins, in part by drawing a line between superstitions and the miracles credibly testified to by the New Testament writers (Butler 1961, 150). If Jesus had not turned loaves into fishes for five thousand, surely someone would have testified against it. But that hadn't happened. Hence the presumption that a miracle had taken place stood. Not only the observably functional order of nature, then, but credited miracles that occasionally violated that order counted as traces (*vestigia*) of the presence of a personal, caring God in His lawfully ordered world. Accordingly, in the most revered apologetic work of the eighteenth century, Joseph Butler's *Analogy of Religion*, we read, "I find no appearance of a presumption from the analogy of nature against the general scheme of Christianity that God created and invisibly governs the world.... There is no presumption against the operation which we should now call miraculous" (Butler 1961, 144–5).

Special creation was "science in a theistic context" (Brooke et al. 2001) because it made sense against the background of a lawful natural order occasionally interrupted by divine interventions. Both "the ordinary doctrine of creation" and the background that legitimated it – what the rhetorical scholar John Angus Campbell called Britain's "grammar of culture" and the historian of science Robert Young dubbed its "common context" (Campbell 1986; Young 1985) – had the force of standing presumptions that could be challenged only by accepting high burdens of proof and the real possibility that, in failing to meet those burdens, one's reputation as sound in mind and character might be damaged. In the middle of the eighteenth century, David Hume accepted the challenge with his double-barreled assault on natural theology and the very possibility of miracles. His reputation suffered accordingly, but in the end his effrontery served as little more than an occasion for reaffirming the cultural grammar's commonplaces.

To be sure, signs of weakness were showing in the secularizing era of the Reform Bill in which Darwin came of intellectual age. Biblical criticism began to push even New Testament miracles into the territory of faith alone, and in the Bridgewater Treatises, six volumes commissioned in the early 1830s to buff up natural theology in the light of up-to-date science, arguments from probability and presumption that had been developed to defend the reasonableness of revelation and credited miracles were now being used to prop up the design inference itself. How improbable that so many physical and chemical processes would coincide in just the right way at just the right time to make a functioning eye or, out of cosmic gases, a planetary system that led to our life on Earth (Chalmers 1833, 13–16; Whewell 1833, 29). So the old presumptions stood. Still, Butler's confidence that presumption was enough to justify assent was coming under pressure from the more empirically demanding scientists. It is telling that Whately, who had started his career by defending miracles as no less worthy of belief than the existence of Napoleon Bonaparte – both were based on testimony – made the management of presumption and burden shifting the keystone of *Elements of Rhetoric* (Whately 1961, 112–32). He was shoring up rhetorical norms of argumentation in a situation of perceived danger.

Nothing could more precisely express the burden taken up by the *Origin*'s "I" than Whately's assertion that if a rhetor wishes to shift the burden of proof onto a standing presumption, he must take pains to "enable men to see the improbability and sometimes utter impossibility of what at first glance they will be apt to regard as perfectly natural" (Whately 1963, 51). From his youth, Darwin had been both immersed in the discursive practices of the cultural grammar and skeptically reflective about them. As a result, he was fully aware that British natural theology was a sublime object of ideology and that "the ordinary doctrine of creation" seemed ordinary because it was deeply embedded within that ideology. *Vestiges* had proclaimed that evolution is simply another secondary cause, like gravity, by which God carries out his creative work. His mentors' doubts about how well Chambers had carried this point had taught Darwin that in defending his alternative he should appeal as substantively as possible to the commonplaces and values with which the "ordinary doctrine" had been defended. His strategy was to use the goodwill accrued by anticipating objections to turn the most sacred assumptions of the cultural grammar against what seemed "at first glance perfectly natural" in their light.

The *Origin*'s treatment of the eye shows this strategy at its most acrobatic. Like Paley and the Bridgewater authors, the speaker takes the eye as the paradigmatic example of an "organ of perfection so extreme and complicated" that seemingly only a god could have made it (*Origin*, 86). In attempting to undermine the assumed naturalness of this inference, the speaker produces a thought experiment about how selection might *conceivably* have worked gradually from small bits of light-sensitive tissue to make an eye. He concludes, "We should be extremely cautious in concluding that [such an] organ could not have been formed by transitional gradations of some kind" (*Origin*, 190). Readers of the *Origin* may have taken this argument not as a question-begging sophism, but as a delicate, decorous appeal to standards with which they were familiar and by which they were capable of being moved. Why? Because its actual warrant is not, in fact, the thin thought experiment about light-sensitive tissue, but rather, as the text makes clear, the supposition that natural selection is a power "always intently watching each slight accidental alteration in the transparent layers [of tissue] and carefully selecting each alternative which, under various circumstances, may in any way or in any degree tend to produce a distincter image" (*Origin*, 189). Here we discover natural selection as a creative power rivaling in goodness the God of Britain's cultural grammar.

The *Origin*'s textual antecedents reveal steady development of the image of natural selection as a beneficent scrutinizer from its first occurrence in 1842, where the selector is a personal, if hypothetical, agent, to the *Origin*'s treatment of natural selection as an agency or natural power defined entirely by its effects (Sloan 2001, 262–5). This agency is far more oriented to the good of each organism than human breeders:

Man can act only on external and visible characters. Nature . . . can act on every internal organ, on every shade of constitutional difference. . . . Man selects only for his own good, Nature only for that of the being which she tends. . . . How fleeting are the wishes and efforts of man! How short his time! And consequently how poor will his products be, compared with those accumulated by nature during whole geological periods . . . . Daily and hourly natural selection is scrutinizing, throughout the world, every variation, even the slightest, rejecting that which is bad, preserving and adding up all that is good. (*Origin*, 83–4)

Echoes of the biblical sublime in this passage have the effect of refiguring the *Origin*'s examples. From metonymic (parts standing for wholes) surrogates for the elided full range of inductive evidence, every adaptation is turned into a synecdoche (a part as a concentrated expression of a whole) proclaiming that natural selection cares completely and exclusively for the good of "every being." A "natural theodicy" is at work here. Traditional British theism had figured the Creator as, at best, a cost-benefit maximizer and at worst as making sport of his own creation, as Shakespeare's *King Lear* at its most nihilistic suspects. "The rattlesnake has a poison-fang for its own defense and for the destruction of its prey," says the *Origin*'s

"I" (201). But on the principles of natural theology the Creator also doles out to the snake's natural enemies vulnerability to poison, balancing harm to one species against the good of another. The *Origin*'s "I" caustically notes, "Some authors [even] suppose that at the same time this snake is furnished with a rattle for its own injury, namely, to warn its prey to escape.... [But] I would almost as soon believe that the cat curls the end of its tail when preparing to spring in order to warn the doomed mouse" (*Origin*, 201).

It was to steer clear of just such difficulties that Robert Chalmers conceded in his Bridgewater Treatise that natural history tells less "distinctly and decisively of [God's] attributes" than do moral experiences like the call of conscience. That is because Chalmers concedes that "the serpent's tooth for the obvious infliction of pain and death upon its victims might furnish as clear an indication of design, though a design of cruelty, as an apparatus for the ministration of enjoyment . . . might furnish . . . a design of benevolence" (Chalmers 1833, 38). The *Origin*'s "I" does better on this troubled subject precisely by staying on the terrain of natural history. Natural selection is as competent as the Creator to produce the snake's adaptations and those that enable its competitors to protect themselves. But it does not weigh the vulnerabilities of one species against another for the simple reason that vulnerabilities cannot be objects of its scrutiny at all. It is blind to everything but "the improvement of each being" and so cannot possibly be either malicious or haplessly compromising.[3]

In this respect, the moral power of selection's agency or power actually increases as its early personification as an all-seeing agent recedes. In its most mature formulation, natural selection trumps a God who might consciously choose evils for one species in order to benefit others. This aspect of creationist theodicies, and even more their anthropocentric assurance that humans get the better of every bargain, had always offended Darwin. As early as 1837, he was protesting Whewell's claim that "length of days adapted to duration of sleep of man!!! whole universe so adapted!!!" as an "instance of arrogance" (*Notebooks*, D 49). For the *Origin*'s "I," natural selection is not cruel or morally indifferent, as Darwinians who abandoned

---

[3] I must pass over the possible ambiguity in "each being."

the metaphor of selection for "survival of the fittest" came to think. Because its sole object must be the good, the speaker's claim that "the vigorous, healthy and happy survive and multiply" (*Origin*, 79) echoes Paley's proclamation that "it is a happy world after all" (Lennox 1994).

Is it special creation or evolution by natural selection that now seems "perfectly natural" in the light of the cultural grammar? I leave it to historians and biographers to say whether, or at what periods in his life, Darwin embraced, transformed, or rejected British natural theology.[4] What matters for a rhetorical reading is that the *Origin* paints natural selection's scrutiny as sublimely oriented toward the good of each being in order to enlist the assumed affective and argumentative dispositions of its audience. The beauty of the metaphor's semantic range is that readers may interpret the *Origin*'s value judgments in as materialistic, Romantic, or crypto-theistic a frame as they wish.

IV. THE INDISPENSABILITY OF METAPHOR

Any rhetorical account of the *Origin* must consider whether Darwin's metaphors are or are not indispensable to the discovery, intelligibility, and epistemic worth of his "one long argument." In no sense are they merely decorative. They were certainly indispensable to Darwin's creative theorizing (Kohn 1996; Beer 2000, 89). They are also indispensable to what I have been calling his natural theodicy. Darwin's metaphors – natural selection as analogous to the breeder's art, the struggle for existence, the scrutiny image, the economy of nature, the branching tree of life – are parts of an interactive system (Beer 2000, 89). Their center of gravity is the numinous kinship conferred on all living beings by common descent. This kinship takes the place in Darwin's natural theodicy of the collapse of the Great Chain of Being from whose last tattered remnant – the qualitative

---

4 Whether Darwin was the last of the natural theologians or their materialist gravedigger is a disputed question. Darwin's Romanticism is a third, increasingly compelling view. See Manier (1978), Richards (2002), Sloan (2001, 2005). I agree with Kohn (1989, 1996) that Darwin's aesthetics keeps natural theology and materialism balanced on a knife's edge.

superiority of man to beasts – Lyell, for one, could never free himself. Hence the *Origin*'s natural theodicy could not even be formulated without its system of interacting metaphors. They cannot be "skimmed off" (Beer 2000, xiv, xxv; Levine 2006, 280, n. 3). Whether the circle of metaphors that licenses this theodicy can be "skimmed off" without damaging the coherence, intelligibility, and explanatory power of Darwin's scientific theory itself, however, is a different and more difficult question. I, for one, think not.

Analogies that encode real scientific knowledge must reveal the point at which they become disanalogous. Otherwise, what literary people prize as polysemy and logicians scorn as ambiguity will throw valid reasoning into disarray. In the *Origin*'s image of natural selection as scrutinizing even the slightest differences, inner or outer, that might serve the good of their possessors, the contrast between artificial selection's self-serving superficiality and natural selection's other-orientedness marks just such point of disanalogy. This contrast moves the intentional agent who is so causally prominent in the breeder's art from foreground to background, allowing the *vera causa* to which the metaphor points to ground objective knowledge that would be missed or misconstrued if represented otherwise. It allows the causal mechanism of artificial selection, for example, to be reflexively explained by reference to its natural counterpart, showing "that the two [and sexual selection as well] . . . have a genuine identity or common nature" rooted in the same material causal processes (Gayon 1998, 51).

The speaker affirms the explanatory indispensability of the image of benign and pervasive scrutiny at precisely those points where the text sets out conditions for its own falsification. "If it could be proved that any part of the structure of any one species had been formed for the exclusive good of another species," the speaker concedes, "it would annihilate my theory" (*Origin*, 201).[5] But the good of each species depends on the tight fit between organisms and their niches, which in turn depends on the gradual shaping of indefinitely small individual variations into adaptations. So the theory would also "break down . . . if it could be demonstrated that any complex organ

---

[5] Another point about pronouns. When "I" wants "we" to attend to logical structure and validity, he uses the impersonal pronoun "it" and passive constructions.

existed which could not possibly have been formed by numerous, successive, slight modifications" (*Origin*, 189). Allowing evolutionary leaps will loosen adaptive fit. So leaps, too, will falsify the theory (*Origin*, 194, 471).

That the speaker is right to regard these entailments as strict is supported by the fact that Darwin's closest allies had little use for either his natural theodicy or his image of a benignly scrutinizing selection and that, not coincidentally, all of them denied one or more of its entailments. Huxley believed in leaps. Wallace took selection to work on races more than on individual differences. Spencer never discriminated clearly between natural selection, whether working on races or individual differences, and the direct effect of the environment in molding adaptations. Their differing commitments afforded Darwin's allies no reason to accept the image of selection as a beneficent scrutinizer. Neglecting this image, they changed or misread Darwin's theory. The reason is not hard to find. Because they had little use for his natural theodicy, Darwin's false comforters had no reason to accept what it logically entailed.

The scrutiny image is also necessary if the speaker is to convince the audience that his theory is not the same as that of "the author of *Vestiges*" or vulnerable to objections against him (*Origin*, 3). The standing accusation that *Vestiges* was purely hypothetical tacitly encoded its failure to conform to a scientific grammar that, ever since Newton, had put a premium on *verae causae* showing nature to be full of inertial tendencies that are shaped by external forces into an instant-by-instant equilibrium like planetary motion. It was easy to discount evolutionary theories other than Darwin's because they often stressed the causal power of inner drives rather than the shaping role of external forces. In mentioning *Vestiges* at the outset, the speaker rejects the very idea of "an impulse... advancing... the forms of life" (*Origin*, 3–4; *Variorum*, 64: c). "This assumption," the *Origin*'s "I" dryly remarks, "seems to me to be no explanation [as it] leaves... the co-adaptations of organic beings to each other and to their physical conditions of life untouched and unexplained" (*Origin*, 4). Natural selection spreads out the emergence of one species from another by separating the source of individual variation from its retention in environments made competitive by the force of Malthusian population pressure operating on a web of interacting species.[6] Accordingly, instant-by-instant – hence necessarily gradual – adaptive equilibrium among living beings is no less universal than the force of gravity to which the speaker refers in the *Origin*'s last sentence (*Origin*, 490). The upshot is a branching but ascending tree of life that disavows the idea that higher types emerge by inner impulses. The *Origin* thus seeks to take its place with other characteristic achievements of British science (Depew and Weber 1995).

In view of its influence, it is customary to praise the *Origin* as rhetorically successful. It was generally received, however, as George Eliot's first impression suggests, as new testimony for the "Doctrine of Development," whose enduring image had already been fixed for the British public by Lamarck, Chambers, and incipiently Spencer (Secord 2000). Thus at the indecorous debate of the British Association for the Advancement of Science at Oxford in the summer of 1860, objections long considered effective against *Vestiges* were indiscriminately hurled once again at the *Origin*. Darwin's correspondence in the early 1860s shows him struggling with such misunderstandings and, in attempts to keep his head above water, protracting the objection-reply structure into new editions of the *Origin* that responded to the changed rhetorical situation created by its first edition. These texts give us a speaker who is not anticipating and deflecting trouble, but experiencing it. He can be at times defensive, didactic, overly accommodating, and even a little testy. He steps out of character in a new chapter on "miscellaneous objections" to complain, for example, that "it would be useless to discuss them all, as many have been made by writers who have not taken the trouble to understand the subject" (*Variorum*, 226: d). Yet under pressure from readers who (mis)read natural selection as a mythical agent, he all too obligingly undercuts the necessity of the metaphor, saying that the phrase is "almost necessary for brevity" in reporting facts that might in principle be put otherwise (*Variorum*, 165: c).

---

[6] Hence Darwin's annoyance with himself for his ambiguous shorthand in the *Origin*'s thought experiment about how "a bear" might become "a whale." (*Origin*, 215; *Correspondence*, 7: 176). Still, Darwin took his dynamical model to be consistent with a fairly strong recapitulationism. See Richards (2002 and this volume), Nyhart (this volume).

Still, Darwin largely resisted Wallace's imprecation that he substitute Spencer's "survival of the fittest" for natural selection (*Correspondence*, 14: 227). Perhaps he realized that wide dissemination of that phrase would mark the emergence of the disenchanted Darwinism that forever left behind the rhetorical situation to which the *Origin*'s "I" had addressed himself.[7]

[7] I am grateful to the editors for commissioning this paper and to them and my fellow authors for helping me improve it. David Kohn, Lynn Nyhart, Bob Richards, and Betty Smocovitis provided detailed suggestions. Gillian Beer, John Angus Campbell, Bruce Gronbeck, John Grula, Michael Ruse, and Rachel Avon Whidden offered various sorts of inspiration and correction.

14 "Laws impressed on matter by the Creator"?

The *Origin* and the Question of Religion

In November 1859, on the brink of publication and eagerly antic-ipating the reaction of the naturalists he most respected, Darwin confided to Alfred Russel Wallace: "If I can convert Huxley I shall be content" (*Correspondence*, 7: 375). A month later he had apparently succeeded, reporting that Huxley "says he has nailed his colours to the mast, and I would sooner die than give up, so that we are in as fine a frame of mind... as any two religionists" (*Correspondence*, 7: 432).

The conversion to which Darwin referred was not to an atheis-tic or materialistic worldview. His goal had been more modest: to corroborate the view that species were mutable and that to explain their appearance it was unnecessary to invoke separate acts of cre-ation. The use of religious language is, however, revealing and was not confined to the metaphor of conversion. Belief in the trans-mutation of species was described as heretical, as when Darwin thanked Huxley for being "my good and admirable agent for the promulgation of damnable heresies" (*Correspondence*, 7: 434). After receiving "unmerciful" admonition from his old Cambridge friend Adam Sedgwick, Darwin described himself as a "martyr" (*Corre-spondence*, 7: 430). As in religious communities, so in the scientific: Darwin exploited personal testimony. "Boastful about my book," he informed a French zoologist as early as December 5, 1859, that "Sir C. Lyell, who has been our chief maintainer of the immutability of species, has become an entire convert; as is Hooker, our best & most philosophical Botanist; as is Carpenter, an excellent physiologist, & as is Huxley; & I could name several other names." Reinforcing the rhetoric, he added that these naturalists "intend proclaiming their full acceptance of my views" (*Correspondence*, 7: 415–16.). In a less boastful overture to the American geologist James Dana, he wrote of "difficulties . . . so great that I wonder I have made any converts, though of course I believe in truth of my own dogmas" (*Correspondence*, 7: 462–3).

This stock of religious metaphor might be seen as nothing more than linguistic embellishment; but it points to something deeper. It underlines a difficulty Darwin faced in winning converts. His own conversion, as he explained to the Oxford geologist John Phillips, had taken "many long years" (*Correspondence*, 7: 372). Writing to Lyell he used a phrase that he would later apply to his religious views: "many & many fluctuations I have undergone" (*Correspondence*, 7: 353). Yet he had to hope that by staggering his readers with his alternative vision of the history of life, he could win converts quickly. He pleaded with them to read his text from beginning to end to grasp its force. The religious metaphors also indicate that two competing paradigms were in contention and that cautious, mediating positions would have the air of compromise. When Phillips sought to restrict the scope of "descent with modification" by confining it within each separate "essential type of structure," he adopted a position for which Huxley already had a colourful slur. Darwin propelled it to Lyell: "Huxley says Phillips will go to that part of Hell which Dante tells us is appointed for those who are neither on God's side nor on that of the Devil's" (*Correspondence*, 7: 409–10).

This was a conspiratorial joke, but one that signals deeper layers of controversy. Those who believed that a theory of separate creation was fundamental to their religious faith saw an immediate threat from Darwin's science. The *Origin of Species* was bitterly attacked, often by clerics with an amateur interest in the study of nature. Enduring his metaphorical martyrdom, Darwin was sustained by the belief that a theory that connected and explained so much that was otherwise inexplicable could not be off the rails.

PARADIGMS AND LAWS

The impression of two paradigms in contention has often been reinforced by anecdotes that have spiced popular accounts of religious reaction. Whether it is Disraeli proclaiming himself on the side of the angels rather than the apes, or the bishop of Oxford, Samuel Wilberforce, humiliated after baiting Huxley with the question whether he would prefer to think of himself descended from an ape on his grand-father's or grandmother's side, the image is of a polarity between scientific credibility and religious obscurantism. Huxley's famous retort at the 1860 meeting of the British Association for the Advancement of Science was that he would rather have an ape for an ancestor than a certain person who used his great gifts to obscure the truth.

We shall return to this event, which has assumed mythic status in the annals of scientific professionalism. It must, however, be stressed that Darwin's theory was published at a time when the established Church in England was seeking to cope with multiple crises. These included the challenge from historical criticism of the Bible, on the subject of which Wilberforce expressed an even greater dismay when condemning *Essays and Reviews* (1860). Among contributors to this collection were Oxford clergy who were suggesting that the Bible should be read like any other book, its authors men of their times, inspired but fallible in their understanding. The Anglican Church was also on the defensive because of the vitality of dissenting Christian denominations and a dispiriting loss of members to the Church of Rome, following the lead of John Henry Newman. Most of Wilberforce's close relatives were defectors. Initially, at least, the Darwinian challenge was of lesser moment, though it undoubtedly added to a sense of crisis when serious thinkers were questioning the foundations of the Christian faith.

The impression of one cultural paradigm fading as another, secular in spirit, gained the ascendancy can also be reinforced by the poignant remarks of Darwin's scientific contemporaries. Harvard's Professor of Geology and Zoology Louis Agassiz favoured an idealist account of the history of life as the unfolding of a plan conceived in the mind of God. He once wrote that "there will be no scientific *evidence* of God's working in nature until naturalists have shown that the whole creation is the *expression of thought* and not the *product of physical agents*" (Roberts 1988, 34). This was not what Darwin had shown! Following a scientific meeting in Boston, the British physicist John Tyndall recalled how Agassiz, "earnestly, almost sadly," had confessed that he had not been "prepared to see Darwin's theory received as it has been by the best intellects of our time." Its success, Agassiz conceded, "is greater than I could have thought possible" (Tyndall 1879, 2: 182). It was as if an older, more spiritual understanding of nature was passing away.

Some of Darwin's own observations resonate with a feature of paradigm change that Thomas Kuhn would stress in his influential book *The Structure of Scientific Revolutions* (1962). Kuhn argued that because of a degree of incommensurability between competing scientific paradigms, the triumph of the new over the old may not flow from a clinching disproof of the one from within the framework of the other. Rather, the newer explanatory scheme gradually displaces its predecessor as a younger generation of scientists adopts it and explores its greater possibilities. In the last chapter of the *Origin*, Darwin anticipated such an eventuality: "A few naturalists, endowed with much flexibility of mind, and who have already begun to doubt on the immutability of species, may be influenced by this volume; but I look with confidence to the future, to young and rising naturalists who will be able to view both sides of the question with impartiality" (*Origin*, 482). This was hardly music to the ears of an older naturalist, Adam Sedgwick, who duly chastised his former student for the "tone of triumphant confidence in which you appeal to the rising generation" (*Correspondence*, 7: 397).

Not surprisingly, then, the Darwinian "revolution" is routinely interpreted as the triumph of a secular scientific paradigm over a religiously inspired natural theology, over a philosophy of nature in which Sedgwick had found evidence of "ten thousand creative acts" (Brooke 1997, 54) as new species first appeared in the fossil record. It is certainly possible to streamline the history with such an interpretation; but it is not the whole story. From the pages of the *Origin* itself, it is clear that Darwin could not dismiss questions that were of fundamental importance to anyone who thought deeply about the order of nature and why it should be as it is. Defending his analogy between the birth of individuals and the birth of new species, Darwin introduced a theological reference: "To my mind it accords better with what we know of the laws impressed on matter by the Creator, that the production and extinction of the past and present inhabitants of the world should have been due to secondary causes, like those determining the birth and death of the individual" (*Origin*, 488). This reference to "laws impressed upon matter by the Creator" remained through all six editions. What did it mean?

This is an important question because it meant different things to different thinkers, depending on their particular religious convictions. The image was both conventional and powerful, reflecting what for some pioneers of European science had been an uncontroversial assumption – that one could not have *laws* of nature without an external *legislator*, whose will had been impressed on the world at its inception. For natural philosophers such as Robert Boyle and Isaac Newton, the basic properties of matter were as they were not out of any necessity, but because they had been bestowed by a Deity who, according to Newton, was "very well skilled in mechanics and geometry" (Newton 1692, 49).

This way of understanding physical laws had repeatedly found expression in the literature of natural theology. The law metaphor did not have to be taken literally. During the Enlightenment there had been less God-laden accounts of laws of nature, as when David Hume argued that for an understanding of causal connection one need look no further than the constant conjunction of cause and effect in human experience and expectation. Nevertheless, in the discourse of natural theology, which Darwin had assimilated in Cambridge, William Paley's view that "a law supposes an agent" was often echoed.

For religious thinkers the matter did not end there. This was because several different constructions could be placed on this relationship between God as lawgiver and the order of nature. One could envisage a Deity who, having impressed laws upon matter at creation, ceased to have anything further to do with the world. Nature, on this view, would be autonomous once it had been created and structured. This interpretation is often described as "deistic" in contrast to a "theism" in which the laws continue to be effective only because they are constantly sustained by the power of their Creator. This concept of divine sustenance has been important in the Christian tradition; but it does not dictate whether divine power might intervene in special ways – in the working of a miracle, for example. This means that, conceptually at least, there is space for an even richer theology of nature in which the laws simply express how the Deity *normally* acts in the world but without prejudice to the question whether there might be *additional* forms of divine activity, in miracles or other forms of self-disclosure. For conservative Christians, the view expressed by the eighteenth-century evangelical reformer John Wesley has often been compelling: surely a Deity with the power and wisdom to create such a universe as this cannot be considered incapable of lesser miracles? These possibilities by no means exhaust the range of options. There were semideistic positions in which nature had full autonomy except when there might be intervention by a *deus ex machina* – as, for example, in the origination of new species.

This last position Darwin clearly contested, but in other respects his reference to "laws impressed upon matter by the Creator" had an ambiguity that could lead to alternative deistic or theistic readings. Given that in his early transmutation notebooks he had referred to a Creator who creates "by laws," it becomes important to ask which gloss Darwin would have placed on his formulation in the *Origin*. His old Cambridge mentor John Henslow had no doubt, declaring that Darwin "believed he was exalting & not debasing our views of a Creator, in attributing to him a power of imposing laws on the organic World by which to do his work, as effectually as his laws imposed upon the inorganic had done it in the Mineral Kingdom" (*Correspondence*, 8: 200).

DARWIN BEFORE THE *ORIGIN*

When Darwin had first studied with Henslow, he had been preparing for the life of a priest in the Anglican Church. In the intervening years, that intention had been displaced by the lure of a life in science, which had become all-absorbing during the *Beagle* voyage. Although Darwin scholars have disagreed over the precise date by which he abandoned Christianity, he had certainly done so by the time he composed the *Origin*. The primary reasons did not emanate from his science. True, he saw the sciences contributing to a worldview in which miracles were increasingly incredible. It is also true, as his wife Emma had anticipated before their marriage, that the high standards of evidence required by rigorous science could induce a sceptical frame of mind when appraising evidence for other forms of belief. But the triggers for his loss of faith largely came from elsewhere.

In common with other early Victorian intellectuals, Darwin had experienced a moral aversion to Christian teaching on heaven and hell. The notion that those outside the fold of Christian orthodoxy (and this applied to his atheist brother Erasmus and his freethinking father) were destined for *eternal* damnation he considered a "damnable doctrine" (*Autobiography*, 87). He recounted later that he had always considered the presence of so much suffering in the world one of the strongest arguments against belief in a beneficent deity. Tellingly, he added that pain and suffering were what one would expect on his theory of natural selection. Early in 1851 he had been devastated by the death of his young daughter Annie, finding it incomprehensible how a God of love could allow such an innocent child to suffer. At the same time he had been reading books by religious thinkers – notably Francis Newman's *Phases of Faith* – whose pilgrimage into unbelief had features in common with his own (Desmond and Moore 1991, 376–8). Add to these considerations the fact that, as early as March 1838, he had flirted with a materialist account of the workings of the mind, even suggesting that love of the Deity might simply be an effect of the brain's organisation (Kohn 1989), and it becomes clear that we should not look in the *Origin* for a Christian theism.

THEOLOGICAL LANGUAGE IN THE *ORIGIN*

Darwin's reference to "laws impressed upon matter by the Creator" was one of few theological remarks in the first edition of the *Origin*. There are, however, several reasons why theological implications would be read into his text. Central to his rhetorical strategy was the antithesis between the explanatory power of natural selection and the sterility of what he called the "theory of creation." In his last chapter he reemphasised that the geographical distribution of species could be perfectly understood if there had been migration of species, followed by gradual modification – as in the striking case of the Galápagos fauna, which most closely resembled, but also deviated from, the species of mainland South America. How could one *explain*, on the basis of separate creation, why some oceanic islands were teeming with life similar to that on a nearby continent, while others, equally habitable but distant from major land masses, were dismally populated by a few bats? To invoke the will of a Deity would simply shut down the enquiry. Today evolutionary biologists with Christian convictions would simply say that the Christian doctrine of creation does not require belief in what Darwin called a "theory of creation." And they would be correct in saying that, in its classic formulation, it is a doctrine about the ultimate dependence of all that exists (including evolutionary processes) on a transcendent power. But Darwin certainly had readers who felt that their faith was being challenged by the severity with which he denounced the model of "independent acts of creation."

A second implication of the text was that it was no longer necessary to regard each species as specially designed and adapted for its particular niche. The argument for design based on intricate anatomical contrivances such as the human eye, in which Paley had so delighted, was rendered nugatory if, as Darwin was proposing, nature could counterfeit the design. If natural selection worked for the gradual improvement of each species, it was not surprising that there was the *appearance* of design, albeit illusory.

Readers would, however, find toward the end of Darwin's book some theological remarks that were far from atheistic. He did not just refer to a Creator who had impressed laws on matter. His language when discussing the original primordial forms incorporated a biblical image. This was the breathing of life into them: "I should infer from analogy that probably all the organic beings which have ever lived on this earth have descended from some one primordial form, into which life was first breathed" (*Origin*, 484). And the image was repeated in the last sentence of the book: "There is grandeur in this view of life, with its several powers, having been originally breathed into a few forms or into one; and that, whilst this planet has gone cycling on according to the fixed law of gravity, from so simple a beginning endless forms most beautiful and most wonderful have been, and are being, evolved" (*Origin*, 490).

Although Darwin had rejected Christianity and would eventually become an agnostic, at the time of writing the *Origin* there is still a sense in which he wishes to persuade his readers that their God is too small. Surely, it was beneath the dignity of the Deity to have produced each species separately as if by so many conjuring tricks! In his third transmutation notebook he had addressed precisely this issue. Was not the notion of God's creating by law "far grander" than the "idea from cramped imagination that God created... the Rhinoceros of Java & Sumatra, that since the time of the Silurian he has made a long succession of vile molluscous animals. How beneath the dignity of him, who is supposed to have said let there be light and there was light" (Brooke 1985, 47). In the first "Sketch" of his theory, completed in 1842, he had insisted that the existence of laws governing the origin of new species did not diminish but "should exalt our notion of the power of the omniscient Creator" (Brooke 1985, 47).

This strand in Darwin's thinking, coupled with his reference in the *Origin* to the laws stemming from a Creator, lends plausibility to the view that in 1859 he was a deist – one who had rejected revelation as a source of knowledge but who was unwilling to regard the *laws* of nature as either self-explanatory or accidental. As becomes clear from his correspondence with the American botanist Asa Gray, he had not yet dropped the belief that the laws governing variation and natural selection might themselves be designed. There is no indication in the *Origin* of belief in a Deity providentially supervising every detail of the evolutionary process; but the possibility of a higher purpose behind the order of nature is not excluded. Indeed, as Robert Richards argues in this volume, Darwin's language encourages that construction.

On this interpretation, Darwin held a perfectly consistent position in 1859. Acts of creation from an intervening Deity were unnecessary to explain the sequences of species in the fossil record. At the same time, the word 'creation' still had fundamental meaning when reserved for the ultimate origin of the universe and the laws that had made possible what Darwin saw as the highest good – the production of the higher animals. In the first edition of the *Origin* he did still use the word 'creation' when referring to nature's productivity, even while ascribing it to secondary causes. This practice was particularly striking in the context of the first primordial form. Here he contemplated the vast "ages which have elapsed since the first creature, the progenitor of innumerable extinct and living descendants, was created" (*Origin*, 488). There was clearly an ambiguity here, which Darwin eventually resolved in the fifth edition by replacing "was created" with "appeared on the stage" (*Variorum*, 757). One might see in this change evidence of his growing agnosticism, or simply an attempt to avoid misleading language.

The first changes that he made to his book, however, were to add to, rather than subtract from, references to a Creator. He evidently wished to offer reassurance that his theory did not contravene a sophisticated understanding of what 'creation' might mean. To the sentence that referred to "one primordial form, into which life was first breathed," he added the three words "by the Creator," repeating the same move in the book's last sentence. Nor were these the only additions for his second edition. In the first, he had conspicuously placed quotations from Francis Bacon and William Whewell at the head of his text, passages carefully chosen to legitimate, both scientifically and theologically, his enquiry into the secondary causes of species production. As master of Trinity College Cambridge, Whewell was a deeply respectable figure who had written extensively on natural theology and the history and philosophy of science. From his Bridgewater Treatise (1833), Darwin extracted the statement that "with regard to the material world, we can at least go so far as this – we can perceive that events are brought about not by insulated interpositions of Divine power, exerted in each particular case, but by the establishment of general laws" (*Origin*, ii). There was opportunism in Darwin's appropriation of this passage, since Whewell would have balked, and did so (Snyder 2006, 195), at its application to the creation of human beings. But, from Darwin's perspective, here was a theologically respectable endorsement of his explanatory programme. From Bacon's *Advancement of Learning* (1605), he extracted a statement to the effect that it was not sacrilegious, but rather a religious duty, to improve one's proficiency in understanding the book of nature, the book of God's works (*Origin*, ii). For his second edition, when it was clear he should do still more to pre-empt religious objections, he found the perfect theological precedent in a classic work of Christian apologetics, Joseph Butler's *Analogy of Religion* (1736).

There is irony here, perhaps, in that Butler had been battling with the deists of his day, arguing that obscurities in the meaning of Scripture (on which the deists pounced) might, with further research, be clarified – just as obscurities in nature yielded to scientific research. The passage that worked for Darwin was this: "The only distinct meaning of the word 'natural' is *stated, fixed,* or *settled;* since what is natural as much requires and presupposes an intelligent agent to render it so, *i.e.* to effect it continually or at stated times, as what is supernatural or miraculous does to effect it for once" (*Variorum*, ii). Butler was renowned as a subtle apologist, and here was a subtle self-defence for Darwin: the appeal to natural causes in species production was not ultimately to negate the action of an intelligent agent.

The view of another clergyman entered the second edition. Darwin was grateful to the Revd. Charles Kingsley for a letter in which he stated what Darwin had certainly once believed and was glad to hear: "I have gradually learnt to see that it is just as noble a conception of Deity, to believe that he created primal forms capable of self development into all forms needful..., as to believe that He required a fresh act of intervention to supply the lacunas which he himself had made. I question whether the former be not the loftier thought" (*Correspondence*, 7: 380). With Kingsley's permission, Darwin reproduced this affirmation, which he described to his friends as "capital" (*Variorum*, 748).

That Darwin had to avail himself of such remarks in the hope of placating hostile critics has sometimes led to the supposition that his references to a Creator were *merely* placatory and not to be taken seriously. This is not a straightforward matter. He was certainly anxious to minimise offence. He was also incensed by reviews (of which the very first, in the *Athenaeum*, was an inauspicious omen) that ignored the cogency of his argument, dwelling instead on its supposedly damaging implications for religion. Darwin fulminated against the *Athenaeum* reviewer: "the manner in which he drags in immortality, and sets the priests at me, and leaves me to their mercies is base" (*Correspondence*, 7: 387). Moreover, in a letter to Hooker, written in March 1863, he claimed that he had "long regretted" having "truckled to public opinion, and used [the] Pentateuchal term of creation" (*Correspondence*, 11: 278). This was a reference again to those first primordial forms into which life had been breathed. By 1863 he could tell Hooker that what he had really meant was nothing more than "appeared" by some wholly unknown process. A naturalistic account of the origin of life was, however, perfectly consistent with his deistic outlook, and the reasons for his regret are more complicated than one might first imagine. Because of his biblical language, he elicited sympathy from some religious writers who seized the opportunity to plug the gaps with their gods. But he also created an opening for fellow naturalists to accuse him of a less than fully naturalistic account.

Prominent among them was his critic Richard Owen, who claimed to have advanced an evolutionary theory before Darwin. Prior to Darwin's publication, Owen had hinted at natural causes in species development but in exceedingly veiled language (Owen, 1849). In his review of the *Origin*, he objected that Darwin had prematurely jumped to a mechanism of change (Owen, 1860). But Owen *was* insistent on a complete continuity of secondary causes in the history of life and had been paining Darwin with the insinuation that, because he had failed to achieve that in the context of the origin of life, he was unworthy of the crown that Owen reserved for himself. So when the "Pentateuchal term" was deleted (though not, interestingly, in the last sentence) from the third edition, the motivation in part was to protect himself and to deflate Owen (*Correspondence*, 11: 278).

Darwin's remarks on the role of a Creator are often ambiguous. But statements made in response to social pressures are not necessarily disingenuous. As we have seen, Darwin held a consistent position in 1859 in which there was space both for a naturalistic science of species production and for a Creator whose laws made the process possible. As he confided to Asa Gray in May 1860, "I can see no reason, why a man, or other animal, may not have been aboriginally produced by other laws; & that all these laws may have been expressly designed by an omniscient Creator, who foresaw every future event & consequence" (*Correspondence*, 8: 224). This was a philosophy of nature that, with its emphasis on divine laws, he hoped theists could share with him. He could see "no good reason why the views given in this volume should shock the religious feelings of any one" (*Variorum*, 748). But they did.

RELIGIOUS RESPONSES TO THE *ORIGIN*

To say that Darwin shocked the religious sensibilities of his contemporaries, and in some constituencies still does so today, is almost a truism. To construct a balanced account of the religious responses is, however, a major challenge and one that has given rise to a massive literature, including a recent evaluation of the Darwinian impact on the Jewish tradition (Cantor and Swetlitz 2006). A fuller account than can be given here is available in *The Cambridge Companion to Darwin* (Brooke 2003). It is easy to list the issues that worried religious commentators, particularly those concerning human dignity, the authority of sacred texts, and the erasure of Providence and design. But what for one thinker was of fundamental concern and a reasonable ground for rejection was, for others, of less significance. We have already seen that Kingsley delighted Darwin with his sympathetic response. The same was true of Asa Gray, Harvard's devoutly Presbyterian professor of botany, who not only impressed Darwin with his grasp of natural selection but also suggested that if severe competition for limited resources was the motor of the evolutionary process, without which the highest forms of life would never have developed, then this insight might assist the theologians in their wrestling with the problem of pain and suffering.

Darwin and Gray would part company on whether design was still discernible in the structures and adaptation of living organisms, but this issue was not of universal concern to religious minds. It *was* a concern for Charles Hodge at Princeton, whose Presbyterian theology could accommodate the reality of evolution but not the specifically Darwinian mechanism of natural selection – a position he made clear in his book *What Is Darwinism?* (1874). By contrast, for the Anglican-turned-Catholic John Henry Newman, the fact that Darwin had undermined one of Paley's arguments for God's existence was of little consequence, for Newman had already decided that arguments for design contributed nothing to the basic doctrines of Christianity.

Objections based on Scripture, and on the book of Genesis in particular, were voiced from the outset. The clerical naturalist Leonard Jenyns informed Darwin early in January 1860 that he doubted whether the image of man as an even *"greatly* improved orang" would find general acceptance: "I am not one of those in the habit of mixing up questions of science and scripture, but I can hardly see what sense or meaning is to be attached to *Gen*: 2.7. & yet more to vv. 21. 22, of the same chapter, giving an account of the creation of *wo*man, – if the human species at least has not been created independently of other animals" (*Correspondence*, 8: 14). For Jenyns and for many others, including Darwin's wife, it was difficult to see how the moral sense could have developed from "irrational progenitors." By contrast, one attraction of evolutionary ideas for some Christian theologians was that a process of cultural evolution could be superimposed on the biological, permitting the argument that human understanding of the Deity (and of consequential moral obligations) had been progressively refined and that this trajectory was discernible in successive books of the Bible. To renounce a literalistic reading of Genesis was not necessarily to renounce the belief that human beings ultimately owed their existence to a transcendent God.

However, the sense in which Darwin had seemingly placed the Deity at arms length, deleting the necessity for divine intervention in the natural order, worried those who cherished a more intimate relationship between God and the world and who looked to the sciences, as well as to history, for support. The question was whether, despite Darwin's reserve, evolutionary processes could be combined with belief in a God who had providential control over their consequences. There is irony here, because Darwin's metaphor of natural *selection* was sometimes seized by Christian theists to legitimate a providentialist interpretation. Darwin had chosen the metaphor with reference to the selection practised by breeders when seeking to accentuate particular features of animals and birds in captivity. The analogy played a crucial role in the rhetoric of the *Origin* because it helped him to collapse an absolute distinction between species and varieties. Thus he observed that even well-trained ornithologists would be tempted to regard the different varieties produced by the pigeon fanciers as separate species if they did not already know they all derived from the common rock pigeon. But if human intelligence and purpose were involved in the selection of the most propitious individuals for breeding under domestication, might not the analogy support the mediation of a divine intelligence and purpose in the natural order? This was a sufficiently common response to take the sting out of the theory for those so minded, and for Darwin to feel it necessary to scotch the misunderstanding of his position. In later editions of the *Origin*, he gave even greater emphasis to the unconscious and unwitting features of the breeders' activity. The metaphysical ambiguity associated with the metaphor of selection also encouraged him to follow Wallace's advice and to introduce the "survival of the fittest"as a synonym for natural selection.

Darwin's theory was a divisive but not necessarily destructive force within Christendom. One reason it was valued by Asa Gray was the support it could lend to the ultimate unity of the human species. The subtitle of the *Origin*, "The Preservation of Favoured Races in the Struggle for Life," could easily be used to sanction the persecution of one race by another that considered itself superior. In New Zealand, for example, Darwinism was invoked to justify the suppression and even extermination of the Maori (Stenhouse 1999). For Darwin, however, and indeed for Wallace, the implications of the theory were otherwise. There was no justification for racial subjugation or for the practice of slavery because all humans had ultimately derived from a common ancestor. It was this support for monogenism that Gray found so attractive and a welcome contrast to the polygenism of Agassiz.

In the diversity of religious response there is an important qual-ification to popular treatments of the Wilberforce/Huxley debate, which sometimes suggest that Wilberforce typifies the religious reaction. At the 1860 Oxford meeting were other Christian clergy more open to Darwin's extension of scientific naturalism. One was Frederick Temple, who later became archbishop of Canterbury. The divisiveness of the Darwinian theory is particularly poignant here because Temple had been one of Wilberforce's ordinands, deeply upsetting the bishop with his liberal contribution to *Essays and Reviews*. In his review of Darwin's *Origin*, Wilberforce did, how-ever, put his finger on the most sensitive subject: the question of human uniqueness:

Man's derived supremacy over the earth; man's power of absolute speech; man's gift of reason; man's free will and responsibility; man's fall and... redemption; the incarnation of the Eternal Son; the indwelling of the Eternal Spirit, – all are equally and utterly irreconcilable with the degrading notion of the brute origin of him who was created in the image of God and redeemed by the Eternal Son. (Wilberforce [1860] 1874, 1: 94)

Such a reaction raises many questions that are debated in theolog-ical circles to this day. It was, however, recognised as an overreac-tion at the time – not least because in all such disparagement was the assumption that there could be no special human uniqueness if humans were derived from brutes. Darwin himself relished that continuity between animal and human, believing it to be a humbler and more authentic view than was displayed by those arrogantly claiming knowledge of their Maker. It was nevertheless possible to argue that, irrespective of origins, humans did have capacities that had advanced beyond those of their simian relatives. A heightened responsiveness to other beings, an ability to imagine the thought worlds of others, a sense of moral obligation, an aesthetic response to the beauties of nature, and, for some, a grateful responsiveness to a supposed Creator of that beauty still remained facets of human experience. Darwin had been no stranger to them. In his *Journal of Researches*, he had reflected on the sublimity of the Brazilian forests and the desolation of the Tierra del Fuego, both "temples filled with the varied production of the God of Nature." No one, he wrote, "can stand in these solitudes unmoved, and not feel that there is more in man than the mere breath of his body" (Darwin [1839] 1910, 473).

ST. GEORGE MIVART AND THE SIXTH EDITION

There was one response to Darwin's theory that, despite the protestations of its author, was seen by Darwin's inner circle, and by Huxley in particular, as insidiously and religiously motivated. This was a critique from St. George Jackson Mivart, a convert to Roman Catholicism and, initially at least, a convert to Darwinian evolution. Mivart, who had been a student of Huxley, remained an advocate of evolution; but during the 1860s he began marshalling arguments against the primacy and sufficiency of natural selection. These were published in his book *The Genesis of Species* (1871) and were principally directed against the gradualness of change that Darwin so stressed. Mivart argued that extremely small variations and modifications would be so slight as to confer negligible advantage. He also exploited gaps in the fossil record, suggesting they were more in line with sudden discontinuous change than with Darwin's gradualism. Darwin was angered by what he considered the unfairness of Mivart's attack because, whereas he had tried in the *Origin* to give a balanced account of the arguments for and against natural selection, Mivart dwelled only on the difficulties (*Variorum*, 242). The mechanism Mivart preferred involved an internal drive, a complexifying and progressive force, at work in evolutionary development that operated irrespective of changes in external conditions. Whereas Darwin had stressed repeated divergence from common ancestors in a branching and seemingly undirected manner, Mivart looked for evidence of convergence toward recurrent structural forms that could be interpreted as goal-directed (Desmond 1982, 140, 176–80). Huxley was particularly embarrassed by Mivart's critique because he considered it impossible to be a good soldier for science and a loyal son of the church. Here was one of his own academic progeny purporting to show that it was not impossible – as long as one's interpretation of evolution was different from Darwin's. It was not difficult for Huxley to impugn Mivart's motives because the appeal to an inner drive pushing the evolutionary process in specifiable directions cohered with a belief in divine immanence. Was this not a blatant case of religion interfering with science?

Darwin himself hit back in a new chapter for his sixth edition, which allowed him to assess the main criticisms he had received – Mivart's in particular. It is instructive to examine his reply because Mivart's version of theistic evolution set a precedent for many responses to the *Origin* during a long period when it remained unclear, even to experts in the life sciences, whether natural selection could be the primary or a sufficient mechanism (Bowler 2000). Darwin could not stomach Mivart's sudden mutations and innate tendencies toward perfectibility (*Variorum*, 241), yet he had to address a worry that weighed with "many readers" (*Variorum*, 242–3). This was the alleged impotence of natural selection to account for the incipient stages of useful structures. There were latent theological issues here, because Darwin described Mivart's sudden abrupt changes as verging on the miraculous (*Variorum*, 266–7). His really telling point, however, referred to the power of his own gradualistic mechanism to explain the phenomenon so beloved by earlier natural theologians – the beautiful adaptedness of living organisms to the conditions of their existence (*Variorum*, 266–7). To suppress the perfecting power of natural selection was to sacrifice an explanation for the very thing that natural theologians, such as Paley, had correctly identified as in need of explanation.

Whether Mivart's alternative account of evolution had been motivated by, shaped by, or simply produced congruously with his Catholic spirituality it is difficult to gauge. Human motivation is notoriously difficult to determine. It is, however, noteworthy that his emphasis on convergence in separate evolutionary lines does have contemporary resonance in the discussion of anatomical structures, such as the human eye, the type of which has appeared many times (Conway Morris 2003).

DARWIN'S AGNOSTICISM

Toward the end of his life Darwin became increasingly agnostic about religious claims. His cousin Julia Wedgwood even referred to a "certain hostility" in his attitude toward religion "so far as it was revealed in private life." She told Darwin's son Frank that he felt he was confronting an influence that "adulterated the evidence of fact" (Brooke 1985, 41). The resistance of religious minds and the machinations of Mivart contributed to that feeling. Insofar as his attempts to pacify committed theists were successful, he could still be discomfited by the confessional superstructures they erected on his statements. His correspondence with William Harvey, an expert on South African flora, would constitute an example (*Correspondence*, 8: 322–32, 370–4).

Along with his deistic metaphysics, the roots of his agnosticism were already in place when he wrote the *Origin*. He knew that whatever one postulated as the cause of the universe would invite questions about *its* cause. There were also certain issues, such as the apparent contradiction between necessity and free will, and questions concerning the origin of evil, that he considered "quite beyond the scope of the human intellect" (*Correspondence*, 8: 106). He felt much the same about the question of design in nature, holding the conviction that so wonderful a universe could not be the result of chance alone, yet also finding it impossible to see intelligent design in the details. In 1860, he had still hoped to hold the contending elements together, asking Gray, "Does not Kant say that there are several subjects on which directly opposite conclusions can be proved true?!" (*Correspondence*, 8: 274). In the ensuing years, this Kantian solution ceased to satisfy him. During an absorbing correspondence, in which he assessed Gray's opinion that variations could be under the control of a superintending Providence, Darwin became convinced that this was no way out of his "hopeless muddle." The problem was the lack of evidence to support Gray's intimation that the variations (however caused) appeared with their prospective use ordained. They surely appeared randomly, and many were detrimental to the species. The analogy Darwin used to articulate his position concerned the building of a house by a builder who used stones that happened to be available in the vicinity. Surely, Darwin argued, no one would say that the stones had come to be there in order that the builder could build his house. Gray's revealing concession that the perception of design had to be through the eye of faith was hardly likely to bring Darwin around (Moore 1979, 275–6).

The word "agnostic" is, by itself, not especially informative. Everything depends on the particular beliefs one is agnostic about (Lightman 1987). A certain kind of agnosticism has been an intrinsic feature of Christian theology itself in contexts where discourse about

God is understood to be discourse about the incomprehensible and ineffable. Even during his later years Darwin denied ever having been an atheist "in the sense of denying the existence of a God" (*Letters*, 1: 304). While admitting that his judgement often fluctuated, he was still prepared to say that he deserved to be called a theist. The image of a Creator who "creates by laws" still endured, but by the end he had relinquished the belief that the *order* of nature necessitated an inference to purpose. To the feelings of "wonder, admiration, and devotion" that had once filled his mind, he was now anaesthetised (*Letters*, 1: 311). He could still find it inconceivable that the universe was the result of chance alone. But – and with Darwin there is so often a nuance – if the human mind was itself the product of evolution, what confidence could one place in any metaphysical or theological conviction – even one's own?

15   Lineal Descendants

The *Origin*'s Literary Progeny

DARWIN AS READER AND WRITER

Toward the end of his life, in the autobiography written for his family, Darwin remembers his early enthusiasm for literature and other arts with melancholy regret, regret for a taste apparently extinguished by the oncoming of age and the need for endless empirical investigations:

My mind seems to have become a kind of machine for grinding general laws out of large collections of facts, but why that should have caused the atrophy of that part of the brain alone, on which the higher tastes depend, I cannot conceive. (*Autobiography*, 139)

Darwin distinguishes his continued pleasure in 'books on histories, biographies, and travels ... and essays on all sorts of subjects' from his earlier delight in poetry and music:

Up to the age of thirty, or beyond it, poetry of many kinds, such as the works of Milton, Gray, Byron, Wordsworth, Coleridge, and Shelley, gave me great pleasure, and even as a schoolboy I took intense delight in Shakespeare, especially in the historical plays. ... But now for many years I cannot endure to read a line of poetry; I have tried lately to read Shakespeare, and found it so intolerably dull that it nauseated me. Music generally sets me thinking too energetically on what I have been at work on, instead of giving me pleasure. I retain some taste for fine scenery, but it does not cause me the exquisite delight that it formerly did. (*Autobiography*, 138)

Even as he declares his 'curious and lamentable loss of the higher aesthetic tastes', faint echoes of former joys return: 'great pleasure', 'intense delight', 'exquisite delight' – the expressions evoke experience not quite forgotten. At this point in the passage he turns to the anomaly in his capacity for aesthetic pleasure, as he sees it:

On the other hand, novels, which are works of the imagination, though not of a very high order, have been for years a wonderful relief and pleasure to me, and I often bless all novelists.

With a certain defensive self-mockery he then declares his taste in novels:

A surprising number have been read aloud to me, and I like all if moderately good, and if they do not end unhappily – against which a law ought to be passed. A novel, according to my taste, does not come into the first class unless it contains some person whom one can thoroughly love, and if a pretty woman all the better. (*Autobiography*, 138–9)

So, ruefully looking back, Darwin presents himself here as one who now has left to him only lower aesthetic experiences. Many would dispute his assumption that novels are necessarily not of a high imaginative order: he is writing in the wake of *Middlemarch* (1872) and in the same year (1876) that *Daniel Deronda* had appeared in part publication. George Eliot might, however, have met his embargo on novels with unhappy endings, depending on how the reader responds to the final pages of each of those novels. Essentially, Darwin seems to be telling us that in his later life he looks to literature for comfort rather than elevation ('some person whom one can thoroughly love'), with a little mild titillation thrown in ('if a pretty woman all the better').

But this self-description is too easily read back into Darwin's earlier life, and some critics have used it to deny that he ever had strong aesthetic responses. This runs counter to the elegiac tone of regret in the passage, which yearns a little for those past intensities of response. Even more, it denies the precision and enthusiasm of his descriptive writing, shown at its full in *The Voyage of the Beagle* and persisting in the *Origin* alongside an argued oratory, sometimes biblical, that insists on reviving so many of the lost forms of life, apparently expunged by time:

How fleeting are the wishes and efforts of man! How short his time! And consequently how poor will his products be, compared with those accumulated by nature during whole geological periods. (*Origin*, 84)

His subtle account of the interdependent structures of any organic being points out that they are determined

in the most essential yet often hidden manner, to that of all other organic beings, with which it comes into competition for food or residence, or from which it has to escape, or on which it preys. This is obvious in the structure of the teeth and talons of the tiger; and in that of the legs and claws of the parasite which clings to the hair on the tiger's body. But in the beautifully plumed seed of the dandelion, and in the flattened and fringed legs of the water-beetle, the relation seems at first confined to the elements of air and water. Yet the advantage of plumed seeds no doubt stands in the closest relation to the land being already thickly clothed by other plants; so that the seeds may be widely distributed and fall on unoccupied ground. (*Origin*, 77)

Tiger and parasite, dandelion and water beetle, are shown as adapted not only in relation to their immediate surroundings, but also in their wider natural milieu. They are, moreover, evoked in the writing: 'clinging', 'beautifully plumed', 'flattened and fringed'.

When in the *Autobiography* Darwin deplores the difficulty he has always had in writing 'clearly and concisely', he modestly does not remark that it is the complexity of what he must simultaneously show that is a fundamental reason for this problem. He does, however, acknowledge that such difficulty also has certain intellectual advantages:

I have as much difficulty as ever in expressing myself clearly and concisely; and this difficulty has caused me very great loss of time; but it has had the compensating advantage of forcing me to think long and intently about every sentence, and thus I have been led to see errors in reasoning and in my own observations or those of others. (*Autobiography*, 136–7)

So, writing spontaneously, however ill-sorted the first attempt, and then scrutinising, together make for that temper of enthusiasm and constant reflection that we encounter in his writing. He continues:

Formerly I used to think about my sentences before writing them down; but for several years I have found that it saves time to scribble in a vile hand, whole pages as quickly as I possibly can, contracting half the words; and then correct deliberately. Sentences thus scribbled down are often better ones than I could have written deliberately. (*Autobiography*, 137)

The zest of observation and insight impels him forward. Afterward he can comb through the cluster of his sentences. And he recognises that quite frequently the first free dash at expression survives better than the reasoned enunciation. But both are necessary to test to the full his argument and his observation.

Darwin, then, drew on literature that he loved during the time of the formation, as well as the formulation, of his theory. We can to some extent track that process in the reading lists that are now included as an appendix to volume 4 of the *Correspondence*. These lists begin only on his return from the *Beagle* voyage and do not note all the books he was reading, but they do, fascinatingly, often note *how* he was reading, even down to differing degrees of skimming: 'skimmed', '*Well* skimmed', 'slightly skimmed'. Among the copious lists, which place mainly scientific books on the left-hand page and nonscientific books on the right, he includes, among travel books and volumes on natural history, physiology, and political economy, many works of literature, both classical and newly published. For example, he reads Harriet Martineau's novel *The Hour and the Man*, based on the life of the black rebel leader Toussaint-l'Ouverture, in 1842, soon after its publication in 1841. Later in that same list, he is reading Carlyle's *Sartor Resartus* ('excellent') and his *Hero Worship* ('moderate') as well as the minor poems of Milton, a volume of Wordsworth in March and another in May. He does not much enjoy Dryden or Bacon or Swift; they all get short shrift: 'Bacon's Essays – dull, and crabbed style'; 'Tale of Tub – poor'; 'Failed in reading Dryden's Poems except Absalom and Ach[itophel] wh. I rather liked'. The very last entry in these lists, which stretch from 1838 to 1860, is another famous book, with almost as long a publication future as the *Origin*, 'Smiles Self-Help (goodish)' (*Correspondence*, 4: 462–3, 497).

As has by now often been observed, Darwin carried Milton's *Poems* with him on his long land journeys during the five-year voyage of the *Beagle*. *Paradise Lost* could furnish the imagination with descriptions of the genesis of life, and its stretched and sinewy sentence structure taught much about keeping ideas in suspension alongside each other, avoiding premature closure. Byron's romantic travels in *Childe Harold* give scope to the ego discovering itself. During Darwin's youth Wordsworth's *Prelude* had not yet been published, but his *Excursion* gave entry to the life of sojourning country dwellers and their philosophies, while 'Lines Composed Above Tintern Abbey' sounds the knell that Darwin came to understand only too well in his retrospect on his own early aesthetic pleasures.

> And now, with gleam of half-extinguished thought,
> With many recognitions dim and faint,
> And somewhat of a sad perplexity,
> The picture of the mind revives again.
>
> (Wordsworth 1969, 164)

Any sailor, especially on so long, trying, and fascinating a journey as that of the *Beagle*, could respond to Coleridge's *Ancient Mariner* with its uncovering of the random violence induced by tedium:

> 'God save thee, ancient Mariner
> ' From the fiends that plague thee thus –
> 'Why look'st thou so?' –With my cross bow
> I shot the ALBATROSS.
>
> (Coleridge 1993, 220)

And Shelley is probably the most informed and ambitious in his response to scientific thought of any of the poets that Darwin names.

But this *Companion* is concerned particularly with the *Origin*, written in the period after the full intensity of his aesthetic responses had – according to his late recollections – already slackened. The writing of that work occupied him over thirteen months as he approached his fiftieth birthday. By then, what he had learned from poets and poetry had silted down in his consciousness and was no longer the object of rapturous attention. Yet it had not gone away. It was so taken for granted as to be part of the mass of his mind. I have called this essay 'Lineal Descendants: The *Origin*'s Literary Progeny': progeny demand forebears.

Understanding Darwin's relation to, and enjoyment of, his copious and eclectic reading is essential for a full grasp of his powers of reflection and lateral reach that placed ideas in new relations to each other. Those qualities of mind were the ones that encompassed the mass of available empirical material and condensed it into a major new theory. As George Eliot wrote in *The Mill on the Floss* (1860), the first of her works to respond in any measure to the *Origin*:

In natural science, I have understood, there is nothing petty to the mind that has a large vision of relations, as to which every single object suggests a vast sum of conditions. It is surely the same with the observation of human life. (Eliot 1980: 238)

The warmth and eloquence of Darwin's writing in the *Origin* is part of the persuasive act by which his ideas engage a broad range of readers, scientific and nonscientific. Those readers then work with the materials of the book to form arguments, observations, and, above all, stories of their own. The great fresh idea of 'natural selection' has to struggle within the formulation that Darwin finds for it, a formulation that threatens to reinstate design even as it flouts it: if 'selection', then by what principles? And is there a ghostly 'selector' making the discriminations, haunting the field of the metaphor? Darwin proposes a dualism of the natural and the artificial: that is, unknowing physical processes as opposed to what is contrived by human artifice. But the term 'selection' moves between the two zones, acting as an agent, perhaps a double agent. Indeed, expelling agency from human language is an impossible task. Darwin expressed exasperation at some of the twists his critics brought to the term: for example, in the third edition: 'that the term selection implies conscious choice in the animals which become modified; and it has even been urged that as plants have no volition, natural selection is not applicable to them!' (*Variorum*, 165:14.3–4:c). But in that same paragraph of the third edition he acknowledges that '[i]n the literal sense of the word, no doubt, natural selection is a misnomer' (145:c), and in the fifth edition that becomes 'natural selection is a false term' (14.5:e).

However, he insists on the need for metaphorical language and buttresses his case with the more established example of chemists:

who ever objected to chemists speaking of the elective affinities of the various elements? – and yet an acid cannot strictly be said to elect the base with which it will in preference combine. (*Variorum*, 165:14.5:c)

It is a telling example, and one that, intriguingly, had already provided the ground for fiction in Goethe's great novel *Die Wahlverwandtschaften* (Elective affinities, 1809). That novel opens with a scene in a garden where the hero is grafting newly cut scions onto rootstock, and it concerns adultery and unwilled affinities. Goethe takes the language of chemistry over into human affairs, affairs from which that technical language had itself sprung. The rapid slide to and fro between ordinary human concerns and constrained scientific usage besets Darwin, exacerbated by his wish to write in a language that any educated person could read. It also provides the creative space for later writers who have in so many different ways responded to Darwin, his ideas, his telling of them, and the often incompatible stories that they have come to inspire.

DARWIN IN LITERATURE NOW

Writers do not simply adopt a scientific idea and carry it through in their novels, plays, or poems. Sometimes a phrase is enough to set new responses in motion. Sometimes the general idea is all the writer needs. As often, it is a problem or contradiction in the ideas that sets the creative juices flowing. Occasionally the writer is a close student of all the intricacy of language and theory that the *Origin* sets forth: that is the exception. All the examples I shall give, both from our contemporaries and from earlier writing, show processes of transformation, sometimes poised at so extreme an angle that it cannot be certain how consciously the writer has drawn on the bank of Darwin's presence. I use the term 'presence' advisedly, since over the years Darwin himself has been mythologized and is liable to appear in a variety of guises in works of fiction: at the turn of the twenty-first century, for example, in Jenni Diski's *Monkey's Uncle* (1994), Roger McDonald's *Mr Darwin's Shooter* (1998), and Harry Thompson's *This Thing of Darkness* (2005). Or in a long poem like 'Darwin' by Lorine Niedecker (1985), drawing intimately on his writing and life, from which I quote these stanzas:

> A thousand turtle monsters
>   drive together to the water
>   Blood-bright crabs hunt ticks
> on lizards' backs
>
> Flightless cormorants
>   Cold-sea creatures–
>   penguins, seals
> here in tropical waters
>
> Hell for FitzRoy
>   but for Darwin Paradise Puzzle
>   with the jig-saw gists
> beginning to fit   (Niedecker 1985, 110)

The fascination with the man has grown stronger in recent years so that he has become a cultural figure buffeted by the winds of controversy, and also a remembered human being tenderly imagined at the end of his life, as in Gejtrud Schnackenberg's poem 'Darwin in 1881', which ends with Darwin wandering in the garden and then coming to bed on his last night:

> He lies down on the quilt,
> He lies down like a fabulous-headed
> Fossil in a vanished river-bed,
> In ocean drifts, in canyon floors; in silt,
> In lime, in deepening blue ice,
> In cliffs obscured as clouds gather and float;
> He lies down in his boots and overcoat,
> And shuts his eyes.
>
> (Schnackenberg 1986, 22)

Darwin the man, then, is very much part of literature now. Ian McEwan introduces a sentence from the conclusion of the *Origin* near the beginning of his novel *Saturday* (2005): 'There is grandeur in this view of life'. The quotation becomes an adamant point of reference through all the vicissitudes of the long day that McEwan then explores with his main character, a neurosurgeon. More allusively, Carol Ann Duffy gives a teasing voice to Emma Darwin:

> Mrs. Darwin
> *7 April 1852.*
> Went to the Zoo.
> I said to Him –
> Something about that Chimpanzee over there reminds
>   me of you.   (Duffy 1999, 20)

In this little squib of a poem Duffy is, of course, remembering Darwin's interest in the London zoo orangutan and is drawing on the long Victorian tradition of illustrating Darwin as almost an ape. This can all be compressed into four lines because by now the life is so familiar to a variety of readers.

But it is not just Darwin's person that continues to inhabit writers' imaginations. One of the most compelling later twentieth century novels to consider the losses of evolution, as well as its strengths, is William Golding's *The Inheritors* (1955). This daring and influential work is presented through the minds of Lok and Fa, gentle

Neanderthals, who try to interpret the behaviour of a new kind of being, our ancestors. By reversing the usual assumption that Neanderthals were brutal and stupid, well overcome by *Homo sapiens*, Golding sheds light on the hubris, and the competitive ferocity, of the human. He also by implication reminds the reader of the outrages perpetrated against indigenous people in different parts of the world as part of the imperial enterprise. The work explores the idea that the Earth itself is an organism of a superior sort; thus it presages Gaia and some aspects of recent ecological thinking. Golding's novel provides a salutary corrective to the notion that the 'survival of the fittest' (in Herbert Spencer's phrase, adopted by Darwin in later editions of the *Origin*) guarantees a morally superior inheritor.

Doris Lessing has twice turned to issues of genetic inheritance, and to reversion, to fuel her novels. In *The Fifth Child* (1988) she imagines a modern liberal family burdened with a male child whom she conceives as an atavistic monster. In her 2007 novel *The Cleft*, she reimagines creation as a lost world of semiaquatic females into which by genetic chance the aberration of maleness enters through the birth of a boy. Similarly, though at a further stretch, her five-volume science fiction sequence *Canopus in Argos: Archives* (1979–83) draws by counter-representation on the evolutionary nature of our customary thinking now. It creates a world in which a new form of civilization spreads through the planetary system by means of what she calls 'forced evolution'.

The idea of evolution and the new technological processes that may drive the human and all surrounding life toward new extreme conditions trouble writers of many varying degrees of talent and insight. At its peak, we have writing such as Margaret Atwood's 2003 novel *Oryx and Crake*. She imagines a post-human world of genetically engineered beings, but her imagination does not need to resort to things not already in the world. The disastrous future that she depicts is generated out of the present. Atwood, herself the daughter of an entomologist, insinuates a chill and detailed account of the conditioned behaviours, as well as the feckless choices, that bring about disaster. Evolution and artificial selection are here at the extreme distance from Darwin's implicitly benign figuring of natural selection, which 'can act only through and for the good of each being' (*Origin*, 84). Yet Darwin also knew that the future could not be foreseen or controlled: 'of the species living at any one period, extremely few will transmit descendants to a remote futurity'.

Looking to the future, we can predict that the groups of organic beings which are now large and triumphant, and which are least broken up, that is, which as yet have suffered least extinction, will for a long period continue to increase. But which groups will ultimately prevail, no man can predict; for we well know that many groups, formerly most extensively developed, have now become extinct. (*Origin*, 126)

RESPONSES IN NINETEENTH-CENTURY LITERATURE

That word 'extinct' gathers into it a terror already abroad before the *Origin* was published. On the one hand: will humankind progress? The expected and hoped-for answer among Darwin's contemporaries was 'yes'. Darwin encouraged that hope in the final paragraph of the *Origin*. He there writes of 'the Extinction of less-improved forms':

Thus, from the war of nature, from famine and death, the most exalted object which we are capable of conceiving, namely, the production of the higher animals, directly follows. (*Origin*, 490)

But 'Man', in the Victorian locution, is not distinguished from 'the higher animals' here: what applies to them, applies to us. And what Darwin has shown is that most species fail to send descendants far into the future. So, on the other hand: will humankind survive? Darwin suggested that mankind, so rarely named apart from other animals in the *Origin*, was subject to the same attrition as all present and past species. That was the chilly message that lay beneath the emphasis on 'improvement', 'development', and 'complexity'. In one aspect Darwin's emphasis on relationships, the most important being the relationship of one organism to another, sounds a note of egalitarianism: 'We possess no pedigrees or armorial bearings; and we have to discover and trace the many diverging lines of descent in our natural genealogies, by characters of any kind which have been long inherited' (*Origin*, 486). But equally, the fine-shaved balances that decide mortality give little room for individual optimism.

To understand the rapid impact that the *Origin* had upon in wider Victorian society and literature it is important to recognise that well before its publication in 1859 Robert Chambers's anonymously published *Vestiges of the Natural History of Creation* (1844) was going through a series of editions and revisions that kept the issue of humankind's relation to the animal kingdom in the foreground of thought:

Man, then, considered zoologically, and without regard to the distinct character assigned him by theology, simply takes his place as the type of all types of the animal kingdom, the true and unmistakable head of animated nature upon this earth. (Chambers, ed. Secord, 1994: 272–3)

The second half of Chambers's sentence is upbeat ('true and unmistakable head'), but the first is more neutral ('the type of all types'). So Man is not set aside for privilege in zoology but manifests all the attributes of the animal kingdom. Perhaps the most famous cry of desolation in Victorian literature, and one habitually linked to the advent of evolutionary theory, is Tennyson's in his great elegy *In Memoriam*, particularly in poems 54–6.

> 'So careful of the type?' but no.
> From scarped cliff and quarried stone
> She cries, ' A thousand types are gone:
> I care for nothing, all shall go.

So opens poem 56, in the voice of nature crying from geological formations that harbour so many vestiges of extinct life. The poem continues from line 8 thus, in a prolonged questioning that embeds the famous 'Nature, red in tooth and claw':

> And he, shall he,
>
> Man, her last work, who seemed so fair,
>   Such splendid purpose in his eyes,
>   Who rolled the psalm to wintry skies,
> Who built him fanes of fruitless prayer,
>
> Who trusted God was love indeed
>   And love Creation's final law –
>   Though Nature, red in tooth and claw
> With ravine, shrieked against his creed –
>
> Who loved, who suffered countless ills,
>   Who battled for the True, the Just,
>   Be blown about the desert dust,
> Or sealed within the iron hills?
>   (Tennyson 1969, 911–12)

Man, despite his ideals and hopes, may, like all other forms, be annihilated, present only as mixed dust or geological detritus.

This poem was almost certainly written before the publication of Chambers's work, and about twenty years before the *Origin*. It is responding to Lyell's *Principles of Geology* (1830–33), a work that profoundly affected the young Darwin at that time too. Tennyson's whole sequence of poems was published (first anonymously) in 1850 and rapidly went through a number of editions. Such was its acclaim, and its closeness to Darwin's own concerns, that it is likely to have been present in Darwin's consciousness as he moved toward the writing of the *Origin*. And Darwin, in the chapter on the 'Struggle for Existence', parallels, though with less extremity of language, what Tennyson observed:

We behold the face of nature bright with gladness, we often see superabundance of food; we do not see, or we forget, that the birds which are idly singing around us mostly live on insects or seeds, and are thus constantly destroying life. (*Origin*, 62)

Two pages later the glad face of Nature is subject to assault:

The face of Nature may be compared to a yielding surface, with ten thousand sharp wedges packed close together and driven inwards by incessant blows, sometimes one wedge being struck, and then another with greater force. (*Origin*, 67)

Such melancholic, and violent, insights are the counter within the *Origin* to Darwin's efforts also to construe the complexity of nature as ultimately harmonious, and enabling perfection:

It is interesting to contemplate an entangled bank, clothed with many plants of many kinds, with birds singing on the bushes, with various insects flitting about, and with worms crawling through the damp earth, and to reflect that these elaborately constructed forms, so different from each other, and dependent on each other in so complex a manner, have all been produced by laws acting around us. (*Origin*, 489)

That ecological and finally orderly, law-governed image is one of the multiple and contradictory stories to emerge from readings of the *Origin*. It can be the story of the genealogical descent of man or, in a different reading, his ascent from lowly organic beginnings; it can be an obliterative tale of extinctions past and to come; it can sustain the intricate ecological balances needed for life to thrive and survive; it can underpin imperial predations in which the more advanced seize the space, and are justified in doing so; it can emphasise the fundamental need for collaborative and even cooperative behaviour. Exactly that capacity to sustain quite diverse narratives helps to explain the creativity of the *Origin* and its fruitfulness in those thought experiments called literature.

So within the work a number of competing potentialities occur. Moreover, when we think about the *Origin* and literature it is important (as my examples suggest) to understand it as part of a network of discussion, as well as seeing it as a fulcrum. Certainly it acts as a hinge for fresh connections, and as a support for new ideas. But its impact would have been less great had it simply stood alone. It is its implication in, and conversation with, other writing of the time that generates much of its initial power. That power grew over the ensuing years.

One can track the presence of the *Origin* above and below ground in Victorian literature. Sometimes it is manifest, used as pointer to a character's free thinking, as in Du Maurier's *Trilby* (1894), where the artist hero Little Billee is reading it for the third time. In Elizabeth Gaskell's last novel, *Wives and Daughters* (1866), the character Roger Hamley, an honest and open natural historian, is modelled on Darwin. This portrayal is unusual in Victorian literature, where doctors, rather than scientists, tend to be the heroes. But beyond these direct references, the *Origin* provides problems for writers to engage with and stories for them to challenge.

Some of the most striking early responses were expressed through fantasy. Two contrasting books play with Darwin's ideas to explore alternatives to current society: Charles Kingsley's *The Water Babies* (1863) and Samuel Butler's *Erewhon* (1874). Kingsley's book is addressed to children and combines a fierce satire on the practice of exploiting children – particularly here little boys as chimney-sweeps, forcing them down chimneys – with a tale of evolutionary and moral renewal. The child Tom drowns and then goes through the processes of rebirth in the ocean, following the stages of foetal gestation and drawing on the then-current theory of 'recapitulation', in which the human being was believed to proceed through the forms of other kinds to reach the human. It is also a story about what's possible and what's natural, challenging the assumptions about what can be, in the service of religious faith but also in the service of new scientific thinking. The whole book is a farrago of ideas, insights, powerful storytelling, and an often very moving exploration of a lost child's experiences. It also tells its young reader to beware of denial:

You must not say that this cannot be, or that is contrary to nature. You do not know what nature is, or what she can do; and nobody knows; not even Sir Roderick Murchison, or Professor Owen, or Professor Sedgwick, or Professor Huxley, or Mr. Darwin, or Professor Faraday, or Mr. Grove, or any other of the great men whom good boys are taught to respect. They are very wise men; and you must listen respectfully to all they say: but even if they should say, which I am sure they never would, "That cannot exist. That is contrary to nature," you must wait a little, and see; for perhaps even they might be wrong. (Kingsley, n.d., 53–4)

Kingsley was, in fact, one of Darwin's earliest allies and among the first to understand that there was no necessary contradiction between natural selection and religious belief. Like Chambers, he placed God at the generative heart of creation, but refused to see Him as persistently active. Intriguingly, when he figures this perception he presents it in the form of a woman, Mother Carey, seated in the middle of the Peacepool:

He expected, of course – like some grown people who should know better – to find her snipping, piercing, fitting, stitching, cobbling, basting, filing, planing, hammering, turning, polishing, moulding, measuring, chiselling, clipping, and so forth, as men do when they go to work to make anything.

But, instead of that, she sat quite still with her chin on her hand, looking down into the sea with two great grand blue eyes, as blue as the sea itself. (Kingsley, n.d., 196–7)

The contrast he describes fits the slow, seeming passivity of natural selection as opposed to the driven activity of artificial selection just as much as it fits godhead or the creative principle. That may be Kingsley's point.

Also in 1863, an article headed 'Darwin among the Machines' appeared in the New Zealand *Press* newspaper. This was the core of Samuel Butler's *Erewhon: or Over the Range* (1872), which, as Butler himself acknowledged, took inspiration from the *Origin*. (His later one-sided quarrel with Darwin has little bearing on this text.)

Erewhon (Nowhere in reverse) is a land where current values are apparently inverted: the ill are punished for their condition, the criminal rewarded. But that inversion itself raises questions about the actuality of current declared values. Perhaps the most enthralling part of the book is Butler's exploration, in the Book of the Machines, of the evolutionary potential of machines and their relationship to Man. On one side of the debate within the book, most argue that machines have the potential to self-generate and vary: 'Machines can within certain limits beget machines of any class, no matter how different to themselves' (Butler 1884: 205). They are therefore a future threat to the dominance of mankind: 'May not man himself become a sort of parasite upon the machines? An affectionate machine-tickling aphid?' (197) On the other side it is argued that Man himself 'is a machinate mammal' (218).

The lower animals keep their limbs at home in their own bodies, but many of man's are loose, and lie about detached. (218)

Umbrellas, for instance: 'If it is wet we are furnished with an organ commonly called an umbrella' (220). The rich, it is argued, are more advanced in evolutionary terms because they can own so many external limbs. Butler's satiric fantasy is full of perceptions that have since been realised – the miniaturisation of computers, for example, and artificial intelligence. His quirky brilliance allowed him to perceive many of the implications of the *Origin* that others missed, including the problem of kinds and degrees of consciousness in other life forms.

That perception became the matter of satire as well as resistance in the work of several notable women writers of the later nineteenth century. The large presence of George Eliot in relation to Darwin has been extensively discussed in many places, including in my book *Darwin's Plots: Evolutionary Narrative in Darwin, George Eliot and Nineteenth-Century Fiction* (1983, 2000). Her responses can be tracked in the very form of her fictions, particularly in the later works *Middlemarch* (1871) and *Daniel Deronda* (1878). In *Middlemarch*, the search for a single originary explanation proves delusive for two contrasted characters, Casaubon, the mythographer, and Lydgate, the medic (though in his case the problem is formulating the necessary question). Instead, the book emphasizes the web of relations and affinities, the play of kinship and difference, among the various groups and individuals in the specific environment of an English midlands town of the late 1820s. It demonstrates how interdependent are the fates of those who barely know each other and how far they are determined by the milieu in which they seek to survive. The 'large vision of relations' that she earlier ascribed to the true scientist is here brought into play in terms of human relations. Indeed, Henry James thought *Middlemarch* 'too often an echo of Messrs. Darwin and Huxley' (*Galaxy* 15 [1873]: 424–8). But, despite her partner G. H. Lewes's deep interest in Darwin, particularly from the mid-1860s on, George Eliot does not engage directly with his work on the surface of her novels. Instead, she works with the forms for experience that Darwin's theories imply: Who survives? How does adaptation take place? What shapes can the unknown future take? Is it possible to plan massive tribal change (as Daniel proposes in the proto-Zionist conclusion of *Daniel Deronda*)?

Much more manifestly responsive to the *Origin* is the work of poets such as Mathilde Blind, Emily Pfeiffer, May Kendall, and Constance Naden. Naden and Kendall wittily satirise current society by means of evolutionary ideas. In 'The New Orthodoxy', Naden, herself well educated scientifically, suggests that faith in science has usurped religious faith as a mark of respectability: the young woman speaker in the poem – a Girton girl – berates her suitor for his backsliding from belief (in Darwin, not religion):

> Things with fin, and claw, and hoof
> Join to give us perfect proof
> That our being's warp and woof
> We from near and far win.
> Yet your flippant doubts you vaunt,
> And – to please a maiden aunt –
> You've been heard to say you can't
> Pin your faith on Darwin!
> (Naden 1894, 313)

Mathilde Blind takes with formidable seriousness Darwin's vision of descent. The appearance of *The Descent of Man* in 1871 triggered considerably more immediate literary response than did the initial publication of the *Origin*. It also enhanced the presence of the *Origin* in later nineteenth-century literary controversy. Blind's one-hundred-page-long epic poem *The Ascent of Man* first imagines the primal stirrings of life ('Chaunts of Life') and then turns to the fate of the outcasts, those who do not survive the evolutionary power struggle ('The Pilgrim Soul'). The last section of the poem is 'The Leading of Sorrow', where a veiled woman conducts the poet through all the reaches of life and death subject to the evolutionary law of struggle and survival. It was first published in 1889 and then again in 1899, that time with a preface by Alfred Russel Wallace. He is shocked as well as impressed by her sombre view of 'the destruction and war ever going on in the animal world, from the lowest to the highest forms' (Blind 1899: ix). Blind magnificently evokes the energies of primal growth:

> Enkindled in the mystic dark
> Life built herself a myriad forms,
> And, flashing its electric spark
> Through films and cells and pulps and worms,
> Flew shuttlewise above, beneath,
> Weaving the web of life and death.
>
> (Blind 1899, 17)

Pfeiffer expresses the extreme disturbance of evolutionary experience in pithy sonnets. Evolution, she writes in the poem of that title, is 'Hunger':

> Hunger that strivest in the restless arms
> Of the sea-flower, that drives rooted things
> To break their moorings
>
> (Pfeiffer 1888, 51)

Nature, meanwhile, in another sonnet in that sequence, has changed from 'nursing mother' to 'Dread Force', 'Cold motor of our fervid faith and song': 'Churning the Universe with mindless motion' (30). Mark Pattison responded to Pfeiffer's poems with a shudder at the 'evolutional idea':

I think the most striking and original of your sonnets are inspired by the evolutional idea – an idea or form of universal apprehension, which, like a boa, has infolded all mind in this generation in its inexorable coil. Try as we may, we cannot extricate our thoughts from this serpent's fold. (Pfeiffer 1888, iii)

So by the 1880s Darwin's ideas have become inextricably wrought into the thoughts of a generation. They can move beneath the surface without the need for direct allusion, inescapable. It is in that style that Hardy responds to them in novels like *Tess of the D'Urbervilles* and *Jude the Obscure*. Tess, scion of an antique family, would be happier remaining a lowly Durbeyfield than rediscovering past D'Urberville grandeur: 'From the first growth of the tree, many a limb and branch has decayed and dropped off' (*Origin*, 129). Tess is fruitful yet doomed because circumscribed by current social conventions. Her physical generosity is exploited and misunderstood by contrasting men, Alec and Angel. She seems to offer fecundity and a wholesome future, but she is sacrificed to the demeaning assumptions of her environment. So energy becomes murder becomes sacrifice. *Jude the Obscure* is bleaker yet: here the general will to live is withering, and Jude's struggles end in a death that is set to one side, unnoticed. The future, as imagined in this novel, is blasted by the division between the intellectual Sue and the grossly physical Arabella. Sue recoils from sexual freedom. Arabella proves to be fittest to survive in a degraded world.

Hardy took further Darwin's hint that natural selection may not produce perfection (bees die as they sting, for example); he combined that with the observation in *The Descent of Man and Selection in Relation to Sex* that sexual selection among humans, in contrast to other species, is driven by the social and economic power of the male, to the detriment of the health of future generations because, among other things, such men tend to select heiresses, who are likely to come from low-fertility families. Hardy responded to Darwin's work with all the intensity of a man who, like Darwin, noticed the variant, the aberrant, and who questioned the normative. The tenor of his work is darker than Darwin's may seem to be, but it is faithful to the insights into waste, extinction, individual withering, and the ravishing beauty of the natural world that are fundamental to Darwin's vision, too.

In these later days of the nineteenth century, the *Origin*, with its abstemious resistance to naming humankind, had become blended in many people's minds with the more robust presence of the human in the argument of *The Descent of Man*. And that is how it continued to be for many years. One can observe the effects of this amalgamation in the two works with which I now conclude. One might choose a number of the novels of H. G. Wells to examine, notably *The Time Machine* (1895), with its vision of a future in which the beautiful

Eloi seem to dominate but prove to be food for their degraded servant class, the Morlocks:

But, gradually, the truth dawned on me: that Man had not remained one species, but had differentiated into two distinct animals: that my graceful children of the Upper-world were not the sole descendants of our generation, but that this bleached, obscene, nocturnal Thing, which had flashed before me, was also heir to all the ages. (Wells 1993, 47)

Wells had been a pupil of Thomas Henry Huxley and remained fascinated by the implications of evolutionary theory. He saw, along with his mentor, that the very distant future would see 'the life of the old earth ebb away', leaving only silence, darkness, and one strange thing 'hopping fitfully about' (86). But the work that bears the full brunt of Darwin's arguments is *The Island of Dr. Moreau* (1896).

This novel, much affected also by the late Victorian controversies around vivisection, takes up Darwin's idea of the 'great family' in which all life is kin and brutally demonstrates what happens when surgical experiment replaces the slow process of natural selection. In that sense it is very much on Darwin's side, and against 'artificial selection' and, perhaps, eugenics. On a remote island the narrator finds himself shipwrecked and at the mercy of two other Englishmen, one of whom is carrying out viciously painful experiments that graft species together: Hyaena-Swine and Bear Man. Moreau is a type of the scientist without empathy, preoccupied solely with 'the study of the plasticity of living forms' and refusing to recognise the suffering he imposes on living creatures. Moreau's dread is that the creatures will breed and that they will gain a taste for meat. So the Beast People are constrained by the Law, which they chant ritualistically and which keeps them enthralled. Of course, the inhibitions fail, in a rout of violence and treachery. This grim fable, like *The Time Machine*, is now sometimes seen as having a racist element. But, more powerfully, it can be read against the horrors of irresponsible experiment that benefits only sterile knowledge at the price of diverse life.

A far more cheerful take on Darwin is the popular masterpiece *Tarzan of the Apes* (1914) by Edgar Rice Burroughs. Here the hero Tarzan, son of an aristocratic young couple killed by the king ape,

Kerchak, is brought up as an ape by his adoptive ape mother. Trouble arrives with Jane, the human heroine whom Tarzan loves. The book, with marvellous absurdity, produces a paradoxical ending that affirms Tarzan's gentlemanly genetic inheritance. Tarzan (though knowing by now his true parentage) tells Jane's fiancé (Tarzan's own cousin) that his mother was an ape and that she therefore could not tell him about the family from which he stems. The 'great family', in the sense of those with chivalric credentials, is put back in charge. Jane marries her gentlemanly fiancé, and Tarzan has to be content with his own honourable behaviour in denying his descent as a man, rather than as an ape. It is that disinterested denial that vouches for his aristocratic birth. Indeed, the paradoxes stretch further, since the cousin has discovered Tarzan's true birth and Tarzan comes to know this; thus the behaviour of one brought up within the pale of high human society is unfavourably contrasted with that of one brought up among apes. Of course, this self-abnegation leaves Tarzan without a mate, and so not a good example of survival. The gentleman must bow out of the struggle for existence.

16 The *Origin* and Political Thought

From Liberalism to Marxism

The publication of the *Origin of Species* propelled Darwin to the status of a public figure, and although he himself preferred to remain secluded in his country house at Downe and pursue specialised research, his theory was at the centre of a heated debate on the social and political implications of evolution. The key principles of Darwin's biology – the struggle for existence and natural selection – became subject to a wide spectrum of interpretations ranging from laissez-faire liberalism to Marxism. This state of affairs raises some interesting questions concerning the claims and arguments advanced by proponents of such opposing views in defence of their positions. Simultaneously, one may inquire after Darwin's own political opinions. This chapter proposes to examine these questions through a close study of reactions to the publication of the *Origin* from three different sources: first, the somewhat misleading enthusiasm of Herbert Spencer, the great philosopher of evolution and an advocate of an extreme type of individualism; second, the progressive attitude of Clémence Auguste Royer, the first translator of Darwin's *Origin* into French; and finally, the comments made by the authors of the *Communist Manifesto*, Karl Marx and Friedrich Engels. All three offer particularly interesting case studies, since Darwin expressed his own opinions of their claims regarding his theory, mostly in private correspondence. Thus, they provide not only a window on the multifarious political uses made of Darwin's biological proposals as put forth in the *Origin*, but also a way to probe his own standpoint on matters political. Another advantage of this pointed examination lies in its focus on the period prior to the publication of *The Descent of Man*. Darwin avoided the question of the origins of mankind in his 1859 publication for fear that parading his views on the subject without any evidence would be "useless and injurious to the success of the book" (*Autobiography*, 130–1). But his readers and contemporaries were much less hesitant to venture into the dangerous territory. Thus, by limiting the time frame to the decade of the 1860s I hope to highlight Darwin's concerns and to examine the tensions that the *Origin* had to face upon publication. This approach will also help gauge the political significance and impact of the theory of biological evolution before Darwin decided to tackle directly the topics of man and his social existence.

I begin my investigation with Herbert Spencer and a review of his intellectual production prior to 1859, since it is necessary for understanding his attitude vis-à-vis Darwin's work and vice versa. During his lifetime, Spencer's reputation was as great as that of his illustrious compatriot, and among the list of subscribers to his most important work – the *System of Synthetic Philosophy* – one can find the names of the main figures in Darwin's close circle: the geologist Charles Lyell, the botanist John Hooker, and the biologist Thomas Henry Huxley. Spencer's glory, however, was not destined to last long, and already toward the end of his life his evolutionary theories were viewed as outdated and inaccurate. The goal of Spencer's *Synthetic Philosophy* was to give a comprehensive account of evolution and progress in both nature and society through a ten-volume publication covering the fields of physics, biology, psychology, sociology, and ethics. His main motive in this ambitious enterprise was primarily political and consisted of providing the doctrines of liberal individualism and laissez-faire economics with a scientific basis. This was evident as early as 1842, when Spencer published a series of twelve letters entitled "On the Proper Sphere of Government," in which he expounded the key elements of his evolutionary theory. First, Spencer expressed his belief in a close and specific relation between animate creatures (including man) and the external world in which they lived, whereby each had appropriate organs and instincts adapted for the performance of its life activities. These activities, in their turn, were dependent upon the position in which the creature was placed, since "Nature provides nothing in vain. Instincts and organs are preserved only as long as they are required." Second, Spencer claimed that natural laws were universal and balanced each other; and finally, he stated his strong opposition to any form of government intervention because it hindered the natural course of affairs, viz. the cobalancing action of universal laws. These convictions, formed at the age of twenty-two, were to remain the core of Spencer's subsequent theories, and, as he himself confessed in his autobiography, without the letters "On the Proper Sphere of Government" he would probably never have written the *System of Synthetic Philosophy* (Spencer 1904, 1: 238–42).

Spencer greatly elaborated his political views in a first book, published in 1851 under the title *Social Statics: or The Conditions Essential to Human Happiness Specified and the First of Them Developed*. It was written while Spencer was employed as coeditor for *The Economist*, a journal known for its open defence of laissez-faire economics. As suggested by the subtitle, the principal goal of *Social Statics* was to give an account of the rules appropriate to a purely scientific moral system, one that would be based on the same universal laws that governed nature. Chief among them was the law of equal freedom: "Every man has freedom to do all that he wills, provided he infringes not the equal freedom of any other man" (Spencer 1851, 103). The existence of evil was explained as the result of a lack of liberty in the exercise of faculties. Thus, in Spencer's view, the moral law of equal freedom was simply the development of a physiological truth. The rest of the treatise was dedicated to studying the various corollaries of the law of equal freedom, such as the freedom of speech, and drawing their political implications. These boiled down to the affirmation that the state should interfere as little as possible in the private affairs of its citizens, since whenever it attempted to alleviate social suffering it actually created more misery. According to Spencer, by employing artificial measures the state meddled with nature's mechanism; it assured the survival of that part of the population that could not adapt to the conditions of existence, and in doing so it obstructed progress and favoured instead a general physical and intellectual degeneration. In the name of this strict noninterventionism Spencer discarded all the legislative measures destined to help the underprivileged, such as the poor laws and other initiatives designed to improve the education and health systems.

After *Social Statics*, Spencer turned his attention to biology. Yet it seemed as though his interest in this domain followed from his political engagement and was due mainly to his wish to give a comprehensive account of evolution in nature and society, rather than to a genuine interest in the natural sciences themselves. In 1852, he published a short article entitled "The Development Hypothesis," which was an open attack on creationism and a repudiation of any appeal to supernatural forces in a truly scientific explanation of the development of life. The biological knowledge used in writing this piece came mainly from Charles Lyell's *Principles of Geology*. Paradoxically, it was Lyell's critique of Lamarck's transformism that convinced Spencer of the validity of the French naturalist's theory (Spencer 1904, 2: 7). There soon followed an article entitled "A Theory of Population, Deduced from the General Law of Animal Fertility," published in the same year. In this second article, Spencer expounded on Thomas Malthus's theory concerning the gap between arithmetic growth of natural resources and geometric multiplication of human beings, declaring that the pressure of population increase is the proximate cause of progress. All men are subject to its "discipline . . . they either may or may not advance under it; but in the nature of things, only those who *do* advance under it eventually survive"; they are the "select of their generation." According to Spencer, these statements brought him to the edge of the theory published eight years later in Darwin's *Origin* (Spencer 1904, 1: 450–1). In truth, however, Spencer's belief that evolution was driven by individual competition and direct adaptive responses to changes in life conditions was more in line with Lamarckism. Consequently, he viewed the struggle for existence as the motor of progress, for it prompted individuals to "advance" and become fitter, that is, to develop useful habits and characters in order to survive in their environments. Since only those who survived were able to reproduce, new acquired traits were passed on to future generations, translating the effects of individual competition into general progress. It is no wonder that Spencer's biological theories were particularly conducive to his political convictions, and indeed in an article published in 1857 under the revealing title "Progress: Its Law and Cause," he ventured to incorporate his diverse views under one overarching law. "Progress," he declared in what was to become the working premise for his entire synthetic system, "is not an accident, not a thing within human control, but a beneficent necessity" (Spencer 1892, 1: 60).

When Darwin's *Origin* appeared, Spencer welcomed it enthusiastically and confessed his satisfaction in reading the book (Spencer

1904, 2: 57). In reality, however, Spencer's eagerness to embrace Darwin's theory was the result of his desire to incorporate it into his own system of progress as yet another proof of the validity of the universal law of development, without noticing the differences that it presented with his Lamarckian position. This was manifest in the arguments employed in *Principles of Biology*, a two-volume work published between 1863 and 1867. In this work, Spencer defined life as the continual adjustment of organisms to their milieus in order to guarantee existence. He then proceeded to declare that the main mechanisms governing this process were the laws of use and nonuse and the principle of acquired characters. Given this line of argumentation, it was clear that in Spencer's view Lamarckian functional adaptation was the determining cause of the structure of living things. Spencer saw natural selection not as the key to evolution but as a secondary mechanism of elimination of those unfit for their environment. Furthermore, random variations had no place in his strictly causal explanation. Evolution was solely the result of the responses of the organism to the changing conditions of its habitat, and because these conditions did not necessarily affect all organisms in the same manner, individuals of the same species would differ from one another. Some were more apt to survive the struggle for existence than others, and they were the ones that reproduced. Variations in structure were thus reduced to two categories – good and bad – with natural selection simply having the function of separating the fit from the unfit. Indeed, according to Spencer, this process of evolution was better identified or could "more literally be called survival of the fittest" (Spencer 1904, 2: 115–16). A. R. Wallace suggested to Darwin that he adopt Spencer's expression instead of "natural selection," since the metaphorical character of this latter seemed to personify nature or imply deliberate choosing and "preferring" of the good of the species. "Survival of the fittest," Wallace argued, could not be misunderstood in this way because it was a more "compact and accurate" definition of the natural process of extermination of unfavourable variations. Wallace added that using Spencer's expression would help differentiate between two senses of natural selection that he detected in Darwin's book: first, natural selection as the simple process of sifting between variations, preserving only the favourable ones; and second, natural selection as the accumulated effect or change produced by this preservation. Darwin agreed with Wallace, confessing that the merits of "Spencer's excellent expression" had not occurred to him before, and stated his intent to use it in future editions of the *Origin*.[1] He nonetheless expressed his doubts whether "the use of any term would have made the subject intelligible to some minds" (*More Letters*, 1: 267–71). Spencer, so it seems, was one of those minds, since Lamarckian functional adaptation continued to play the main role in his theory of evolution.[2]

Another essential point of difference between Spencer's and Darwin's biological views concerns sexuality and its role in the evolutionary process. Spencer postulated that there exists a direct relationship between the degree of fertility and the level of mental development of living beings. The more developed and complex an organism, the higher its "expense" in terms of nervous influx spent on perfecting its intelligence and ameliorating its situation. The result is a diminution of its reproductive capacities, since the sexual organs receive less nervous influx. According to this view, the degree of individual development and the aptitude to produce offspring stand in inverse proportion to each other. Nonetheless, Spencer maintained that the advantage acquired in one domain was not entirely offset by the loss in another, since the more developed beings were also better adapted and therefore had a better chance to come out as winners in the struggle for existence (Spencer 1865–67, Part IV). In this way, the problem of a possible contradiction between the efforts of the individual to optimise its own existence – so important to Spencer's political credo – and the perpetuation of the species was avoided. In fact, this view of sexuality combined with the principle of "survival of the fittest" guaranteed that the species would be maintained in its highest possible form, for only the "best," or those most mentally developed and physically adapted, could reproduce. Evolution was therefore the promise of progress, and indeed, for Spencer, these two expressions were completely interchangeable:

[1] Darwin started using the expression "survival of the fittest" in the fifth edition of the *Origin*, where Chapter 4 is titled "Natural Selection or the Survival of the Fittest."

[2] Spencer continued to advocate adaptive-progressive evolution even after this biological explanation was strongly contested by neo-Darwinians like August Weismann, who sought to do away with the Lamarckian vestiges in Darwin's doctrine. See the *Contemporary Review*, nos. 63–4, 66 (1893–94).

"I say progress, but I ought to say evolution; for now the word is introduced and begins to be used instead of progress" (Spencer 1904, 1: 589).

The similarity between Spencer's views and those of the political economist Adam Smith and the utilitarian thinker Jeremy Bentham, both of whom are mentioned in *Social Statics*, is manifest. Spencer also made explicit the importance of his family background in forming his convictions, especially the education given by his uncle, a figure known for his political support of free trade and minimal state interference (Spencer 1904, 1: Part I; Duncan, 1: Chapters 1–2). Throughout his life Spencer remained faithful to his early and rather extreme interpretation of laissez-faire liberalism, even when a shift in the stance of his own political party seemed to be taking place in the 1880s, with the liberals accepting the necessity of certain social reforms (Spencer 1884). Some have claimed that Darwin shared many of Spencer's views on human and social progress and that he also believed that evolution is a process of improvement, though its main mechanism is natural selection rather than environmentally and functionally induced modifications (Greene 1981). It is also noteworthy that Darwin came from a strong Whig tradition and was a member of the prosperous bourgeoisie. But can it truthfully be said that his description of natural selection as "daily and hourly scrutinising..., rejecting that which is bad, preserving and adding up all that is good" and acting "only through and for the good of each being" (*Origin*, 84) is equivalent to Spencer's theory of progress? Can it be said that Darwin's biological proposals were also derived from or subjected to his political inclinations? The claims advanced in the *Origin* seem to indicate otherwise.

Clearly, for Darwin, evolution was not a teleological process tending upward in a linear mode toward a specific goal. Though he claimed that it led "to the production of the higher animals," that is, to those with a better-adapted organisation and thus having an advantage in the struggle for life (*Origin*, 336–7, 490), it was nonetheless an open-ended process and did not entail progress as Spencer understood it. Operating through random variations and divergence, Darwin's evolutionary development could not have a specific and predicted outcome or a political application like the one that dictated the structure of Spencer's theory. Although Darwin referred to Spencer a number of times in his works and confessed a certain esteem for his efforts, in a telling passage in his autobiography he specified that his biological reasoning was far removed from that of the English philosopher and professed his mistrust of the rather hasty generalisations made by Spencer.

After reading any of his books, I generally feel enthusiastic admiration for his transcendent talents, and have often wondered whether in the distant future he would rank with such great men as Descartes, Leibnitz, etc. about whom, however, I know very little. Nevertheless I am not conscious of having profited in my own work by Spencer's writings. His deductive manner of treating every subject is wholly opposed to my own frame of mind. His conclusions never convince me: and over and over again I have said to myself after reading one of his discussions, – "Here would be a fine subject for half-a-dozen years' work." His fundamental generalisations (which have been compared in importance by some to Newton's laws!) – which I daresay may be very valuable under a philosophical point of view, are of such a nature that they do not seem to me to be of any strictly scientific use. They partake more of the nature of definitions than of laws of nature.... Anyhow they have not been of any use to me. (*Autobiography*, 108–9)

Further light may be shed on Darwin's political views and the political significance of the *Origin* through an analysis of the preface to the first French translation by Clémence Auguste Royer, and Darwin's reaction to it. Royer was a woman of independent mind and one of the early advocates of feminism. She had been trained as a secondary school teacher, but her great interest in science had led her to read Lyell and Lamarck, as well as Malthus and other economists. It is unclear how the arrangement for Royer to translate Darwin's work was made; however, she obviously had an agenda of her own in taking on this task. According to her biographer, J. Harvey, Royer never saw herself as a science popularizer; her mission was to construct a philosophy that would make the world comprehensible using science as a tool. (Harvey 1997). As a result, Royer did not simply translate Darwin's *Origin*, but added extensive notes, offered some of her own corrections (Miles 1988), and changed its title from *On the Origin of Species by Means of Natural Selection* to *De l'Origine des espèces, ou des Lois du progrès chez les êtres organisés* (On the origin of species, or the law of progress of organic beings). More importantly, Royer added a long preface explaining her own interpretation of Darwin's theory and detailing the implications of his ideas for human evolution. The preface begins with an attack both on religion and on the church's dogma and is followed by a declaration of faith in reason and scientific progress, which set the tone for the whole piece. Royer then proceeds to situate Darwin's book as the continuation of Lamarck's theory and Enlightenment philosophy. This affirmation appears to be particularly important given Royer's own position, which shows a strong attachment to Enlightenment ideals such as the belief that human progress is an inevitability that follows naturally from the progress of science and reason. The question remains: was this in line with Darwin's evolutionary thinking?

Royer maintained that there was a "solidarity" between her views and Darwin's and that, in fact, she had arrived at similar conclusions to his on the succession and progressive evolution of living beings independently, while giving a course on the philosophy of nature and history in Lausanne during the winter of 1859. She further claimed that the notes added in her translation of the *Origin* simply offered developments of Darwin's theory or general summaries that recapitulated its main claims in a more synthetic form. Yet Royer also confessed that she may have "dared hypotheses" more than the English naturalist. These hypotheses were clearly explicated toward the end of the preface and concerned the political and moral consequences in which "Mr. Darwin's theory is fecund." In a passage that echoes Spencer's laissez-faire individualism, Royer argued against "exaggerated pity and charity" and the "unintelligent protection of the weak," since they lead to an increase in suffering and a degeneration of the race. This is because the weak slow down the strong who need to support them, thus preventing the powerful elements of society from developing and multiplying, which in turn causes a serious obstacle to progress. Since, as she believed, men were unequal by nature, Royer argued that in the political realm all doctrines that aim to achieve equality were unrealistic and harmfully utopian. Instead, she claimed that evolutionary biology gave support to the most unlimited individual liberty and free competition, which had already proved themselves to be the means of progress in the biological domain, leading to the evolution of man from very simple organisms. By the same token, Darwin's theory offered, according to Royer, an absolute criterion by which to judge between what is good and what is bad from a moral point of view, since the moral

law of any species is that which leads to its conservation, multiplication, and progress. Royer concluded by cautioning the reader that
she could only hint cursorily at the political and moral consequences
of Darwin's theory in her preface, since these last would fill a whole
book (Darwin 1862, xii–xxxvii). She also announced her desire to
write such a book, a task she accomplished in 1869 with a study
entitled *The Origin of Man and Societes*.

When Darwin received the French version of his book he was
unpleasantly surprised by the liberties taken by his translator. He
wrote to J. D. Hooker, complaining: "Almost everywhere in the
Origin when I express great doubts she appends a note explaining
the difficulty or saying there is none whatever! It is really curious
to know what conceited people there are in the world." Darwin
was also displeased with Royer's preface and her sweeping declarations that "natural selection and the struggle for life will explain
all morality, nature of man, politics etc etc!!!" And he continued:
"She makes some very curious and good hits and says she will publish a book on these subjects and a strange production it will be"
(Harvey 1997, 67–8). Darwin worried that Royer's preface and translation were affecting acceptance of his ideas in France, and he had
good reason to think so. Royer, for example, had translated "natural selection" as "*élection naturelle*," which she claimed was better
suited to illustrate Darwin's meaning in French, but also seemed to
indicate an intelligent and deliberate choice. She defended her word-
ing by invoking the term "elective affinities" in chemistry, which
refers to "inorganic, dead and inert nature." She agreed, however,
to adopt the more widely diffused expression "natural selection" in
the second edition of her translation, which appeared in 1865.[3] The
addendum "laws of progress" was also removed from the title. The
1865 edition was accompanied by a new foreword that reiterated
many of Royer's views, namely, her attacks against the church and
religion and her admiration for eighteenth-century Enlightenment
philosophy with its celebration of reason. To counter criticism of
her mélange of philosophy and science in the first preface, Royer
replied that she was a synthetic spirit who believed that all phenomena in the world are connected. She did not retract her developments
of Darwin's ideas and in fact highlighted the liberties she had taken

---

[3] For a different view on the subject see S. Miles (1988, 1: 42–5, 78–82).

in the translation. As for the political consequences she deduced from Darwin's biology, Royer affirmed that she did not pretend that Darwin accepted everything she dared to say and that it was not her intention to jeopardise his theory. Finally, to quiet the rumours that Darwin himself was displeased with her work, she claimed to have letters in her possession stating the opposite (Darwin 1866, viii–xiii, 95).

Darwin was dissatisfied with Royer's modifications and supplements, especially the comments concerning the political and moral implications of the theory of natural selection. Indeed, Royer's views on evolution and progress appear to be closer to Spencer's than to Darwin's, as does her defence of free competition and limited state intervention. Like Spencer, she remained faithful to Lamarck's transformism throughout her life, describing the French naturalist as Darwin's forerunner and dedicating articles to his life and work (Royer, 1868–69). Furthermore, in the new foreword written for the third French edition of the *Origin*, Royer criticised Darwin severely for his pangenesis theory, claiming that its archaic character and clearly false assumptions endangered the theory of natural selection (Darwin 1870, vi, xviii–xxvi). This attempt to sabotage his work was the last straw for Darwin, and he decided to dissociate himself totally from Royer and her opinions. He was also extremely displeased with Royer for not having asked his permission to issue another edition of her translation and for overlooking the corrections and notes that he had included in the previous two English versions of the *Origin*. He was determined, therefore, to ask for a new translator or withdraw his authorisation. A solution was quickly found with another Parisian editor, who published a translation of the fifth edition of the *Origin* in competition with Royer's third edition. This was the end of the correspondence between Darwin and Royer. It is nonetheless worthwhile mentioning, as Harvey has pointed out, that Darwin seemed interested enough in Royer's comments to include her book on *The Origin of Man and Society* in a list of possible sources for his own study of human evolution. As for Royer, she confessed to having been unlucky with her book on human evolution: some copies were published in 1869 and some in 1870, but the outbreak of the Franco-Prussian War diverted public attention (Harvey 1997, 100–1).

Darwin's reaction to Royer's attempt to portray evolution as a progressive process, as well as his comments on Spencer's philosophy, are revealing. Far from cautionary, his attitude was one of rejection of any direct application of notions like the struggle for existence and natural selection to the socioeconomic realm, a move he deemed to be hasty and unwarranted. In this respect it is no coincidence that both Spencer and Royer were inspired by Lamarck's transformism and that the French naturalist's theory remained the chief influence over their evolutionary thinking long after the publication of the *Origin*. In reality, they were not adepts of Darwin's theory at all and perhaps even ignored its most original aspects that did not sit well with their political agendas. Simply put, Darwin's kind of evolution could not be equated with progress in the sense of development in a specific direction or toward a specific goal. Natural selection can lead to radically different outcomes in different conditions; there is nothing linear in its modus operandi, and as a result there can be no hierarchy of weak and strong, good and bad. This was a major difference between Darwin's science and Spencer's and Royer's evolutionary conceptions.

It appears as though what eluded Spencer and Royer was specifically what attracted their counterparts on the other side of the political spectrum: Karl Marx and Friedrich Engels. At first, the authors of the *Communist Manifesto* seemed particularly impressed with the lack of a predetermined *telos* in Darwin's theory of evolution. However, this too quickly changed, a sign that there is a more profound difference between Darwin's enterprise and the underlying logic of political discourses, as an analysis of Marx and Engels' reaction to Darwin's theory will demonstrate.

A short time after the publication of the *Origin*, Engels wrote to Marx: "Darwin . . . whom I'm reading just now, is absolutely splendid. There was one aspect of teleology that had yet to be demolished, and that has now been done. Never before has so grandiose an attempt been made to demonstrate historical evolution in Nature, and certainly never to such good effect. One does, of course, have to put up with the crude English method" (*Collected Works*, 40: 551). Engels's letter is dated less than three weeks after the *Origin* appeared in bookshops in England, a fact that indicates his great interest in Darwin's theory. Marx, on the other hand, waited a year before replying to Engels on the topic. He agreed that although Darwin's treatise on natural selection is developed "in the crude

English fashion, this is the book which, in the field of natural history, provides the basis for our views." In fact, Marx drew an even closer connection between Darwin's theory and his own doctrine in a letter to Ferdinand Lassalle in 1861, declaring: "Darwin's work is most important and suits my purpose in that it provides a basis in the natural science for the historical class struggle" (*Collected Works*, 41: 232, 246–7). A closer examination shows, however, that Marx and Engels did not share exactly the same views on Darwin's theory and that their acceptance of the biological claims put forward in the *Origin* was conditioned by powerful political motives, as noted by Yves Christen (Christen 1987). In 1866, for example, Marx wrote to Engels:

A very important work which I shall send on to you . . . as soon as I have made the necessary notes, is: *P. Trémaux's Origine et transformations de l'homme et des autres êtres, Paris, 1865*. In spite of all the shortcomings that I have noted, it represents a *very significant* advance over Darwin. . . . Progress, which Darwin regards as purely accidental, is essential here on the basis of the stages of the earth's development.

This last remark refers to one of the main theses in the work of French ethnologist, archaeologist, painter, and photographer Pierre Trémaux, who claimed that the physical features and evolution of the Earth determine the structure and modifications of living beings and are the main cause of differentiation. Marx was particularly impressed with Trémaux's declarations regarding man's total subjection to the laws of nature. He confirmed that "in its historical and political applications [the work of Pierre Trémaux is] far more significant and pregnant than Darwin." Engels's opinion was quite different. He wrote back to Marx that though he had not yet finished reading Trémaux's work, he had "come to the conclusion that there is nothing to his whole theory because he knows nothing of geology, and is incapable of even the most common-or-garden literary-historical critique. . . . In addition, it is another pretty notion of his to ascribe the differences between a Basque, a Frenchman, a Breton, and an Alsatian to the surface-structure." Engels concluded with the categorical verdict: "The book is utterly worthless, pure theorising in defiance of all the facts, and for each piece of evidence it cites it should itself first provide evidence in turn" (*Collected Works*, 42: 304–5, 320).

Engels's uncompromising attack did not cause Marx to alter his view. In his response Marx evoked Cuvier, arguing that Engels's judgment echoed almost word for word the French naturalist's criticism of German theories on the variability of species. Yet scientific progress had proved Cuvier to be wrong, and it turned out to be "German nature-worshippers, amongst others who *formulated* Darwin's basic idea in its entirety, however far they were from being able to *prove* it." Engels conceded that Darwin might not have insisted enough on the influence of the soil in the evolution of organisms, perhaps for lack of data. He nonetheless continued to maintain that when Trémaux "goes on to declare that the effect of the soil's greater or lesser age, modified by crossing, is the *sole* cause of change in organic species or races, I see absolutely no reason to go along with the man thus far, on the contrary, I see numerous objections to so doing," namely, that 9/10 of his "ridiculous evidence" is based on "erroneous or distorted fact" (*Collected Works*, 42: 322–4). This was the end of the exchange between Marx and Engels on the subject, and it seems as though they remained divided in their opinions, for only a few days after Engels's last letter Marx sent a letter to his friend Ludwig Kugelmann recommending Trémaux's book and insisting that "although written in a slovenly way, full of geological howlers and seriously deficient in literary-historical criticism, it represents – WITH ALL THAT AND ALL THAT – an advance over Darwin" (*Collected Works*, 42: 327). Marx's appraisal of Trémaux is interesting if only because it enables us to better understand his position vis-à-vis Darwin's evolutionary theory. His insistence on the all-importance of the soil in explaining evolutionary changes reveals a strong attachment to a causal determinism of environmental conditions that is more in tune with Lamarck's views than with Darwin's biology. After all, Lamarck's view of evolution was better suited to Marx's claims concerning the overriding importance of material conditions of existence in determining the nature of individuals as well as their intellectual and social existence, professed two decades earlier in his critique of German ideology. These claims subsequently became the building blocks for Marx's materialist conception of history and therefore could not easily have been modified or reworked.

Marx may have also leaned toward a favourable appreciation of Trémaux's theory because of the latter's claim that the Darwinian struggle for life can have only negative results for both survivors and losers. According to Trémaux, when animals and plants fight with each other for the same living space they are all weakened by the fierce competition and as a result degenerate rather than develop and progress (Christen 1981, 58). Marx himself condemned the mechanism of the struggle for existence when he wrote to Engels in an often-quoted letter from 1862:

I'm amused that Darwin, at whom I've been taking another look, should say that he *also* applies the 'Malthusian' theory to plants and animals, as though in Mr Malthus's case the whole thing didn't lie in its *not* being applied to plants and animals, but only – with its geometric progression – to humans as against plants and animals. It is remarkable how Darwin rediscovers, among the beasts and plants, the society of England with its division of labour, competition, opening up of new markets, 'inventions' and Malthusian 'struggle for existence'. It is Hobbes' *bellum omnium contra omnes* and is reminiscent of Hegel's *Phenomenology*, in which civil society figures as an 'intellectual animal kingdom', whereas, in Darwin, the animal kingdom figures as civil society. (*Collected Works*, 41: 381)

Engels, on the other hand, tried to overcome the problematic aspect of Darwin's Malthusianism by arguing for a discrepancy between the evolution of man and the evolution of animals. "I, too," he wrote in 1865,

was immediately struck on first reading Darwin by the remarkable similarity between his description of the vegetable and animal life and the Malthusian theory. Only my conclusion was ... that it is to the everlasting disgrace of modern bourgeois development that it has not yet progressed beyond the economic forms of the animal kingdom. The so-called 'economic laws' are not eternal laws of nature but historical laws that appear and disappear, and the code of modern political economy, insofar as the economists have drawn it up correctly and objectively, is for us merely a summary of the laws and conditions in which modern bourgeois society can exist. (*Collected Works*, 42: 136)

When the first volume of Marx's *Das Kapital* appeared Engels was able to reconcile his admiration for Darwin's science – minus the latter's adherence to Malthus – with his loyalty to Marx's ideas by proclaiming his good friend as the first thinker to truly grasp the meaning of evolution in the human domain. He heralded Marx's contribution to the study of economics as the most important scientific innovation in the field, indeed as comparable to Darwin's achievement in the realm of biology: "In so far as he [Marx] endeavours to show that present-day society, economically considered, is pregnant with another, higher form of society, he merely strives to present as law in the social sphere the same process which Darwin traced in natural history, a process of gradual evolution" (*Collected Works*, 20: 224–5). This statement was intended to establish Marx's independent scientific status as Darwin's equal, rather than to portray him as a follower or disciple. In other words, according to Engels, whereas Darwin was able to reveal to us the truth about the organic world, he could not take us further. Marx was the one who finally uncovered the truth about human evolution in all its complex economic, social, and political relations. In this manner Engels was able to assuage his dissatisfaction with the political views that inspired Darwin's biology and at the same time to affirm Marx's revolutionary conclusions. In short, his strategy amounted to putting each thinker in charge of a different sphere of knowledge: "Just as Darwin discovered the law of development of organic nature, so Marx discovered the law of development of human history.... Such was the man of science. But this was not even half the man. Science was for Marx a historically dynamic, revolutionary force..." (*Collected Works*, 24: 467–8).

These words, pronounced at Marx's graveside, seem to summarise well the difference between the two thinkers. Marx's objective could not have been further from Darwin's. His goal was to change the world, not to interpret it in a different way, while Darwin wanted to understand nature through careful observation. This is perhaps why Darwin was not particularly interested in Marx's study of capitalism. Only the first 102 pages of the 822 that comprise the German copy of *Das Kapital*, which Marx sent Darwin as his "sincere admirer," were cut, and they do not contain Darwin's customary comments (Gruber 1961, 582). Darwin's polite response was very short, yet under the humble mask of his supposed lack of education he made a point of highlighting the distance that separated him from Marx: "I heartily wish that I was more worthy to receive it [Marx's "great work on Capital"], by understanding more of the deep and important subject of political Economy. Though our studies have been so different, I

believe that we both Earnestly [sic] desire the extension of Knowledge ... " (Feuer 1975, 2). That Marx allegedly proposed to dedicate
Das Kapital to Darwin as a sign of respect and appreciation has
been proven false. Darwin's letter of refusal was actually destined
for Edward Aveling, Marx's son-in-law and the translator of the first
volume of *Capital*, who asked Darwin to endorse his secular, atheistic views as expressed in a compilation of his tracts bearing the
general title *The Student's Darwin* (Feuer 1975, 1–12; Carroll 1976,
384–94).

Marx quoted Darwin twice in *Capital*, but he did so without giving any real weight to the principle of natural selection, focusing
instead on adaptation. The two footnotes in Chapters 14 and 15
constitute the only public mentions Marx made of Darwin's work
(Christen 1981, 66–7). These notes refer to Darwin's comments in
the *Origin* on the formation of organs in plants and animals as instru-
ments that may be adapted to very specific functions through natural
selection (*Origin*, 149). For Marx, these comments were of particular
significance because he claimed that the "history of Nature's Technology," as expounded by Darwin, should be complemented by the
"history of the productive organs of man, of organs that are the material basis of all social organisation" (*Collected Works*, 35: 346, 375).
But this type of evolutionary thinking (i.e., that organs are adapted
to certain ways of life) was not particular to Darwin, nor was it
the source of his originality. It seems that Marx referred to Darwin
mainly because his was the most recent scientific work on evolution,
not because he judged it more favourably than other, lesser-known
treatises, like Trémaux's book. Engels's comments in the 1870s indicate the same attitude: "Of Darwin's doctrine, I accept the *theory of
evolution*, but assume Darwin's method of verification (STRUGGLE
FOR LIFE, NATURAL SELECTION) to be merely a first, provisional,
incomplete expression of a newly discovered fact." He continued to
specify in the same letter from 1875 that "if a self-styled naturalist
takes it upon himself to subsume all the manifold wealth of historical development under the one-sided and meagre axiom 'struggle
for existence', a phrase which, even in the field of nature, can only
be accepted *cum grano salis*, his method damns itself from the outset" (*Collected Works*, 45: 106–7). Engels also seemed to tilt toward
a more Lamarckian view of biological evolution when he claimed
in his 1878 publication *Anti-Dühring* that Darwin attributed to his

discovery of natural selection "too wide a field of action," making it the sole agent in the alteration of species and neglecting the causes of "repeated individual variations" (*Collected Works*, 25: 65–6).

Engels's change of heart and disavowal of the main mechanisms of Darwin's theory, together with Marx's reluctance to acknowledge their role in evolution, show that in the case of Marxism, as in the previously examined laissez-faire views of Spencer and Royer, the similarities between Darwin's ideas and their ideological discourse were only superficial. In fact, this seems to have been Darwin's own opinion. In a letter from 1879 to E. Haeckel, his most prominent follower in Germany, Darwin expressed his full support of Haeckel's rejection of any direct transfer of the theory of evolution into the realm of practical politics, specifically in support of socialism (Richards 2008). Thus, a difference in kind between science and politics becomes apparent, one that Darwin was aware of. Political theories, by their very nature, have to be normative, since they aim to convince the listener or reader that the solution offered is beneficial, or at least more desirable than the existing alternatives. Therefore, by definition, they need to have a specific objective, be it a more equitable or prosperous society, a freer society, or some other goal. Darwin's biology was constructed in response to a different set of rules, those of scientific investigation and explanation. In this sense it was not normative but descriptive. Darwin was as far removed from Marx's claim that we must change the world as he was from Spencer's and Royer's advocacy of laissez-faire liberalism as the best social structure. His theory of evolution could not be said to postulate progress toward a particular goal, for it depicted a process of constant change involving numerous factors. He looked to the past and the present as a way to understand the workings of nature, not as a means for controlling it or predicting future outcomes. Humbly, he declared in the concluding paragraph of the *Origin*: "There is grandeur in this view of life, with its several powers," whereby "from so simple a beginning endless forms most beautiful and most wonderful have been, and are being, evolved" (*Origin*, 490). This is why he was reluctant to draw the political conclusions of his findings; quite simply, he may have felt that he lacked the tools and the data to judge one way or the other. In any case, his biological theory as developed in the *Origin* could not be of any help to him in this domain. As for the proponents of laissez-faire liberalism and Marxism we have examined here, it appears as though Darwin's theory fulfilled for them only the function of a pretext and was not in reality connected with their views, nor could these latter be inferred from it.

17   The *Origin* and Philosophy

I. DARWIN AND DARWINISM

Darwinism is all the rage in philosophy these days. Evolutionary thinking of one kind or another is frequently used to illuminate such areas as ethics (e.g., Joyce 2006), epistemology (e.g., Hull 1988), and the philosophy of mind (e.g., Sterelny 2003). But it is one thing to examine how evolutionary work in a broadly Darwinian style has influenced philosophy, another to ask what form of philosophical insight is present in Darwin's own oeuvre. And it is something else yet again to ask what the specific relationship might be between philosophy and the *Origin of Species*. This last question can be broken down into an analysis of the work's philosophical legacy and an analysis of its philosophical content.

Since the *Origin* has so often been taken by later thinkers as the canonical statement of a Darwinian worldview, any project of assessing that book's philosophical legacy risks sliding toward a hopelessly ambitious attempt to embrace all those subsequent forms of philosophical Darwinism that the *Origin* has inspired. This essay consequently focuses on the *Origin*'s own philosophical content. In the next section I will catalogue some of the philosophical themes that arise in it. The third section contains some brief reflections on Darwin's philosophical method. Section four addresses the *Origin*'s major philosophical *input*, namely, the scientific methodology of

This chapter draws heavily on scattered material from Lewens (2007). I am grateful to the editors for inviting me to contribute to this Companion, and I am especially grateful to Robert Richards, whose acute comments on an earlier draft saved me from some embarrassing mistakes.

John Herschel. And the final two sections examine what, in the eminent biologist Ernst Mayr's eyes, was the *Origin*'s major philosophical *output* – the shift from so-called typological thinking to 'population' thinking.

II. DARWIN AS PHILOSOPHER

Darwin's assessment of his own philosophical abilities is unpromising. In his autobiographical reminiscences, he claims that '[m]y power to follow a long and purely abstract train of thought is very limited; I should, moreover, never have succeeded with metaphysics or mathematics' (*Autobiography*, 140). In spite of this alleged handicap, Darwin took an active interest in philosophy. In the years immediately after the return of the *Beagle* to England, he filled notebooks with ideas whose descendants would appear in modified form in the *Origin* and other works. During this period he read books by the philosophers David Hume and Adam Smith, by the leading methodologists of science John Herschel and William Whewell, and by other philosophers better known then than they are now, such as James Mackintosh.

Darwin's notebooks are consequently rich in observation and speculation regarding many of the same topics that today's philosophers have attempted to view in an evolutionary light – the formation of our moral sense, the nature of emotional expression, the origins of innate knowledge, and so forth. Darwin's thoughts on these topics appeared in published form in his later works, written well after the *Origin*. It seems likely that Darwin wished to establish the credentials of his basic transmutationist views prior to developing in print any extrapolation toward philosophical matters. The value of this form of caution is illustrated by the hostile reception that greeted another transmutationist work, the anonymously published *Vestiges of the Natural History of Creation*, which appeared in 1844 (fifteen years before the *Origin*). *Vestiges* was exceptionally ambitious, covering everything from the origins of the universe as a whole to the differences between men and women. The men of science whom Darwin held in the highest esteem dismissed it as speculative nonsense, and dangerous nonsense to boot. The *Origin* was altogether a more serious affair, and Darwin worked hard at presenting its theory in a sober manner. He offered a wide variety of data to support transmutationism, and he largely restricted his discussion of the theory's application to the topics of speciation and adaptation in plants and animals.

In spite of this, it would be a mistake to regard the *Origin* as wholly silent on issues of broad philosophical significance such as divine design, the mind, and morality. Most obviously, the book provides a defence of transmutationism against special creation. Although Darwin was no atheist when the *Origin* was written, he nonetheless tackles directly the issue of whether it is necessary to invoke a Creator to explain the immediate origin of species, concluding that such explanations are vacuous: 'It is so easy to hide our ignorance under such expressions as the "plan of creation," "unity of design," &c., and to think we give an explanation when we only restate a fact' (*Origin*, 481–2).

The *Origin*'s exploration of the action of natural selection on animal instincts also demonstrates Darwin's view that his theory can shed light on the mind. Although these ideas were not applied to our own species until the publication of *The Descent of Man*, Darwin leaves his reader in no doubt about the potential of his theory in this domain:

In the distant future I see open fields for far more important researches. Psychology will be based on a new foundation, that of the necessary acquirement of each mental power and capacity by gradation. Light will be thrown on the origin of man and his history. (*Origin*, 488)

Darwin's remarks on the evolution of instincts in animals also hint at his later work in *Descent* on the origins of the moral sense. In *Descent*, Darwin argues that the good of the community should be identified as the 'criterion' of morality (for discussion, see Richards 1987). That is to say, Darwin claims that what makes an action right is its contributing to the good of the community, defined as 'the means by which the greatest possible number of individuals can be reared in full vigour and health, with all their faculties perfect, under the conditions to which they are exposed' (*Descent*, 98). The case he makes rests in part on his view that natural selection at the community level can explain how social instincts come to exist. These instincts produce behaviour that promotes the community good.

This set of views is partially prefigured in the *Origin* – for example, when Darwin suggests that:

It may be difficult, but we ought to admire the savage instinctive hatred of the queen-bee, which urges her instantly to destroy the young queens her daughters as soon as born, or to perish herself in the combat; for undoubtedly this is for the good of the community; and maternal love or maternal hatred, though the latter fortunately is most rare, is all the same to the inexorable principle of natural selection. (*Origin*, 202–3)

We ought to admire the queen bee's hatred, says Darwin, because that hatred is for the good of the bees' community.

III. THE *ORIGIN* AND PHILOSOPHICAL METHOD

So far I have tried to show only that Darwin was interested in philosophical topics, and that he regarded them as important enough to introduce various hints in the *Origin* regarding the philosophical promise of his theory. The *Origin* is also a useful text for demonstrating a little about Darwin's views regarding proper philosophical method. Remember that Darwin claimed to have been no good at following 'a long and purely abstract train of thought'. I suggest that this comment was somewhat disingenuous. Darwin proudly described the *Origin* itself as 'one long argument' (*Autobiography*, 140): he was perfectly capable of keeping a long train of thought on the rails. The *Origin*'s argument is, of course, peppered with hard-won empirical data, and Darwin's other remarks suggest that any difficulty he may have suffered in coping with lines of thinking that are 'purely abstract' rested as much on his scepticism regarding the value of fact-free enquiry as on his cognitive limitations. Consider, for example, that Darwin was disappointed when he reread his own grandfather's proto-evolutionary work *Zoonomia*, 'the proportion of speculation being so large to the facts given' (*Autobiography*, 49). Of Herbert Spencer he said:

After reading any of his books, I generally feel enthusiastic admiration for his transcendent talents, and have often wondered whether in the distant future he would rank with such great men as Descartes, Leibnitz, etc., about whom, however, I know very little. Nevertheless I am not conscious of having profited in my own work by Spencer's writings. His deductive manner

of treating every subject is wholly opposed to my frame of mind. His conclusions never convince me: and over and over again I have said to myself,
after reading one of his discussions, – 'Here would be a fine subject for
half-a-dozen years' work.' (*Autobiography*, 108–9)

The high regard in which Darwin is held among modern philosophers is undoubtedly due in large part to the multipurpose explanatory power many attribute to the idea of natural selection. But,
as these comments suggest, Darwin also shares a methodological
stance with modern-day 'philosophical naturalists' who stress the
need for close attention to the results of the natural sciences in
order to discipline and inspire philosophical inquiry.

Darwin's empiricism does not make him wholly sceptical of the
value of armchair pastimes beloved of more traditionally minded
philosophers, such as the fabrication of thought experiments, and
abstract reflection on the meanings of terms. The *Origin* includes
several highly idealised scenarios intended to illustrate how, in broad
terms, aspects of Darwin's theory are meant to work. He requests
our indulgence, for example, as he explains the workings of natural
selection:

In order to make it clear how, as I believe, natural selection acts, I must beg
permission to give one or two imaginary illustrations. (*Origin*, p. 90)

Darwin then shows us how natural selection accounts for adaptation
by helping himself to a stylised example of a population of wolves
that lives by hunting deer. Darwin is aware, however, that a combination of further reasoning and further evidence concerning such
matters as the character of variation in nature, the extent of resemblance between parents and offspring, and the nature of the struggle
for existence is required in order to show that natural selection is
in fact responsible for organic change, either generally or on specific
occasions. A famous paragraph at the end of Chapter 4 of the *Origin*
(*Origin*, 126–7), which summarises the argument for natural selection's existence and efficacy in largely abstract terms, is immediately
followed by a reminder that the case in favour of this principle is not
yet complete:

Whether natural selection has really thus acted in nature, in modifying and
adapting the various forms of life to their several conditions and stations,

must be judged of by the general tenour and balance of evidence given in the following chapters. (*Origin*, 127)

The importance of pinning down the scope and form of natural selection also drives Darwin to clarify the meanings of the terms he uses. He takes pains to point out, for example, that the 'struggle' for existence is meant in a 'large and metaphorical sense' that does not require literal battles over resources, and that a plant 'may truly be said to struggle with the plants of the same and other kinds which already clothe the ground' (*Origin*, 62–3). In these cases, Darwin's use of imaginary illustration and conceptual clarification is essential for the articulation of the central explanatory schemata of his new theory. When discussing other topics, he sometimes regards excursions into similarly abstract matters as unnecessary, or distracting. At various points in the *Origin*, and throughout its later editions, Darwin makes an effort to clarify the meanings of 'progress'. But consider these remarks on what it might mean to say that one organism is 'higher' than another:

The embryo in the course of development generally rises in organisation: I use this expression, though I am aware that it is hardly possible to define clearly what is meant by the organisation being higher or lower. But no one will probably dispute that the butterfly is higher than the caterpillar. (*Origin*, 441)

Here, Darwin seems content with the thought that even if we cannot say with clarity what is meant by 'height' of organisation, we can at least recognise rising organisation when we see it. Or consider his views on classification. Darwin argues that a genealogical classification is the best one. On Darwin's view, organisms with shared ancestry tend to resemble each other in many respects, both trivial and important. This means that a genealogical classification will also group together organisms that resemble each other most closely:

In classing varieties, I apprehend that if we had a real pedigree, a genealogical classification would be universally preferred. . . . For we might feel sure, whether there had been more or less modification, the principle of inheritance would keep the forms together which were allied in the greatest number of points. (*Origin*, 423)

This leaves open the question of whether the ultimate goal of classification is to give an accurate genealogy or to group organisms by resemblance. So long as Darwin is right to say that genealogy and resemblance ('filiation' and 'affinity', as he sometimes says) reliably go together, practical taxonomic projects do not require that we answer this question. But one can ask various 'what if?' questions. What would happen if genealogy and resemblance were to come apart? Isn't it logically possible that two organisms could resemble each other very closely in many trivial respects without this being attributable to inheritance from a common ancestor? What if it turned out, for example, that just one of the many species of kangaroo were descended from bears? How should that species be classified? Darwin's genealogical system would demand that it be grouped with the bears. But shouldn't we then violate Darwin's genealogical rule and classify all the other kangaroo species with bears, too? Or are we to say, absurdly, that only one kangaroo species should be classed with the bears, no matter how much it resembles the other species of kangaroo? Darwin has no truck with this sort of objection:

> I might answer by the *argumentum ad hominem*, and ask what should be done if a perfect kangaroo were seen to come out of the womb of a bear?...The whole case is preposterous; for where there has been close descent in common, there will certainly be close resemblance or affinity. (*Origin*, 425)

This counterexample, Darwin thinks, is too silly to be worth considering, and because of that it gives us no reason to revise his view of classification (Lewens 2007, 80). Here, again, Darwin has allies among some modern philosophers who refuse to let far-fetched imaginings undermine a serviceable theory.

IV. THE *ORIGIN* AND SCIENTIFIC METHOD

One of the most important philosophical *inputs* to Darwin's work lies in the contribution of a particular view about scientific method to Darwin's structuring of the *Origin*. It is tempting to cast the *Origin*'s argument as a form of 'inference to the best explanation'. There is no doubt that in day-to-day reasoning we often infer that theories are true by virtue of the fact that were they true, they would best explain some body of data (Lipton 2004). This is typically the case in forensic reasoning, where the identity of the murderer is established by pointing out how the evidence is most neatly tied together on the assumption that it was Professor Plum who did it. In the final chapter of the *Origin*, Darwin summarises the varied facts his theory is able to explain – facts about anatomy, embryology, the distribution of species around the globe, even about the characteristic arguments among natural historians – and he notes how ill-equipped are rival theories to explain the same facts. In the *Origin*'s sixth edition he adds that:

It can hardly be supposed that a false theory would explain, in so satisfactory a manner as does the theory of natural selection, the several large classes of facts above specified. It has recently been objected that this is an unsafe method of arguing; but it is a method used in judging of the common events of life, and has often been used by the greatest natural philosophers. (*Variorum*, 748: *183.I:f*)

In other words, the fact that a theory is able to explain diverse phenomena successfully is, Darwin thinks, strongly indicative of the theory's truth.

'Inference to the best explanation' (IBE) is an attractive slogan, and it has considerable intuitive appeal. Even so, there are plenty of hypotheses that have many of the marks of a good explanation, which we are inclined to regard as false. Conspiracy theories often neatly tie up bodies of diverse data, but they are nonetheless greeted with scepticism. The defender of IBE would need, ideally, to give some account of what features make an explanation good, and an account of why the 'best' explanation in this sense is likely to be true. Any such defence will face an uphill struggle. Suppose, for example, IBE's defender claims that good explanations tend to be simple explanations. In addition to answering the tricky question of what is meant by 'simplicity', he would also need to say why the simplest explanation is likely to be the right one. There is, after all, no obvious a priori reason to think the universe must be a simple place, and plenty of a posteriori reasons to think it is not.

I do not propose to give any general defence of the reliability of IBE here. Instead, I will try to show that some versions of IBE are obvious nonstarters, and that John Herschel's methodology of science – the methodology followed by Darwin in the *Origin* – can be understood as a more sophisticated, more defensible form of IBE (Lewens 2007, Chapter 4).

One might think that a good explanation is one that, were it true, would make our data probable. Yet IBE is obviously hopeless if it asserts that a theory need do nothing more than make our data probable in order for us to be warranted in believing it. One problem with this view is that it leads to absurd results. If Gordon Brown were a Martian, then (since he would also be a living organism) it is probable that he would have a metabolism. However, if we observe that Gordon Brown has a metabolism, this clearly does not warrant our inferring that he is a Martian. This version of IBE is also incomplete. Different, incompatible theories often entail, and thereby make probable, many of the same observational consequences. The hypothesis that Gordon Brown is a human also makes it probable that he has a metabolism, but not even Brown can be both a Martian and a human at once. So a plausible version of IBE needs to offer some resources for choosing between competing theories when the observations we are seeking to explain are common consequences of all of them.

John Herschel was an astronomer and mathematician, and one of the leading intellectual figures in Cambridge during the time Darwin spent there as a student. Herschel also wrote on scientific method – a topic that is often regarded today as part of philosophy. The two men got to know each other after Darwin graduated from Cambridge, meeting for the first time when the *Beagle* stopped off in South Africa. Darwin respected Herschel enormously:

I felt a high reverence for Sir J. Herschel, and was delighted to dine with him at his charming house at the C. of Good Hope and afterwards at his London house. I saw him, also, on a few other occasions. He never talked much, but every word which he uttered was worth listening to. (*Autobiography*, 107)

Darwin read Herschel's major methodological work during his student days, and he credits it with having a great impact on his thought (for detailed discussion of Herschel and Darwin, see Ruse 1975, Hodge 1977, and Waters 2003):

During my last year at Cambridge I read with care and profound interest Humboldt's *Personal Narrative*. This work and Sir J. Herschel's *Introduction to the Study of Natural Philosophy* stirred up in me a burning zeal to add even the most humble contribution to the noble structure of Natural Science. No one or a dozen other books influenced me nearly so much as these two. (*Autobiography*, 68–9)

Herschel argues that a respectable scientific theory should appeal only to 'true causes', or *verae causae*. Herschel understands that if we are to place confidence in an explanatory theory, it is not enough for that theory to posit causes that make sense of a single, narrow range of phenomena. Further conditions must be satisfied. One of these involves an extension of the explanatory range of a theory. Herschel argues that our confidence should increase if a theory is able to explain several distinct classes of phenomena. This aspect of Herschel's methodology is reflected in Darwin's efforts to show that his theory of descent with modification can account for diverse facts in domains as varied as embryology, morphology, and biogeography. Herschel also argues that we should have direct experience, either of the posited cause itself or, failing that, of something closely analogous to it. This feature of Herschel's thinking is reflected in Darwin's efforts to demonstrate both that natural selection is real – that offspring do resemble parents, that variation is plentiful, that there is a struggle for life – and that the analogous force of artificial selection has known efficacy in modifying species.

I do not mean to show that Herschel's methodological principles are justified, but they do have a reasonable descriptive fit with the criteria we use for theory choice. The hypothesis that Gordon Brown is a Martian is not one that we should take seriously, if it does nothing more than explain his possessing a metabolism. But we might take it more seriously if it explained many diverse facts about Brown's constitution and behaviour, and we would take it more seriously still if we had direct experience of the existence of Martians.

The *Origin*'s argument can be understood as an inference to the best explanation. Darwin's adherence to Herschel's *vera causa* standard means that Darwin's sophisticated version of IBE avoids the most obvious objections to this mode of argument. This is important, in part because modern-day Intelligent Design (ID) theorists sometimes appeal to IBE to ground their inferences from the structure of particular organic traits to a creative intelligence. But ID theorists are typically far less sensitive than Herschel to the perils of simplistic versions of IBE. A powerful designing intelligence is sometimes posited merely on the grounds that were there to be such a thing, it would make probable the existence of an otherwise puzzling phenomenon, namely, the phenomenon of adaptation. This, we have seen, is an indefensible mode of argument.

V. FROM TYPOLOGICAL THINKING
TO POPULATION THINKING

The eminent biologist and historian of biology Ernst Mayr argues, in the introduction he wrote in 1964 to a facsimile edition of the *Origin*, that contained within that book was a new and revolutionary philosophical conception of nature. It represented nothing less than the rejection of a thereto dominant Platonism:

> The *Origin of Species* was far more than a mere accumulation of facts proving evolution.... Darwin started from a new basis by completely eliminating the last remnants of Platonism, by refusing to admit the *eidos* (Idea; type, essence) in any guise whatever. (Mayr 1964, xi)

The specific philosophical breakthrough for which Darwin was responsible is, says Mayr, the transition from a form of Platonism called 'typological thinking' to a new way of thinking about nature, which Mayr calls 'population thinking'. Mayr regards the introduction of population thinking into biology as Darwin's third great conceptual innovation, at least as important as the hypothesis of the evolutionary relatedness of species, and the articulation of natural selection as the primary mechanism of organic change. Mayr described the typological/population distinction in different ways through his career (Chung 2003), but by far the best-known formulation comes from a paper originally published to mark the *Origin*'s centennial in 1959. The typologist is one who, in a manner reminiscent of Plato, posits an ideal specimen or 'form' for each biological species:

> According to [typological thinking], there are a limited number of fixed, unchangeable 'ideas' underlying the observed variability, with the *eidos* (idea) being the only thing that is fixed and real, while the observed variability has no more reality than the shadows of an object on a cave wall, as it is stated in Plato's allegory. The discontinuities between these natural 'ideas' (types), it was believed, account for the frequency of gaps in nature... Since there is no gradation between types, gradual evolution is basically a logical

impossibility for the typologist. Evolution, if it occurs at all, has to proceed in steps or jumps. (Mayr 1976, 27)

Darwin, by contrast, is a population thinker:

The assumptions of population thinking are diametrically opposed to those of the typologist. The populationist stresses the uniqueness of everything in the organic world.... All organisms and organic phenomena are composed of unique features and can be described collectively only in statistical terms. Individuals, or any kind of organic entities, form populations of which we can determine only the arithmetic mean and the statistics of variation. Averages are mere abstractions; only the individuals of which populations are composed have reality. (ibid.)

Mayr argues that the two positions express fundamentally different philosophical stances:

The ultimate conclusions of the population thinker and the typologist are precisely the opposite. For the typologist, the type (*eidos*) is real and the variation an illusion, while for the populationist the type (average) is an abstraction and only the variation is real. No two ways of looking at nature could be more different. (ibid.)

It is important to say a little in defence of 'typological thinking', lest we make the position appear absurd. As Mayr says, there are 'gaps in nature'. Organic forms that we might think possible are never observed – there are no six-legged elephants. Moreover, some organic structures are encountered with great regularity. Many individual organisms, and even organisms of different species, appear to be variations on a common underlying theme. The 'vertebrate archetype' of Richard Owen, for example, was an effort to represent the common structural plan, modified to various degrees in particular species, that underlies all vertebrates. We can think of 'types' as explanatory posits: some forms are seen rarely or not at all because there is no corresponding type; others are seen frequently because they are variations on an underlying type. Note that for these reasons Mayr's claim that for the typologist variation is 'an illusion' is rather misleading (Sober 1980). A typologist need not claim that all dogs are identical. But a typologist might claim that by positing a dog type, we can explain why there are so many individual organisms with 'doggy' characteristics.

Mayr's equation of types with Platonic 'ideas' makes them sound unaccountably mystical – the sorts of posits that have no business in respectable science. It is true that some of Darwin's contemporaries, Owen included, made positive references to Plato. Yet the question of how important Platonism was in the dominant strands of typological thinking at the time the *Origin* was published has been contested (Amundson 2005; see also Winsor 2006 for scepticism regarding Mayr's historiography). The point that I wish to stress here is that we can make 'types' more respectable by understanding them as stable configurations of organic matter (Lewens 2007, 85–6). Moving away from biology for a moment, something like a typological explanation seems appropriate when we try to understand why some crystal structures are seen frequently, for example, while others are not seen at all. We can take reference to types here to be shorthand for sets of physical facts that make some crystalline forms stable, others unstable. Perhaps we can think of organic types in a similar way. The typologist claims that only a few basic organic configurations are stable. These stable configurations then explain the diversity of forms manifested by individual organisms.

Mayr characterises typological thinking as a sterile product of a scientifically uninformed philosophical relic. But typologists were pursuing a productive research program, one that the philosopher Ronald Amundson (2005) convincingly argues had a positive impact on Darwin's own work. Darwin's reading of naturalists of the typological school (people like Geoffroy St-Hilaire in France and Richard Owen in England) pushed him to recognise the existence of an important phenomenon in nature – the existence of underlying commonalities in structure among species whose conditions of life were quite different – that demanded explanation. So Mayr's characterisation of typological thinking is misleading. Even so, we will see that Mayr is right to say that Darwin is opposed to typological thinking.

Darwin says that species are formed by the action of natural selection on slight variation. His position demands, then, that these small variations, if they can be added up to produce new species, are themselves stable. Hence the reason why we do not observe forms that are intermediate between existing species cannot be that these forms are unstable. This presents Darwin with a dilemma. On the face of things, if he is right about common ancestry and the stability of slight variations, there should be no gaps between existing organic forms. On the other hand, Darwin learns from the typologists that this is not the pattern we observe. So Darwin needs to give a non-typological explanation for apparently typological phenomena:

... why, if species are descended from other species by insensibly fine gradations, do we not everywhere see innumerable transitional forms? Why is not all nature in confusion instead of the species being, as we see them, well defined? (*Origin*, 171)

Darwin reinterprets Owen's archetypes as ancestors: the diverse vertebrate species appear to be variations on a common theme not because they are manifestations of a single timeless ground plan, but because they have retained the characteristics of a common ancestor (*Origin*, 435; see also Amundson 2005, Chapter 4). But Darwin's way of thinking about shared history does not guarantee that we should expect the world to contain species that are what he calls 'tolerably well-defined objects' (*Origin*, 177). We still need some explanation for the coherence of species, and for the gaps between them.

In response, Darwin claims that many stable forms fail to be observed alive today because, while they may have been suited to earlier conditions of existence, they have been driven out of populations by shifting competitive demands. He supplements this assertion with geological reasoning that explains why the known fossil record consists of a highly impoverished collection of these intermediate forms (*Origin*, 279–311). Darwin, in other words, gives us an alternative way of thinking about the modality of species. In this respect, Mayr is right about the philosophical novelty of Darwin's view. While the typologist stresses the instability of unobserved forms given constant natural law, Darwin stresses the improbability of the continued survival of unobserved forms given shifting competitive demands. One of Darwin's primary explanatory tools for this task is an offshoot of the more general principle of natural selection, which Darwin calls the 'principle of divergence of character'. Darwin had learned from Adam Smith that competition will be most intense between individuals in the same line of business. Darwin argues that in the economy of nature, no less than in human affairs, competitive advantage will come to those who find new ways of making a living:

... the more diversified the descendants from any one species become in structure, constitution, and habits, by so much will they be better enabled to seize on many and widely diversified places in the polity of nature, and so be enabled to increase in numbers. (*Origin*, 112)

By coupling principles such as this one to his hypothesis of common ancestry, Darwin is able to explain the existence of discrete species while also accounting for their underlying commonalities, and he is able to do so in a non-typological way. This much is good news for Mayr. What is not such good news for Mayr is that the primary resources Darwin uses to replace typological explanation are natural selection and the 'tree of life' hypothesis. This makes it hard to characterise 'population thinking' as a third conceptual innovation wholly distinct from Darwin's better-known ideas.

On Darwin's view, species are 'tolerably well-defined objects' by virtue of the corralling forces of local environments, which discipline the tendencies of individuals to vary, and thereby maintain coherence over time at the level of the population. This population-level coherence is achieved in spite of differences constantly being introduced among individuals, not because of something shared by all individuals. Philosophers and scientists have frequently latched onto the significance of this sort of 'population thinking' and claimed diverse consequences for topics such as race, systematics, progress, health, and even cultural evolution. There is no sense trying to evaluate all of these claims here, but it is worth pointing out that when Darwin himself wrote on many of these topics, he found ways to reconcile his own population thinking with claims that will sound distinctly typological to many.

The variability of species in every respect is something that Darwin learned from his work on barnacles, and he constantly draws our attention to it: 'I am convinced that the most experienced naturalist would be surprised at the number of cases of variability, even in important parts of structure . . .' (*Origin*, 45). In spite of this, some of Darwin's observational practices show scant regard for the ubiquity of variation. Consider Darwin's work on domesticated pigeons. This work seeks to show the power of artificial selection in modifying domesticated species, and in doing so it points to the potential for natural selection to modify wild species. Darwin illustrates artificial selection's power by demonstrating the degree to which distinct domesticated varieties of pigeon differ from each other, and the degree to which these varieties differ from wild rock pigeons. In order to do this, Darwin needed to make measurements of various traits in these different varieties. However, as James Secord points out:

As Darwin confessed in the *Variation under Domestication*, his measurements on the wild rock pigeon were based on only two birds, and originally he planned to have but one. Without knowing the variation of his standard, readers must have been highly suspicious of his comparisons with other birds. One can hardly call Darwin's work in this instance 'population thinking', although he certainly was aware of a problem. (Secord 1981, 179)

Darwin's views on race are also noteworthy. His 'population thinking' made him sceptical of any exceptionless generalisation regarding the members of particular human races. Even so, he was happy to endorse fairly strong rules of thumb, including generalisations relating to racial psychology:

Their mental characteristics are likewise very distinct.... Every one who has had the opportunity of comparison, must have been struck with the contrast between the taciturn, even morose, aborigines of S. America and the light-hearted, talkative negroes. (*Descent*, 216)

And while it is true that Darwin rejected some conceptions of evolutionary progress, distancing himself in the *Origin*'s third edition from Lamarck's belief in 'an innate and inevitable tendency towards perfection in all organic beings' *(Variorum, 223: 382.14–16:c)*, this does not mean that he eschewed talk of progress altogether. Even in the domain of race, Darwin appeared to think some form of progress, while not inevitable, was highly likely. He offers an evolutionary explanation for why the current gap in advancement between man and his nearest living relative will in time become even greater:

At some future period, not very distant as measured by centuries, the civilised races of man will almost certainly exterminate and replace throughout the world the savage races. At the same time the anthropomorphous apes, as Professor Schaaffhausen has remarked, will no doubt be exterminated. The break will then be rendered wider, for it will intervene between man in a more civilised state, as we may hope, even than the Caucasian, and some ape as low as a baboon, instead of as at present between the negro or Australian and the gorilla. (*Descent*, 201)

One of the reasons for Darwin's continued adherence to many typological pronouncements in spite of his population thinking is that the shift to an explanation of species composition in terms of external ecological forces, rather than internal organismic tendencies, leaves open the issue of how homogeneous we should expect species to be. Modern evolutionary game theory has shown how selection can actively promote the existence of mixed populations. But this view is not entailed by the more basic Darwinian thought that species coherence is a function of natural selection acting to shape the variation that is constantly arising in populations.

VI. DARWIN AND PHILOSOPHY

Let us return to Mayr's assessment of the *Origin* in his 1964 introduction to that work. He casts Darwin's conceptual innovation not merely as a rejection of a dominant philosophy, but as a rejection triumphantly effected without engagement with the great philosophers, hence a rejection that also demonstrates the general lack of effectiveness of traditional philosophical methods in answering broad philosophical questions:

No one resented Darwin's independence of thought more than the philosophers. How could anyone dare to change our concept of the universe and man's position in it without arguing for or against Plato, for or against Descartes, for or against Kant? Darwin had violated all the rules of the game by placing his argument entirely outside the traditional framework of classical philosophical concepts and terminologies.... No other work advertised to the world the emancipation of science from philosophy as blatantly as did Darwin's *Origin*. For this he has not been forgiven to this day by some traditional schools of philosophy. To them, Darwin is still incomprehensible, 'unphilosophical,' and a bête noire. (Mayr 1964, xi–xii)

There are problems for Mayr's interpretation of Darwin's relationship with philosophy. Herschel was famously unimpressed by the *Origin*, apparently describing the theory defended therein as the 'law of higgledy piggledy' (quoted in Browne 2003, 107). This remark supports Mayr's view: one might say that Herschel's failure to appreciate Darwin's theory betrays him as one of those philosophical reactionaries whose views Darwin made obsolete. But we need to remember that Darwin modelled the *Origin*'s argument on Herschel's principles. If Herschel's methodological work counts as philosophical in nature, then while it remains reasonable to regard

Darwin's views as being in tension with some important elements of Victorian philosophy, one cannot claim that Darwin 'had violated all the rules of the game.'

In other ways, Mayr is right. Darwin, as we have seen, was sceptical of the ability of abstract thought alone – the 'deductive method' of Herbert Spencer – to deliver factual conclusions regarding the natural world. Because of this, Darwin looked to scientific observation to shed light on matters that some philosophers might still be tempted to regard as the preserve of reason alone. But Darwin's views put him in an established philosophical tradition – Mayr exaggerates Darwin's claim to be regarded as a philosophical revolutionary. Philosophers of the British empiricist school insisted on the necessity of experience for the possession of knowledge, unless that knowledge was itself restricted to abstract logical or mathematical matters. David Hume, whose works Darwin had read, famously closed his *Enquiry Concerning Human Understanding* by recommending that:

If we take in our hand any volume; of divinity or school metaphysics, for instance; let us ask, *Does it contain any abstract reasoning concerning quantity or number?* No. *Does it contain any experimental reasoning concerning matter of fact and existence?* No. Commit it then to the flames: for it can contain nothing but sophistry and illusion. (Hume 1978, 165)

While Darwin saw considerable philosophical promise in the evolutionary perspective, he was also aware that his views were in tune with some existing themes in the history of philosophy. He wrote in his Notebook N: 'I suspect the endless round of doubts & scepticisms might be solved by considering the origin of reason. as gradually developed. see Hume on Sceptical Philosophy.' (*Notebooks*, 348: N, 101) The Humean sceptic gives up on the possibility of finding a rational justification for such things as our expectation that the future will resemble the past, for our belief that causes in some sense necessitate their effects, and for our inclination to regard some acts as wrong, others as right. The Humean sceptic instead rests content with an account of human thought and human nature that explains why we regard these expectations and inclinations as compelling. In this way, Hume aims to give 'sceptical solutions' to the sceptical problems he raises. They are sceptical solutions because they concede to the sceptic the impossibility of justifying our fundamental practices of moral judgement, reasoning about the future, and so forth. It would be a stretch to call Darwin a Humean sceptic, not because Darwin is at odds with Hume, but because Darwin does not say enough about his general philosophical stance for their views to be closely aligned. Yet Darwin holds in common with Hume a willingness to settle for description and explanation of the characteristic habits of human thought, in lieu of watertight justifications of those habits.

18   The *Origin of Species* as a Book

Varieties and variation were the keys to Darwin's theory of evolution by means of natural selection – barnacles and pigeons – and varieties and variations are the keys to our understanding of the *Origin of Species* as a book.

The three largest collections of editions of the *Origin* are the Kohler Collection held by the Natural History Museum in London; a collection assembled by R. B. Freeman at the Thomas Fisher Library of the University of Toronto; and the books collected by Warren Mohr, Jr., now in the Henry E. Huntington Library in San Marino. Anyone wanting to study the *Origin* as a book needs to have close by a copy of R. B. Freeman's *The Works of Charles Darwin: An Annotated Bibliographical Handlist* in its second edition. With all its faults and quirks, it is indispensable.

On November 24, 1859, *On the Origin of Species by Means of Natural Selection, or the Preservation of Favoured Races in the Struggle for Life* was published by John Murray on heavy cream-coloured stock from Spalding, printed by W. Clowes and Sons and bound by Edmonds & Remnant, London, in green cloth with gilt blocking on the spine. John Murray had held a trade sale on November 22 when orders were taken for copies from booksellers, wholesalers, and circulating libraries. Twelve hundred and fifty copies had been printed and bound. Of those, 5 were sent to Stationers' Hall; 12 were "allowed Author"; and 41 were "Presented Reviews" leaving 1,192 for sale (National Library of Scotland (NLS) Acc 12604/570/158). This number was oversubscribed by 250 (*Correspondence*, 7: 394). Darwin required ninety copies to be sent as presentations to friends, family, and scientists (*Correspondence*, 8: 554–6). It is not clear if that number included the 12 author's copies, but it would seem

333

that about 1,100 copies were available for sale to distributors on the day of the trade sale. Of those, apparently 500 copies were taken by Mudie's Subscription Library – the largest commercial lending library in the country, which also supplied book societies and village libraries (*Correspondence*, 7: 395 n.1), which means that only 600 copies were available for sale to the general public on November 24. Darwin has been blamed for the canard that the book sold out completely on the first day of sale, although he actually wrote to Lyell, "I heard, also, from Murray that he sold whole Edition the first day to the trade" (*Correspondence*, 7: 394). It is quite likely that the entire edition was sold out at bookshops soon after the twenty-fourth.

Murray proposed a new edition immediately, and for Darwin this was an opportunity to make emendations. Morse Peckham, in his variorum edition, traced these and other corrections that Darwin made to Murray editions over the next seventeen years. The *Origin* went through six editions during Darwin's lifetime, and for each one he made alterations and corrections. Peckham and Freeman note that a few further small corrections were made to the 1876 printing, and it is this version that was reprinted again and again by Murray. The sixth edition differed from the first in many ways – Darwin had been bombarded by criticisms and had made alterations as a result. Freeman and Peckham both believed and found it highly significant that the second through fifth editions of the *Origin* were printed from type that was left standing after the printing of the first edition. They both cited this as the reason for what they mistakenly presumed was a reduction in price of the second edition to fourteen shillings. Freeman also thought that the "Fifth thousand" was not really a second edition because it was made from standing type. Recent research by Peter L. Shillingsburg has shown that, except for the inner form of gathering X and all of gathering Y of the 1869 edition, each new edition was completely reset (Shillingsburg 2006). For such a well-studied title, this is a surprising and slightly disorienting discovery. For example, some errors arising in the text that were previously attributed to Darwin are probably not his at all. This new research has implications for our whole understanding of the early publication of the book.

In November 1901, the first edition came out of copyright in Britain and was reprinted widely. Murray printed the following note in the 1902 "Popular Impression (copies 40,000 through 47,000) of the corrected copyright edition issued with the author's executors" that

> Darwin's "Origin of Species" has now passed out of Copyright. It should, however, be clearly understood that the edition which thus loses its legal protection is the imperfect edition which the author subsequently revised, and which was accordingly superseded. The complete and authorized edition of the work will not lose copyright for some years.
>
> The only complete editions authorized by Mr. Darwin and his representatives are those published by Mr. Murray.

Today another opinion is widely held – that to approach more nearly Darwin's original and revolutionary ideas it is best not to read the sixth edition but rather the first or its corrected version, usually referred to as the second but actually called the "fifth thousand" on the title page.

Freeman assigns numbers to fifty-eight editions – not all of which he handled – although some of the listings include variant bindings. The Natural History Museum has sixty-five in its collections; the Huntington Library and Toronto can account for a further seven not in the museum's collection. These are divided among three series: the standard edition published in dark green cloth, the library edition in two volumes, and the popular cheap edition usually bound in light green cloth or in paper wrappers.

Murray had a deep commitment to the book – he accepted it sight unseen based on Darwin's reputation and his own experience as the publisher of Darwin's *Journal of Researches* (*Correspondence*, 7: 275). He had a good working relationship with his author, who was grateful for the publisher's attentions, agreeableness, and prompt payment of royalties. After Darwin's death, his sons, as his executors, tried Murray's patience. Murray set out his own position in a letter of May 18, 1882, to William M. Hacon, the Darwin family solicitor.

> The very peremptory order which Mr. Darwin's executors have sent me through you, not to reprint their Father's works without their express sanction, was scarcely required since it is my unvarying rule, observed with all my Authors, except in cases of emergency, to apprize them when new Editions are called for.

It has been followed in the case of Mr. Darwin ever since I became his publisher. I need scarcely add that the orders of the Executors shall be strictly carried out in future." (NLS Acc 12604/735/135)

The executors accepted, and the relationship continued.

The Murray Archive has recently been deposited at the National Library of Scotland in Edinburgh. Although no new specific information about the 1859 edition has come to light, there are lists for the trade sales of the fifth edition on June 23, 1869 (NLS Acc 12604 Sales Book 1869/1870) and the sixth edition on February 14, 1872 (NLS Acc 12604 Sales Book 1871/1872).

The other major nineteenth-century English-language publisher of the *Origin* was the firm of D. Appleton & Co. in New York. Asa Gray, the Harvard botanist and Darwin's long-time correspondent, had offered to arrange an American edition. Although his first choice was Ticknor and Fields in Boston, Appleton was already preparing its text from the 1859 edition and had made stereo plates. By the time Gray approached Appleton the second edition was on sale in Britain, and Darwin hoped they would reprint this. Corrections and additions were sent by ship. By May 1860 Appleton had sold most of the original 2,250 copies from three separate printings of its first edition (*Correspondence*, 8: 571–2) and proceeded to produce a fourth printing incorporating the corrections. Appleton was not obliged to do so, but the company agreed to pay Darwin a royalty.

In 1872, Murray sent Appleton a set of stereo plates for the sixth English edition of the *Origin* (NLS Acc 12604/566/118). Appleton reprinted from it, which meant that the few changes made by Darwin in 1876 – including, for example, changing "Cape de Verde" in the text to "Cape Verde" – were not made to the Appleton one-volume edition. Murray later supplied Appleton with stereo plates for the two-volume edition. Gerard Wolfe claims that "the house reprinted Darwin's classic 38 times," although he also claims that the first Appleton edition appeared in 1859 (Wolfe 1981, 42, 112). Freeman lists forty-seven different printings, bind-ups, and editions, ending with the Appleton edition in two volumes in 1937. The Natural History Museum holds twenty-eight, the last of which is dated 1927. Toronto and the Huntington Library between them hold a further twenty-two not housed in the museum, making a total of at least fifty-editions, printings, and bindings. Murray used bindings that can be generally described as green; Appleton bound volumes with less consistency. Although the main binding colour for Appleton editions was brick red, the Natural History Museum copies are bound in eleven different colours and cloths, including publisher's half-maroon morocco with marbled boards. Appleton published the *Origin* in one-volume, two-volume, and two-volumes-in-one formats, including an Authorized Edition that they intended to be part of a "collected edition" – owing to the Authorized Edition's being reprinted, when actually "collected" together the resulting assemblages tended to vary one from the other – and a luxury Westminster Edition.

While Murray and Appleton were dominant in the English-language market in the nineteenth century, there were other players. Murray sold Routledge 274 sets of sheets in 1894 (NLS Acc 12604/658/75) and 4,004 sets of sheets in 1898 (NLS Acc 12604/570/105), at least in part for their Sir John Lubbock Hundred Books series. There was more competition in the American market, with a Humboldt Library of Popular Science edition in wrappers and one from John B. Alden in 1886 that included *The Descent of Man*.

The United States Congress passed a copyright law in 1891 that included an exemption for any book first published before July 1891. This legal clarification coincides with a rash of new American editions of the *Origin*. Excluding Appleton editions, for the period covering the next nine years the Natural History Museum holds twenty-six different editions and variants published by fourteen different American publishers in New York, Chicago, and Akron, Ohio. These versions appeared in one or two volumes, mostly from the sixth edition, although the Hennebery Company of Chicago included preliminary material with its version of the sixth edition – material last seen in the fifth thousand of 1860.

The twentieth century saw many changes and experiments in the publication of the *Origin*. In England, cheap editions joined Murray's popular edition early in the century: Ward, Lock & Co., Grant Richards, the Unit Library, Hutchinson, Oxford World Classics, Watts and Co. issuing for the Rationalist Press Association, Cassell and Company, Collins Clear Type, and Dent Everyman. In America, early twentieth-century editions were published by P. F. Collier, Hurst, A. L. Burt, Thomas Y. Crowell, J. A. Hill, and Rand McNally. A leading player from the 1930s onward was the Modern Library.

Although Freeman cites only three issues – including a British one – of which he himself had two, twenty-two different variants can be found in the Natural History Museum. The *Origin* was reissued in various colours and with the different styles of the Modern Library torchbearer on the bindings. Using the lists of publications on the dust wrappers and the known dates for the different torchbearers, it is possible to date the different issues throughout the century.

The second half of the twentieth century was a rich time for cheap editions in both Britain and America. Paperback editions include those published by Penguin, Everyman, Oxford World Classics, Mentor, Doubleday, Bantam, and Harvard University Press, using variously the first, second, and sixth editions. The first microform edition was part of the Readex Microprint Landmarks of Science II, published on six cards from the fifth thousand of 1860 in the 1960s. Electronic online editions started to appear from Gutenberg and Electronic Scholarly Publishing in the 1990s. In 1997, Lightbinder Inc. in San Francisco issued a reprint of the sixth edition on computer optical disk, probably a CD-ROM. A version on audio cassette was produced by Audio Scholar of Mendocino, California, in 1990. In 1991, the shortest edition of the book was published: a children's illustrated board book, reduced to ninety-one words with illustrations enabling children "to learn from imaginative adult readers how Darwin understood the evolving world" (Karlinsky 1991).

Throughout the twentieth century publishers vied with each other for the imprimatur and introductory material of leading scholars, scientists, and pundits: Grant Allen, Sir Arthur Keith, Leonard Darwin, Edmund B. Wilson, C. D. Darlington, Julian Huxley, Sir Gavin de Beer, Charles G. Darwin, Ernst Mayr, J. W. Burrow, L. Harrison Matthews, George Gaylord Simpson, Patricia Horan, Richard Leakey, Jeff Wallace, Dame Gillian Beer, and, most surprising of all, the anti-Darwinian W. R. Thompson, an entomologist and director of the Commonwealth Institute of Biological Control, Ottawa, Canada, who wrote the introduction to the 1956 Dent Everyman edition. This was reprinted in 1967 by the Evolution Protest Movement as *New Challenging "Introduction" to The Origin of Species Everyman Library No. 811 (1956)*. An edition aimed at a more popular audience included material by Walter Cronkite and James Michener.

In Britain in the nineteenth century there were neither condensed nor illustrated editions – a sign of continued family interest in and control of the publication of the *Origin*. At the turn of the century, Francis Darwin wrote to Murray of his brother William's disapproval of "an illustrated *Origin* as very difficult to do well – even if so done of no real use," while also conveying his own thoughts: "I don't see that one can do better than produce a really cheap well printed *Origin* – it is much better than an annotated or illustrated edition, and it would be difficult to get these done in my opinion" (NLS Acc 12604 Folder 109). As for condensed versions, he noted: "My own impression is that the modern student is fed on lecture notes & selections until he is forgetting how to read a stiff book altogether. And I should be sorry to let the Origin help this process – I daresay the book would sell, but that is not the only question" (NLS Acc 12604 Folder 108). When the texts finally came out of copyright, things changed. There have been a few deluxe editions. The Heritage Press of New York and Norwalk and the Griffin Press of Adelaide published finely bound editions with wood engravings by Paul Landacre. Less beautiful but probably more frequently read are the various condensed versions – most notably *What Mr. Darwin Really Said*, published by George Routledge in 1929 with an introduction by Julian Huxley. Fleetway House published extracts with a précis by C. W. Saleeby in 1926, and Quarter Books in Pasadena published its own condensed version in 1936. Regenery of Chicago, better known today for its anti-Darwinian material, published a volume of extracts in the 1940s. Charlotte and William Irvine abridged the *Origin* for Frederick Ungar, and Richard Leakey's *Illustrated Origin of Species* is also abridged – and translated into many languages. In 1996, Orion Books published Chapter 3 and the first part of Chapter 4 of the *Origin* in fifty-nine pages for sixty pence. Extracts from the *Origin* appear widely in collections of selections for biology students and general readers and also occasionally in translation.

Darwin is certainly the most translated scientific author of all time and probably one of the most translated of any author originally published in English. Before the publication of the *Origin*, Darwin wrote to Murray: "I am *extremely* anxious for the subject sake (& God knows not for mere fame) to have my Book translated; & indirectly its being known abroad will do good to English Sale ... " (*Correspondence*, 7: 376).

The German edition was first translated by Heinrich Bronn in 1860. Darwin had written to Bronn in February asking if he could offer suggestions for a translator and if he could, perhaps, proofread the result before publication, "for I am most anxious that the great & intellectual German people should know something about my work" (*Correspondence*, 8: 70). Bronn did not agree with Darwin about natural selection but produced a prompt translation himself, with its own quirks and oddities, published by E. Schweizerbart in Stuttgart. Darwin's greatest concern was obviously the translation of the term "Natural Selection":

Several scientific men have thought the term "natural Selection" good, because its meaning is <u>not</u> obvious, & each man could not put on it his own interpretation, & because it at once connects variation under domestication & nature. – Is there any analogous term used by German Breeders of animals? – "Adelung" – ennobling – would perhaps be too metaphorical. It is folly in me, but I cannot help doubting, whether "Wahl der Lebensweise" expresses my notion. – It leaves the impression on my mind of the Lamarckian doctrine (which I reject) of habits of life being all-important. Man has altered & thus improved the English Race-Horse by <u>selecting</u> successive fleeter individuals; & I believe, owing to the struggle for existence, that similar <u>slight</u> variations in a wild Horse, <u>if advantageous to it</u>, would be <u>selected</u> or <u>preserved</u> by nature: Hence natural Selection. But I apologise for troubling you with these remarks on the importance of choosing good German terms for "natural Selection." (*Correspondence*, 8: 82–3)

Bronn finally settled on "*natürliche Züchtung*" ("natural breeding") for his translation, although it was still not exactly what Darwin thought correct. Contrary to Freeman's note, Bronn remained the sole translator of the second German edition, completing the revisions just before his own sudden death. The third edition of 1867, with the change from *Züchtung* to *Zuchtwahl* or selection, was a revision of Bronn's translation by Julius Victor Carus incorporating Darwin's changes to the fourth Murray edition. The 1872 edition was Carus's new German translation based on the sixth Murray edition. The last edition to have Bronn's name on the title page is the one dated 1876. There have been a total of eight translators of the *Origin* into German: Heinrich Bronn, Julius Victor Carus, Georg Gärtner, David Hack, Paul Seliger, Dr. Richard Böhme, Dr. Heinrich Schmidt of Jena, and Carl W. Neumann. Further editorial matter is attached to several Reclam editions by Georg Suchmann, Gerhard Heberer, Rolf Lothar, and Gerhard H. Müller, while Walter Domann has edited a translation for Diesterweg.

The first Dutch translation, by Tiberius Cornelius Winkler, was published in 1860 – Darwin knew nothing about it until he received a copy in 1861 (*Correspondence*, 9: 195). Freeman calls him T. E. Winkler in his handlist and claims that the first edition was not published until 1864 (Freeman 1978, 221). Other Dutch translators have included J. Klerkx in 1983, Ludo Hellemans in 2000, and Ruud Rook in 2001. In 1958, A. Schierbeek edited a condensed version of the Winkler translation in *Darwin's werk en personlijkheid* – which was mistakenly described by Freeman as a Flemish translation. Freeman had mistranscribed the imprint of the Wereld-Bibliotheek edition as published in the Belgian city of Antwerp, a subsidiary place of publication for this Amsterdam firm. In fact, there is no Flemish language – the formal language spoken and written by the Flemish community of Belgium is Dutch.

Even before the *Origin* was published Darwin had an offer from Louise Swanton Belloc – an Irish-born French writer – to translate his work, but she ultimately decided it was too scientific for her. She was a friend of Darwin's friend Mary Butler (*Correspondence*, 7: 376–7). Pierre Théodore Alfred Talandier, a professor of French at Sandhurst, also offered to translate the book but failed to find a publisher (*Correspondence*, 8: 135). Eventually, the first French translation was prepared by Clémence Royer, a French naturalist living in Lausanne, Switzerland, and published in 1862. It was characterised by a long preface and an abundance of explanatory texts. Darwin described her as probably "one of the cleverest & oddest women in Europe," though lacking in experience as a naturalist. Darwin tried but failed to get her to remove or modify her commentary (*Correspondence*, 10: xx). Her translation went through at least five editions, including a two-volume reprint of 1918. In 1873, a translation by J. J. Moulinié appeared, "Traduit sur l'invitation et avec l'autorisation de l'auteur sur les cinquième et sixième éditions anglaises ... " This was followed in 1876 by Edmond Barbier's translation of the sixth or "l'édition Anglaise définitive," which remains, modified by Daniel Becquemont, the standard French translation.

In September 1874, Milan Radovanovitch wrote to Darwin asking for his authorisation to publish a Serbian translation of the *Origin*

(*Calendar*, 414). It was ready for the press in 1876 but was not published until 1878 owing to war between the Serbs and the Turks.

The *Origin* was translated into a further seven languages during Darwin's lifetime: Russian (1864), Italian (1864), Swedish (1869), Danish (1872), Polish (1873), Hungarian (1873–74), and Spanish (1877). These translations were all either initiated by their authors or arranged by Darwin himself. It has since appeared in a further thirty-three languages: Arabic, Armenian, Bahasa Indonesian, Basque, Bulgarian, Catalan, Chinese, Croatian, Czech, Finnish, Gallegan, Greek, Hebrew, Hindi, Icelandic, Japanese, Korean, Latvian, Lithuanian, Malayalam, Macedonian, Marathi, Norwegian, Persian, Portuguese, Punjabi, Romanian, Slovakian, Slovenian, Tibetan, Turkish, Ukrainian, and Vietnamese.

As we approach the sesquicentenary of the publication of the *Origin*, what can we say about the book today? It remains in print in English from Bantam Books, Barnes & Noble, Broadview Press, Castle Books, Collector's Library, Dover, Everyman's Library, Harvard University Press, Modern Library, New York University Press, W. W. Norton, Oxford University Press, Penguin, Pickering and Chatto, Prometheus, Running Press, Signet, and Wordsworth. Print-on-demand editions are available from Bibliobazaar, Cosimo, Echo Library, Elibron Classics, IndyPublish.com, Kessinger, University of Pennsylvania Press, ReadHowYouWant.com, Standard Publications, and Wildside Press. Editions with new introductions by leading Darwinians, including Richard Dawkins, James Watson, and E. O. Wilson, have recently appeared. Due to be published in 2009 are an edition for Cambridge University Press edited by Jim Endersby, a new edition for Penguin edited by William Bynum, and a new edition for John Murray edited by Olivia Judson. It remains available in translation in Chinese, Danish, Dutch, French, Gallegan, German, Hebrew, Hungarian, Icelandic, Bahassa Indonesian, Italian, Japanese, Korean, Norwegian, Polish, Portuguese, Punjabi, Russian, Slovakian, Spanish, Swedish, Tibetan, and Vietnamese, and probably in several other languages that we have not been able to verify definitively.

There is a clear distinction between the "traditional" publishers and the print-on-demand publishers. The majority of the traditional publishers reprint the first edition 1859 text, and most of these explain why they do so. Oxford World's Classics reprints the second edition for reasons given by Dame Gillian Beer in her note on the text (Beer 1996, xxix), and the Pickering & Chatto/New York University Press edition reprints both the first and the sixth editions for completeness. The remaining few either use the sixth edition with no explanation or do not cite which edition they are reprinting. These "traditional" publishers usually provide introductory material of varying quality. The print-on-demand publishers tend to use whatever digital version they find on the Web without further editorial comment. The exception is the University of Pennsylvania Press, which scanned in its own 1959 variorum text and prints on demand to satisfy distributors' orders. This is not a "photographic" reprint as an alteration has been made to the dedicatory leaf, for no reason that the publisher can provide. The Australian company ReadHowYouWant.com prints on demand to order in a variety of print formats: large print, print for those with reading disabilities such as dyslexia, audio versions, and Braille.

The Everyman's Library edition of the *Origin* also includes *The Voyage of the Beagle* in the same volume, while both W. W. Norton and Running Press produce omnibus volumes printing *The Voyage of the Beagle*, *The Descent of Man*, and *The Expression of the Emotions in Man and Animals* along with the *Origin*.

There is a print-on-demand condensed version of the first edition published by Trafford – the editor has "taken the first edition and removed the superfluous words" (Sheldon 2004, vii). The version published by Broadview and edited by Joseph Carroll is the kind of edition that Francis Darwin claimed to abhor but even he might agree it is an interesting one. Carroll gives his account of the volume: "The basic text for the present edition is that of the first edition. Necessary corrections have been incorporated from the relevant revisions in the second edition. Neither the trivial changes of punctuation and phrasing nor the substantive revisions in the second edition have been incorporated into the present text" (Carroll 2003, 76). The volume includes the historical sketch and the glossary from the corrected sixth edition of 1878. Carroll also provides a full introductory essay together with a bibliographic review of books about Darwin and evolution, adding excerpts from the *Journal of Researches*, *Autobiography*, *Notebooks*, the 1844 *Manuscript*, and *The Descent of Man*. This is followed by excerpts from other texts and authors: Genesis, Paley, Lamarck, Spencer, Malthus, Lyell,

Wallace, and Huxley (Carroll 2003, 76). The *Origin* appears online or can be downloaded in its first through sixth Murray editions and in Danish, French, German, Italian, and Russian translations. There are electronic books from xlibris.com, Ebooks.lib, Digireads.com, Nuvision, and E-classics. There are audio CDs from Tanton and Pisces Conservation Limited and even "free audio mp3 files...for downloading onto a computer or portable mp3 player" (www.darwin-online.org.uk/audio_darwin.html). In the audio version that he has edited and read for CSA, Richard Dawkins's "priority was to cut those passages that are now known to be wrong, notably those concerned with genetics" (Dawkins 2006, insert). An interesting new project for 2009 is a genetic electronic edition of the *Origin* in progress at the University of Birmingham under the direction of Barbara Bordalejo and her colleagues. It is designed to show the creative process behind the text. Like a variorum, it will show the changes through the first six editions, but it "will also include manuscript materials going back, at least, to 1842. We are developing a view that will allow the reader to see the text develop in front of her eyes" (personal e-mail to the authors from Barbara Bordalejo, June 6, 2007).

How many copies of the *Origin* are now sold each year? We asked the publishers of current editions for their latest annual sales figures, promising them anonymity. Some cooperated, some did not. Our best informed estimate is that approximately 75,000 to 100,000 copies of the *Origin* are now sold in many different languages throughout the world each year.

Until this point we have been considering the publishing history of the book. However, it is worth noting that the *Origin* has had a long and interesting life in the antiquarian book trade. From 1859 readers have recognised the importance of the *Origin* and tended to keep their copies, with the result that the first edition has never been a rare book and has been regularly offered for sale at auction and by antiquarian booksellers.

Bernard Quaritch have been antiquarian booksellers in London since 1847 and remain one of the world's greatest firms. They sold early editions of the *Origin* throughout the nineteenth century, often at a discount to the published price, but never offered it as a first edition aimed at the private collector. In 1862, they offered the 1861 third edition for sale at 11s.8d., calling it the "last edition of this remarkable book (pub. at 14s)," and in 1865 they offered the same edition, describing it as "one of the most important publications of our times." An 1881 catalogue entry says "*Origin of Species* . . . cloth, 5s. 1859."

The change came in 1903 when Quaritch offered an 1859 *Origin*, stating that it was the first edition, presenting Darwin's theories in their original form, and consequently would be of interest to "the collector." The price was £2–10-0, a premium on the price of a new copy, not a discount.

The first auction record that we have found for the first edition is for a copy bound in calf and including a Darwin autograph letter, which sold at Sothebys for £2 in 1905. In the years since, some three hundred copies have appeared at auction. All prices mentioned hereafter are for good copies of the first edition in the original cloth, unless otherwise stated. We have, in the main, excluded copies in bad condition and copies presented by Darwin to colleagues or friends and copies containing autograph letters. The prices of copies sold outside Britain have been converted to pounds sterling, at the rates prevailing at the time of the sale.

Book auctioneers were primarily wholesalers to the antiquarian trade until recent times, and most copies of the *Origin* were sold to booksellers, mainly buying for their stock. Nowadays auctioneers court the private collector and hope sometimes to sell important and expensive books and manuscripts directly to collectors.

Before the First World War, three further copies sold for between one guinea and two pounds. Although Quaritch had promoted the first edition in 1903, the book was not yet sought after by private collections. Seymour de Ricci's influential *The Book Collectors Guide*, published in 1921, does not mention Darwin.

Auction prices of the *Origin* started to rise in the late 1920s, and in November 1929 the innovative booksellers Elkin Mathews paid the highest price yet, £80, for a first edition that contained a four-page Darwin autograph letter. A. W. Evans, senior director of Elkin Mathews, was keen to push those books that he thought were not sufficiently appreciated by collectors and priced the book at £130, noting in the catalogue entry of December 1929: "one of the epoch-making books of the nineteenth century, or, indeed of all time, for Darwin's evolution theory not only revolutionised biology but 'changed the whole intellectual outlook of mankind'. . . . Darwin has influenced the whole range of human thought, and his book is comparable, on this account, with Galileo's *Dialogues* or Newton's *Principia*" (Elkin Mathews 1929, item 15; thanks to the Lilly Library, Indiana University, Bloomington, Indiana, for permission to quote from their copy in the Elkin Mathews archive). In the eighty years since, no bookseller or auctioneer has put it better. Evans's judgement was right, but the timing was poor. The Wall Street crash ended the boom years, and depression followed. Elkin Mathews finally sold this copy for £45 in 1932. Some fifty more copies were sold at auction during the 1930s, more than in any other decade, and the average price was down to £12 by 1939.

By the late 1950s the average auction price was £36, increasing to £100 during the first half of the 1960s and then to £250 by the end of that decade. Ten years later copies were going for almost £1,000, and in 1978 Quaritch offered a copy, "the finest we have seen," for almost £3,000. And so it went – up. By the middle of the 1980s auction prices were averaging almost £4,000, and between 1989 and 1990 three copies fetched over £10,000 each. By the late 1990s the price had risen to an average of £17,500, with one copy selling for £49,000 in 1999. The record auction price for a good copy of the first edition of the *Origin* in original cloth was achieved in New York in 2001, when it fetched £101,000. Prices came down after that, with the average price for the eleven copies sold at auction in the twenty-first century being about half that record figure.

What were the factors driving the price up? The importance of the book has always been acknowledged, but in recent years more and more collectors have recognised this – partly through the publication in 1967 of *Printing and the Mind of Man*, which has become the bible for collectors of important and influential books in the development of human thought; partly as a result of the marketing skills of the auction houses and antiquarian booksellers; and partly through investment buying.

A generation of private collectors is now buying famous and important science books. They want fine copies of the great books and see the *Origin* as an icon, as a landmark, as a trophy. A lot of these buyers have made their money in business, often in America and often in businesses in the science and technology fields.

The growth of interest in this area has outstripped all other collecting fields in recent years. There is a difference between the words "famous" and "important." The pricing of significant science books in the rare book market tends to favour fame over importance. Euclid's *Elements (Elementa Geometriae*, 1482), Copernicus's *De Revolutionibus (De Revolutionibus Orbium Coelestium*, 1543), Newton's *Principia (Philosophiae Naturalis Principia Mathematica*, 1687), and Darwin's *On the Origin of Species* are all very famous and very important books. Their values are: Euclid, £100,000/£225,000; Copernicus, £250,000/£500,000; Newton, £125,000/£225,000; Darwin, £50,000/£100,000. There are other science books that are important but that are not famous: Darwin took the first edition of Lyell's *Principles of Geology*, three volumes, 1830–33, with him on the *Beagle* – its value now is £3,000/£5,000.

Dick Whittington went to London to see if the streets were indeed paved with gold. We went to London to see if we could find Darwin gold. We did. We visited antiquarian bookshops and saw twelve copies of the *Origin* that had been published in Darwin's lifetime – three first editions, one each of the second, third, and fifth editions, and six of the sixth edition. We know that Quaritch bought twenty-five copies of the sixth edition at Murray's trade sale in February 1872; currently (summer 2007) they have for sale, for £200,000, a remarkable copy of the first edition. This is the presentation copy to William Carpenter, one of the earliest and most influential reviewers of the book, inscribed in ink "From the author" by one of Murray's clerks (as usual) and with Carpenter's pencil annotations in the text. (We are grateful to a number of antiquarian booksellers for their help with this section, in particular to Anthony Payne for allowing us to look at Quaritch's records and to Julian Wilson of Maggs for alerting us to Elkin Mathews's pushing of the *Origin* in the late 1920s.)

The *Origin of Species* was originally published at a price of fourteen shillings, but there is a long history of claiming that the price for the first edition was fifteen shillings. On November 2, 1859, John Murray wrote to Darwin: "Now as to price – the book from its bulk & size will not be dear at 14/-. & this is the price I propose.... The only alternative is 12/-.... At 14/- the lowest trade allowance will be 9/6" (*Correspondence*, 7: 364–5). Darwin replied the next day: "I have received your kind note & the copy: I am *infinitely* pleased & proud at the appearance of my child. – I quite agree to all you propose about price..." (*Correspondence*, 7: 365). The book was listed in *The English Catalogue* for 1859 at 14s. (thanks to Professor Simon Eliot for this reference). Simple enough, but . . . George Paston, in *At John Murray's: Records of a Literary Circle 1843–1892*, wrote that "*The Origin of Species by means of Natural Selection, or The Preservation of Favoured Races in the Struggle for Existence* – to give the book its full title [sic] – appeared in November. Much to the astonishment of the author, the whole edition of 1,250 copies at 15s. was sold at Murray's annual sale . . . " (Paston 1932, 173). A copy of the relevant letter was held at Murray's, so there is no excuse for the error. All calculations for the edition in Murray's accounts are based on a price of 9s.6d. to the trade (NLS Acc. 12604/158/158). Paston seems to have had no understanding of the different editions published by Murray, conflating titles and also referring to the Historical Introduction as if it were part of the first edition when it did not appear until the third edition of 1861. In his introductory material Peckham cites the letter from Darwin to Murray of November 3 saying that "he was in agreement about the price which was to be 15s" (Peckham, 16). He repeats that "the retail price was 15/0" on the following page, where he also states that Darwin got £90 when the figure in the Murray accounts that he records is £180. Freeman comments about the fifth thousand that "[t]he price fell to 14s." (Freeman 1977, 78). Desmond and Moore, in their biography of Darwin, state that the price was fifteen shillings (Desmond and Moore 1991, 476). Richard Dawkins, in the insert for his CD reading of the *Origin*, claims that the book was "priced at fifteen shillings (the equivalent of 75p, but at a time when it was possible to live on a pound a week)" (Dawkins 2006, insert). James Secord concurs: "The *Origin* looked like the standard run of books published by John Murray, with green cloth casing, fifteen shilling price, and octavo size" (Secord 2000, 508). The book was in Murray green cloth; the book was octavo size, though in gatherings of twelves – which accounts for the book frequently being described as duodecimo – but the book was not priced at fifteen shillings, nor was that a standard price for Murray books; there are only a handful of titles priced at fifteen shillings in Murray's own inserted advertisements. Janet Browne in her *Darwin's* Origin of Species: *A Biography* gets it right at fourteen shillings (Browne 2006, 6).

How did print runs of the *Origin* compare with those of other revolutionary scientific books of its time? Evolutionary theories and transmutation were part of the common intellectual life in 1859, in large part as a result of the anonymous publication of Robert

Chambers's *Vestiges of the Natural History of Creation*. James Secord has studied the print runs for this title and found that the book went through eleven editions between 1844 and 1860, for a total of 23,350 copies printed (Chambers 1994, xxvii). He provides a comparative chart in his *Victorian Sensation* (Secord 2000, 526) and shows that by 1882, 25,500 copies of both *Vestiges* and the *Origin* had been printed in Britain. His chart shows that in 1890 about 39,000 copies of *Vestiges* had been printed and, using Peckham's data, indicates that this equalled the Murray production of the *Origin*, whereas the actual total for the *Origin* should be 40,750, including the more sumptuous two-volume edition (NLS Acc.12604, various ledgers). In 1921, when Murray last reprinted the sheets of the *Origin*, the total figure for Darwin's book produced from Albemarle Street was 175,144 copies, and although 1,706 copies were washed (i.e., pulped) sometime between 1932 and 1943, the book remained in print with Murray through the 1940s, with the last entry in the ledger showing three copies on hand on December 31, 1948 (John Murray Albemarle Street Archive, Copies Ledgers B1/Cr. 154 and M1/296).

The other revolutionary title that makes an interesting comparison is Sir Charles Lyell's *Principles of Geology*, the first volume of which was published by John Murray in 1830 and taken by Darwin on HMS *Beagle*. It was a multivolume work – volumes two and three were sent out to Darwin in South America on publication. The print run reached 29,500 sets in 1875 (Baldwin 1998, 114). Lyell's book was much more expensive than Darwin's. Surprisingly, there really is not much difference in the figures for the three titles during Darwin's lifetime.

There is no full-scale Darwin bibliography. In an appendix, Peckham gives full bibliographic descriptions of some of the editions published by Murray but does not take into account the variant bindings. He would appear to have looked at single copies of many of the editions he describes – but this is not a bibliography. R. B. Freeman first published *The Works of Charles Darwin: An Annotated Bibliographical Handlist* in 1965, but it is the "second edition, revised and enlarged" of 1977 that is usually cited. Freeman called it a handlist, not a bibliography, which has not prevented the widespread belief that it is the Darwin bibliography. He himself notes that "[t]here is no full bibliographical work even of the first editions of CD's books" (Freeman 1978, 79–80). The Freeman handlist has rightly been important in framing our understanding of the publishing history of Darwin's various titles – collectors, booksellers, librarians, and scholars refer to Freeman.

There are several problems with the handlist. Freeman was a scientist who collected books. He had a deep interest in Darwin, but he was not a bibliographer. He uses the digressive method for describing reprints, and this leads, sometimes, to a mess, especially when he combines variants in one entry and then on other occasions gives them separate entries. Entries refer back to previous ones that they appear to replicate, and when this means going back a page or more it can be maddening, especially if there is a typo or an error. Freeman included items in his list that he had never handled; he was dependent on descriptions produced by others and had no way of knowing whether or not they were correct. His dating is problematic. His cavalier use or nonuse of square brackets for dates causes problems when the date on the title page is the issue point. There are a number of typographical errors and transposed numbers, which can be exasperating for the user.

None of this would seem to matter so much as the book is in limited supply. However, as part of his project of mounting all of Darwin on the Web, John van Wyhe, in an admirable effort to make bibliographical information widely available, has put the handlist, together with updates, on his website. Freeman's own update is known in the book trade in the form of a badly mimeographed samizdat publication handed around like a secret code but of little real use or interest. Considerable effort has been made by van Wyhe to deal with the problems of the digressive nature of Freeman, but his not having seen the original volumes themselves has again led to errors that will mislead and confuse collectors and scholars who use the website without handling the books. For example, as already noted, when Heinrich Bronn published his German translation of the *Origin* he used the expression "*natürliche Züchtung*" on the title page. By the third edition this had become "*natürliche Zuchtwahl*," but in the online version it remains in its original form through the 1882 edition, when the online version no longer lists the title in German.

As we go to press we have learned that the American Philosophical Society in Philadelphia has acquired the Darwin collection of Professor James W. Valentine, Professor Emeritus in the Department of Integrative Biology at the University of California, Berkeley.

Collecting for fifty years, Professor Valentine has accumulated nearly one thousand different copies of the *Origin*, including items not in the Natural History Museum, Toronto, or Huntington collections. Starting with these major collections of the *Origin* together with the Murray Archive at the National Library of Scotland and the Darwin manuscript material at the Cambridge University Library, it should be possible to compile a full-scale bibliography of Darwin's works. The task, though, is enormous. Perhaps such a book will be available for the bicentenary of the *Origin*.

Varieties and variation – it is clear that from 1860 onward there were various editions available to anyone seeking to read the *Origin*. They vary in price, printing, binding, advertisements, language, format (including electronic and audio), and indeed even in the actual text. Once the first edition came out of copyright, the introductory material also varied. It is time now for a modern critical edition of the *Origin*. Today it is possible to study some 1,000 different copies in one place, fulfilling the Romantics scholar Charles Robinson's editorial requirement that "any editing that is going on must have complete runs of books, early and late, good and bad" (personal e-mail to the authors, June 2, 2002). The introductions in editions in English and in translation reflect changing knowledge about and attitudes toward evolution by natural selection and provide a history of how Darwin's classic has been presented to readers throughout the last 150 years. Such a critical edition would help to resolve many of the issues raised by Peckham fifty years ago in his own introduction. "Here is a task for a dozen maids with a dozen mops for more than a dozen years. Such an edition, once completed, would be a foundation for future studies" (Peckham, 10).

BIBLIOGRAPHY

Alter, S. G. 1999. *Darwinism and the Linguistic Image: Language, Race and Natural Theology in the 19th Century*. Baltimore: Johns Hopkins University Press.

Amundson, R. 2005. *The Changing Role of the Embryo in Evolutionary Thought*. Cambridge: Cambridge University Press.

Arnold, Michael. 1997. *Natural Hybridization and Evolution*. Oxford: Oxford University Press.

Babbage, C. [1838] 1967. *The Ninth Bridgewater Treatise: A Fragment*. 2nd ed. London: Frank Cass.

Bachmann, H. 1906. "Der Speziesbegiff." *Verhandlungen der Schweizerischen Naturforschenden Gesellschaft* 88: 161–208.

Baldwin, Stuart A. 1998. "Charles Lyell and the Extraordinary Publishing History of His Works." *Geology Today* 14 (3): 113–15.

Balme, D. M. 1962. "*Genos* and *Eidos* in Aristotle's Biology." *Classical Quarterly* 12: 81–98.

Barrett, P. H., P. J. Gautrey, S. Herbert, D. Kohn, and S. Smith, editors. 1987. *Charles Darwin's Notebooks, 1836–1844*. Ithaca, NY: Cornell University Press.

Barrett, P. H., D. Weinshank, and T. T. Gottleber, editors. 1981. *A Concordance to Darwin's* Origin of Species. Ithaca, NY: Cornell University Press.

Barsanti, G. 1992. *La Scala, la Mappa, l'Albero: Immagini e classifacazione della natura fra Sei e Ottocento*. Florence: Sansoni.

Beatty, J. 1985. "Speaking of Species: Darwin's Strategy." In *The Darwinian Heritage*, edited by D. Kohn, 265–81. Princeton, NJ: Princeton University Press.

Beer, Gillian. 1983. *Darwin's Plots: Evolutionary Narrative in Darwin, George Eliot and Nineteenth-Century Fiction* (2nd ed. published 2000). London: Routledge and Kegan Paul.

1996. *Open Fields: Science in Cultural Encounter*. Oxford: Clarendon Press.

Beer, Gillian, editor. 1996. *The Origin of Species by Charles Darwin*. Edited with an introduction by Gillian Beer. Oxford and New York: Oxford University Press.

Bernhardi, J. J. 1834. *Über den Begriff der Pflanzenart und Seine Anwendung*. Erfurt: Otto.

Blind, M. *The Ascent of Man*. 1899. London: Fisher, Unwin.

Boitard, Pierre and Corbié. 1824. *Les pigeons de Volière et de Colombier*. Paris: Audut.

Boreau, Alexandre. 1840. *Flore du centre de la France et du bassin de la Loire, ou, Description des plantes qui croissent spontanément, ou qui sont cultivées en grand, dans les départements arrosés par la Loire et par ses affluents, avec l'analyse des genres et des espèces*. Paris: Roret.

Bowler, Peter J. 1992. *The Fontana/Norton History of the Environmental Sciences*. London: Fontana; New York: Norton.

1996. *Life's Splendid Drama: Evolutionary Biology and the Reconstruction of Life's Ancestry, 1860–1940*. Chicago: University of Chicago Press.

2000. *Reconciling Science and Religion*. Chicago: University of Chicago Press.

2003. *Evolution: The History of an Idea*. 3rd ed. Berkeley and Los Angeles: University of California Press.

2008. "Geographical Distribution in the Origin of Species." In *The Cambridge Companion to the "Origin of Species,"* edited by Michael Ruse and Robert J. Richards, 153–72. Cambridge: Cambridge University Press.

Bredekamp, H. 2006. *Darwins Korallen: Die frühe Evolutionsdiagramme und die Tradition der Naturgeschichte*. Berlin: Verlag Klaus Wagenbach.

Brewster, D. 1844. "Vestiges." *North British Review* 3: 470–515.

Briggs, D., and S. M. Walters. 1997. *Plant Variation and Evolution*. 3rd ed. Cambridge: Cambridge University Press.

Brooke, J. H. 1985. "The Relations between Darwin's Science and His Religion." In *Darwinism and Divinity*, edited by J. R. Durant, 40–75. Oxford: Blackwell.

1997. "The Natural Theology of the Geologists." In *Images of the Earth*, edited by L. J. Jordanova and R. Porter, 53–74. Oxford: Alden Press.

2003. "Darwin and Victorian Christianity." In *The Cambridge Companion to Darwin*, edited by J. Hodge and G. Radick, 192–213. Cambridge: Cambridge University Press.

Brooke, J. H., M. Osler, and J. van der Meer, editors. 2001. *Science in Theistic Contexts. Osiris* 16.

Brougham, Henry, Lord. 1839. *Dissertations on Subjects of Science Connected with Natural Theology*. London: C. Knight and Co.

Browne, Janet. 1980. "Darwin's Botanical Arithmetic and the 'Principle of Divergence', 1854–1858." *Journal of the History of Biology* 13: 53–89.

1983. *The Secular Ark: Studies in the History of Biogeography*. New Haven, CT: Yale University Press.

1995. *Charles Darwin: Voyaging. Volume 1 of a Biography*. New York: Knopf.

2002. *Charles Darwin: The Power of Place. Volume II of a Biography*. New York: Knopf.

2003. *Charles Darwin: The Power of Place*. London: Pimlico.

2006. *Darwin's* Origin of Species: *A Biography*. London: Atlantic Books.

Buck, R., and D. L. Hull. 1966. "The Logical Structure of the Linnaean Hierarchy." *Systematic Zoology* 15: 97–111.

Buckland, W. 1836. *Geology and Mineralogy considered with reference to Natural Theology*. Vol. 6 of the Bridgewater Treatises. London: William Pickering.

Buckle, Thomas Henry. 1857–61. *History of Civilization in England*. 2 vols. London: J. W. Parker and Son.

Bulmer, Michael. 2004. "Did Jenkin's Swamping Argument Invalidate Darwin's Theory of Natural Selection?" *British Journal for the History of Science* 37: 281–97.

Burkhardt, Frederick, et al., editors. 1985–. *The Correspondence of Charles Darwin*. Cambridge: Cambridge University Press.

Burkhardt, Frederick, and Sydney Smith, editors. 1994. *A Calendar of the Correspondence of Charles Darwin, 1821–1882: With Supplement*. Cambridge: Cambridge University Press.

Burroughs, E. R. 1914. *Tarzan of the Apes*. New York: A. L. Burt.

Butler, J. 1961. *The Analogy of Religion* (1st ed. 1736). New York: Ungar.

Butler, Samuel. 1884. *Erewhon; or, Over the Range*. London: Trubner.

Campbell, J. A. 1986. "Scientific Revolution and the Grammar of Culture: The Case of Darwin's *Origin*." *The Quarterly Journal of Speech* 72: 351–76.

2003. "Why Was Darwin Believed? Darwin's *Origin* and the Problem of Intellectual Revolution." *Configurations* 11: 203–37.

Candolle, Alphonse de. 1855. *Géographie biologique raisonnée*. 2 Vols. Paris: Victor Masson.

Cantor, G., and M. Swetlitz, editors. 2006. *Jewish Tradition and the Challenge of Darwinism*. Chicago: University of Chicago Press.

Carroll, Joseph, editor. 2003. *On the Origin of Species by Means of Natural Selection. By Charles Darwin*. Peterborough: Broadview Texts.

Carroll, P. T. 1976. "On the Utility of Collating the Darwin Correspondence." *Annals of Science* 33: 383–94.

Chalmers, T. 1833. *On the Power, Wisdom and Goodness of God as Manifested in the Adaptation of External Nature to the Moral and Intellectual Constitution of Man.* London: William Pickering.

[Chambers, R.] 1844. *Vestiges of the Natural History of Creation.* London: Churchill.

1994. *The Vestiges of the Natural History of Creation and Other Evolutionary Writings.* Edited with a new introduction by James A. Secord. Chicago: University of Chicago Press. (Originally published anonymously, London: Churchill, 1844.)

Christen, Y. 1981. *Marx et Darwin.* Paris: Alban Michel.

Chung, C. 2003. "On the Origin of the Typological/Population Distinction in Ernst Mayr's Changing Views of Species, 1942-1959." *Studies in History and Philosophy of Biological and Biomedical Sciences* 34: 277–96.

Clark, J. F. M. 1997a. "'A Little People but Exceedingly Wise'? Taming the Ant and the Savage in Nineteenth-Century England." *La Lettre de la Maison Française* 7: 65–83.

1997b. "'The Ants Were Duly Visited': Making Sense of John Lubbock, Scientific Naturalism and the Senses of Social Insects." *British Journal of the History of Science* 30: 151–76.

2006. "History from the Ground Up: Bugs, Political Economy, and God in Kirby and Spence's *Introduction to Entomology* (1815-1856)." *Isis* 97: 28–55.

Coleridge, S. T. 1993. *Poems.* Edited by J. B. Beer. London: Heinemann.

Conway Morris, S. 2003. *Life's Solution.* Cambridge: Cambridge University Press.

Cornell, J. F. 1984. "Analogy and Technology in Darwin's Vision of Nature." *Journal of the History of Biology* 17: 303–44.

Coyne, Jerry A., and H. Allen Orr. 2004. *Speciation.* Sunderland, MA: Sinauer.

Croll, James. 1864. "On the Physical Cause of the Change of Climate during Geological Epochs." *Philosophical Magazine* 27: 121–37.

1875. *Climate and Time in Their Geological Relations.* London: Daldy, Isbister.

Darwin, Charles. 1839. *Journal of Researches into the Geology and Natural History of the Various Countries Visited by HMS Beagle.* London: Henry Colburn.

[1839] 1910. *A Journal of Researches.* London: Ward Lock.

1839. "Observations on the Parallel Roads of Glen Roy, and of Other Parts of Lochaber in Scotland, with an Attempt to Prove That They Are of Marine Origin." *Philosophical Transactions of the Royal Society of London* 129, 1: 39–81.

1840. "On the Connexion of Certain Volcanic Phenomena in South America; and on the Formation of Mountain Chains and Volcanos, as the Effect of the Same Power by which Continents are Elevated". *Transactions of the Geological Society of London*, second series (part 3): 601–3.

1842. *The Structure and Distribution of Coral Reefs*. London: Smith, Elder and Co.

1844. *Geological Observations on the Volcanic Islands Visited during the Voyage of HMS Beagle, together with some Brief Notices of the Geology of Australia and the Cape of Good Hope*. London: Smith, Elder and Co.

1846. *Geological Observations on South America. Being the third part of the Geology of the Voyage of the Beagle*. London: Smith, Elder and Co.

1851a. *A Monograph of the Fossil Lepadidae; or, Pedunculated Cirripedes of Great Britain*. London: Palaeontographical Society.

1851b. *A Monograph of the Sub-Class Cirripedia, with Figures of all the Species. The Lepadidae; or Pedunculated Cirripedes*. London: Ray Society.

1854a. *A Monograph of the Fossil Balanidae and Verrucidae of Great Britain*. London: Palaeontographical Society.

1854b. *A Monograph of the Sub-Class Cirripedia, with Figures of all the Species. The Balanidge (or Sessile Cirripedes); the Verrucidae, and C*. London: Ray Society.

1859. *On the Origin of Species by Means of Natural Selection, or the Preservation of Favoured Races in the Struggle for Life*. London: John Murray.

1862. *De l'Origine des espèces par sélection naturelle, ou des lois du progrès des êtres organisés*. Paris: Guillaumin et V. Masson.

1862. *On the Various Contrivances by which British and Foreign Orchids are Fertilized by Insects, and On the Good Effects of Intercrossing*. London: John Murray.

1866, 1870. *De l'origine des espèces*. Paris: Guillaumin et V. Masson.

1868. *The Variation of Animals and Plants Under Domestication*. London: John Murray.

1871. *The Descent of Man, and Selection in Relation to Sex*. Edited by J. T. Bonner and R. M. May. Princeton, NJ: Princeton University Press.

1878. *The Effects of Cross and Self Fertilization in the Vegetable Kingdom*. London: John Murray.

1880. *The Power of Movement in Plants*. London: John Murray.

1881. *The Formation of Vegetable Mould, Through the Action of Worms, with Observations on their Habits*. London: Murray.

1909. *The Foundations of the Origin of Species: Two Essays Written in 1842 and 1844*. Edited by F. Darwin. Cambridge: Cambridge University Press.

1958. *The Autobiography of Charles Darwin, 1809–1882*. Edited by N. Barlow. London: Collins.

1959. *The Origin of Species by Charles Darwin: A Variorum Text*. Edited by M. Peckham. Philadelphia: University of Pennsylvania Press.

1964. *On the Origin of Species: A Facsimile of the First Edition*. Introduction by Ernst Mayr. Cambridge, MA: Harvard University Press.

1975. *Charles Darwin's Natural Selection, Being the Second Part of His Big Species Book Written from 1856 to 1858*. Edited by R. C. Stauffer. Cambridge: Cambridge University Press.

1985-. *The Correspondence of Charles Darwin*. Cambridge: Cambridge University Press.

Darwin, Charles, editor. 1840. *The Zoology of the Voyage of HMS Beagle, under the Command of Captain Fitzroy, RN, during the Years 1832–1836*. Vol. 1. London: Smith, Elder and Co.

Darwin, Charles, and A. R. Wallace. 1958. *Evolution by Natural Selection*. Foreword by Gavin de Beer. Cambridge: Cambridge University Press.

Darwin, Erasmus. 1794–96. *Zoonomia or the Laws of Organic Life*. 2d ed. 2 vols. London: Johnson.

[1794–96] 1801. *Zoonomia; or, The Laws of Organic Life*. 3rd ed., 2 vols, London: J. Johnson.

Darwin, F. 1887. *The Life and Letters of Charles Darwin, Including an Autobiographical Chapter*. London: John Murray.

Darwin, F., and A. C. Seward, editors. 1903. *More Letters of Charles Darwin*. London: John Murray.

Daston, Lorraine. 2004. "Attention and the Values of Nature in the Enlightenment." In *The Moral Authority of Nature*, edited by Lorraine Daston and Fernando Vidal, 100–26. Chicago: University of Chicago Press.

Dawkins, Richard. 2006 *On the Origin of Species. Charles Darwin. Abridged and Read by Richard Dawkins*. Five CDs. London: CSA Word.

de Beer, G. R. 1954. *Archaeopteryx lithographica: A Study Based upon the British Museum Specimen*. London: Trustees of the British Museum (Natural History).

de Beer, G. R., editor. 1959–69. "Darwin's Notebooks on Transmutation of Species." *Bulletin of the British Museum Historical Series*, 2 and 3.

De Chadarevian, Soraya. 1996. "Laboratory Science versus Country-house Experiments. The Controversy between Julius Sachs and Charles Darwin." *The British Journal for the History of Science* 29: 17–41.

Delair, J. B., and W. A. S. Sarjeant. 1975. "The Earliest Discoveries of Dinosaurs." *Isis* 66: 5–25.

Depew, D., and B. Weber. 1995. *Darwinism Evolving*. Cambridge, MA: MIT Press.

Desmond, Adrian J. 1975. *The Hot-blooded Dinosaurs: A Revolution in Palaeontology*. London: Blond and Briggs.

1979. "Designing the Dinosaur: Richard Owen's Response to Robert Edmond Grant." *Isis* 70: 224–34.

1982. *Archetypes and Ancestors*. London: Blond and Briggs.

1985. "The making of Institutional Zoology in London, 1822–1836." *History of Science* 23: 153–85, 223–50.

1989. *The Politics of Evolution: Morphology, Medicine and Reform in Radical London*. Chicago and London: University of Chicago Press.

Desmond, A., and J. R. Moore. 1991. *Darwin*. London: Michael Joseph.

di Gregorio, Mario A., and N. W. Gill, editors. 1990. *Charles Darwin's Marginalia. Volume I*. New York: Garland.

Drouin, Jean-Marc. 2005. "Ants and Bees: Between the French and the Darwinian Revolution." *Ludus Vitalis* 13: 3–14.

Duffy, C. A. *The World's Wife*. 1999. London: Picador.

Duncan, D. 1908. *Life and Letters of Herbert Spencer*. 2 vols. New York: Appleton.

Eaton, J. M. 1851. *A New and Compleat Treatise on the Art of Breeding and Managing the Almond Tumbler, Calculated for the Information and Amusement of the Young Pigeon Fancier, Containing the Whole Natural History of Pigeons, also Observations on the Complaints and Disorders They are Liable to and Instructions for Treating Them when Labouring under Such Complaints, with Many Other Observations and Instructions Necessary for the Young Fancier and Adapted to Pigeons of All Kinds*. London: Alexander Hogg.

1858. *A Treatise on the Art of Breeding and Managing Tame, Domesticated, Foreign, and Fancy Pigeons, Carefully Compiled from the Best Authors, with Observations and Reflections, Containing All That is Necessary to be Known of Tame, Domesticated, Foreign and Fancy Pigeons, in Health, Disease, and Their Cures*. London: John Matthews Eaton.

Eliot, G. 1980. *The Mill on the Floss*. Edited by G. Haight. Oxford: Oxford University Press.

Elkin Mathews Ltd. 1929. *No. 27. A Catalogue of Fifty Famous First Editions. December 1929*. London: Elkin Mathews Ltd.

Ellegård, A. 1958. *Darwin and the General Reader*. Stockholm: Almqvist and Wiksell.

Endersby, James John. 2008. *Imperial Nature: Joseph Hooker and the Practices of Victorian Science*. Chicago: University of Chicago Press.

Ereshefsky, M. 2001. *The Poverty of the Linnaean Hierarchy: A Philosophical Study of Biological Taxonomy*, Cambridge: Cambridge University Press.

Evans, L. T. 1984. "Darwin's Use of the Analogy between Artificial and Natural Selection." *Journal of the History of Biology* 17: 113–40.

Farber, P. 1972. "Buffon and the Concept of Species." *Journal of the History of Biology* 5: 259–84.

Feuer L. S. 1975. "Is the 'Darwin-Marx Correspondence' Authentic?" *Annals of Science* 32: 1–12.

Freeman, R. B. 1977. *The Works of Charles Darwin: An Annotated Bibliographical Handlist.* Second edition revised and enlarged. Folkestone: Dawson and Hamden: Archon Books.

    1978. *Charles Darwin: A Companion.* Folkestone: Dawson and Hamden: Archon Books.

Gale, B. G. 1982. *Evolution without Evidence: Charles Darwin and the Origin of Species.* Albuquerque: University of New Mexico Press.

Gardiner, B. G. 2004. "Darwin and South American Fossils." *The Linnean* 20: 16–22.

Gayon, J. 1998. *Darwinism's Struggle for Survival: Heredity and the Hypothesis of Natural Selection.* Cambridge: Cambridge University Press.

Ghiselin, M. T. 1984. *The Triumph of the Darwinian Method.* Chicago: University of Chicago Press.

Gliboff, Sander. 2008. *Translation and Transformation: H. G. Bronn, Ernst Haeckel, and the Origins of German Darwinism.* Cambridge, MA: MIT Press.

Gould, Stephen Jay. 1977. *Ontogeny and Phylogeny.* Cambridge, MA: The Belknap Press of Harvard University Press.

Gray, Asa. 1856–57. "Statistics of the Flora of the Northern United States." *American Journal of Science and Arts* 22 (1856): 204–32; 23 (1857): 62–84, 369–403.

Grene, M., and D. Depew. 2004. *The Philosophy of Biology: An Episodic History,* Cambridge: Cambridge University Press.

Gruber, H. E. 1961. "Marx and das Kapital." *Isis* 52: 582–6.

Haeckel E. 1866. *Generelle Morphologie der Organismen.* Berlin: Georg Reimer.

    1876. *The History of Creation: Or the Development of the Earth and Its Inhabitants by the Action of Natural Causes.* 2 vols. New York: Appleton.

Hamilton, W. D. 1964. "The Genetical Evolution of Social Behaviour." *Journal of Theoretical Biology* 7: 1–52.

Harper, J. L. 1967. "A Darwinian Approach to Plant Ecology." *The Journal of Ecology* 55: 247–70.

Harvey J. 1997. *"Almost a Man of Genius"*: *Clémence Royer, Feminism, and Nineteenth Century Science*. New Brunswick, NJ: Rutgers University Press.

Hayden, Sarah, and Peter S. White. 2003. "Invasion Biology: An Emerging Field of Study." *Annals of the Missouri Botanical Garden* 90: 64–6.

Hennig, W. 1977. *Phylogenetic Systematics*. Urbana: University of Illinois Press.

Herbert, S. 1971. "Darwin, Malthus, and Selection." *Journal of the History of Biology* 4: 209–17.

2005. *Charles Darwin, Geologist*. Ithaca, NY: Cornell University Press.

Herschel, J. F. W. 1827. "Light." In *Encylopaedia Metropolitana*, edited by E. Smedley et al. London: J. Griffin.

1830. *Preliminary Discourse on the Study of Natural Philosophy*. London: Longman, Rees, Orme, Brown, Green, and Longman.

1841. Review of Whewell's *History and Philosophy*. *Quarterly Review* 135: 177–238.

Heslop-Harrison, J. 1979. "Darwin and the Movement of Plants: A Retrospect." In *Plant Growth Substances*, edited by F. Skoog. Berlin and Heidelberg: Springer-Verlag.

Himmelfarb, G. 1959. *Darwin and the Darwinian Revolution*. New York: Norton.

Hitchcock, E. 1858. *Ichnology of New England: A Report on the Sandstone of the Connecticut Valley Especially its Fossil Footmarks*. Boston: W. White.

Hodge, M. J. S. 1977. "The Structure and Strategy of Darwin's 'Long Argument'." *British Journal for the History of Science* 10: 237–46.

1985. "Darwin as a Lifelong Generation Theorist." In *The Darwinian Heritage*, edited by David Kohn, 207–43. Princeton, NJ: Princeton University Press.

Hooker, Joseph Dalton. 1861. "Outlines of the Distribution of Arctic Plants." *Transactions of the Linnean Society* 23: 251–348.

Huber, François. 1814. *Nouvelles observations sur les abeilles*. Paris-Genève: J. J. Paschoud.

Huber, Pierre. 1810. *Recherches sur les moeurs des fourmis indigènes*. Paris: J. J. Paschoud.

Hull, David, editor. 1973. *Darwin and His Critics. The Reception of Darwin's Theory of Evolution by the Scientific Community*. Cambridge, MA: Harvard University Press.

1988. *Science as Process: An Evolutionary Account of the Social and Conceptual Development of Science*. Chicago: University of Chicago Press.

Hume, D. 1978. *Enquiries Concerning Human Understanding and Concerning the Principles of Morals.* Edited by L. A. Selby-Bigge and P. H. Nidditch. Oxford: Clarendon Press.

Huxley, Leonard. 1900. *Life and Letters of Thomas Henry Huxley.* 2 vols. New York: Appleton.

Huxley, Thomas H. 1868. "On the Animals which Are Most Nearly Intermediate between Birds and Reptiles." *Geological Magazine* 5: 357–65.

——— 1893. *Darwiniana: Collected Essays*, vol. 2. London: Macmillan.

Joyce, R. 2006. *The Evolution of Morality.* Cambridge, MA: MIT Press.

Karlinsky, Harry. 1991. *The Origin of Species. By Charles Darwin. A Special Board Book for Very Young Children.* Adapted by Harry Karlinsky. Illustrated by Brock Irwin. Toronto: Davis Press.

Kingsley, C. n.d. *The Water Babies, A Fairy Tale for a Land-baby.* London and Glasgow: Collins. (First edition London: Macmillan, 1863.)

Kirby, William, and William Spence, editors. 1860. *An Introduction to Entomology; or, Elements of the Natural History of Insects: Comprising an Account of Noxious and Useful Insects, of their Metamorphoses, Food, Stratagems, Habitations, Societies, Motions, Noises, Hybernation, Instinct, etc. etc.* 7th ed. London: Longman, Green, Longman, and Roberts.

Kohn, David. 1980. "Theories to Work By: Rejected Theories, Reproduction and Darwin's Path to Natural Selection." *Studies in History of Biology* 4: 67–170.

——— 1985. "Darwin's Principle of Divergence as Internal Dialogue." In *The Darwinian Heritage*, edited by D. Kohn, 245–57. Princeton, NJ: Princeton University Press.

——— 1989. "Darwin's Ambiguity: The Secularization of Biological Meaning." *British Journal for the History of Science* 22: 215–39.

——— 1996. "The Aesthetic Construction of Darwin's Theory." In *Aesthetics and Science: The Elusive Synthesis*, edited by A. Tauber, 13–48. Dordrecht: Kluwer.

——— 2006. *Natural Selection Portfolios.* <http://darwinlibrary.amnh.org>.

——— 2008a. "Divergence in the *Origin of Species.*" In *The Cambridge Companion to the "Origin of Species,"* edited by Robert J. Richards and Michael Ruse, 87–108. Cambridge: Cambridge University Press.

——— 2008b. *Darwin's Garden: An Evolutionary Adventure.* Catalog for an exibit at the New York Botanical Garden, April–June 2008.

Kohn, David, Gina Murrell, John Parker, and Mark Whitehorn. 2005. "What Henslow Taught Darwin." *Nature* 436: 643–5.

Larson, J. L. 1971. *Reason and Experience: The Representation of Natural Order in the Work of Carl von Linné.* Berkeley: University of California Press.

Latreille, Pierre-André. 1802. *Histoire naturelle des fourmis*. Paris: Théophile Barrois père.

Lennox, J. 1994. "Darwin *Was* a Teleologist." *Biology and Philosophy* 8: 405–21.

2006. "Aristotle's Biology." In *The Stanford Encyclopedia of Philosophy (Fall 2006 Edition)*, edited by Edward N. Zalta, available at <http://plato.stanford.edu/archives/fall2006/entries/aristotle-biology/>.

Lessing, D. 1988. *The Fifth Child*. London: Cape.

2007. *The Cleft*. London: Fourth Estate.

Levine, G. 1988. *Darwin and the Novelists: Patterns of Science in Victorian Fiction*. Cambridge, MA: Harvard University Press.

2006. *Darwin Loves You*. Princeton, NJ: Princeton University Press.

Lewens, T. 2007. *Darwin*. London: Routledge.

Lewontin, Richard. 1978. "Adaptation." *Scientific American* 239: 220.

Lightman, B. 1987. *The Origins of Agnosticism*. Baltimore: Johns Hopkins University Press.

Limoges, C. 1970. *La sélection naturelle*. Paris: Presses Universitaires de France.

Lindroth, S. 1983. *Linnaeus, the Man and His Work*. Edited by T. Frängsmyr. Berkeley: University of California Press.

Lipton, P. 2004. *Inference to the Best Explanation*. 2nd ed. London: Routledge.

Lloyd, E. A. 1983. "The Nature of Darwin's Support for the Theory of Natural Selection." *Philosophy of Science* 50: 112–29.

Lustig, A. J. 2004. "Ants and the Nature of Nature in Auguste Forel, Erich Wasmann, and William Morton Wheeler." In *The Moral Authority of Nature*, edited by Lorraine Daston and Fernando Vidal, 282–307. Chicago: University of Chicago Press.

Lyell, C. [1830–33] 1987. *Principles of Geology: Being an Attempt to Explain the Former Changes in the Earth's Surface by Reference to Causes now in Operation*. London: John Murray. Facsimile reprint, Chicago: University of Chicago Press.

Mallet, James. 2004. "Poulton, Wallace and Jordan: How Discoveries in Papilio Butterflies Led to a New Species Concept 100 Years Ago." *Systematics and Biodiversity* 1 (4): 441–52.

2005. "Speciation in the 21st Century." *Heredity* 95: 105–9.

Malthus, T. R. [1798] 1959. *Population: The First Essay*. Ann Arbor: University of Michigan Press.

[1826] 1914. *An Essay on the Principle of Population*. 6th ed. London: Everyman.

Manier, E. 1978. *The Young Darwin and His Cultural Circle*. Dordrecht: Reidel.

Marx, K., and F. Engels. 1975–. *Collected Works of Karl Marx and Friedrich Engels*. New York: International Publishers.

Mayr, Ernst. 1942. *Systematics and the Origin of Species from the Viewpoint of a Zoologist*. New York: Columbia University Press.

1968. "Illiger and the Biological Species Concept." *Journal of the History of Biology* 1: 163–78.

1976. "Typological versus Population Thinking." In *Evolution and the Diversity of Life*, edited by E. Mayr, 26–90. Cambridge, MA: Harvard University Press.

1982. *The Growth of Biological Thought*. Cambridge, MA: Belknap Press.

1991. *One Long Argument: Charles Darwin and the Genesis of Modern Evolutionary Thought*. Cambridge, MA: Harvard University Press.

1992. "Darwin's Principle of Divergence." *Journal of the History of Biology* 25 (3): 343–59.

Mendel, Gregor. 1865. *"Experiments in Hybridization."* Available at <http://www.mendelweb.org/mendel.html>.

Miles, S. J. 1988. "Evolution and Natural Law in the Synthetic Science of Clémence Royer." Ph.D. dissertation, University of Chicago.

Milne-Edwards, Henri. 1844. "Considérations sur quelques principes relatifs à la classification naturelle des animaux." *Annales des sciences naturelles*, 3rd ser., 1: 65–99.

Moore, J. R. 1979. *The Post-Darwinian Controversies*. Cambridge: Cambridge University Press.

Morrell, J. 2005. *John Phillips and the Business of Victorian Science*. Aldershot: Ashgate.

Morris, Solene, L. Wilson, and David Kohn. 1987. *Charles Darwin at Down House*. London: English Heritage.

Mullen, P. M. 1964. "The Preconditions and Reception of Darwinian Biology in Germany, 1800–1865." Unpublished Ph.D. dissertation, University of California, Berkeley.

Müller, Fritz. 1969. *Facts and Arguments for Darwin*. Translated by W. S. Dallas. London: John Murray. (Originally published as *Für Darwin*. Leipzig: Engelmann, 1864.)

Naden, C. 1894. *The Complete Poetical Works of Constance Naden*. London: Bickers and Son.

Nägeli, K. von. 1865. *Entstehung und Begriff der naturhistorischen Art*. Munich: Königliche Akademie.

Newton, I. 1692. "Letter to Richard Bentley." In *Newton's Philosophy Of Nature*, edited by H. S. Thayer, 46–58. New York: Hafner, 1953.

Niedecker, L. 1985. *The Granite Pail: Selected Poems*. San Francisco: North Point Press.

Nyhart, Lynn K. 1995. *Biology Takes Form: Animal Morphology and the German Universities, 1800–1900*. Chicago: University of Chicago Press.

Olby, R. 1993. "Constitutional Diseases." In *Companion Encyclopedia of the History of Medicine*, edited by William Bynum and Roy Porter, 412–37. London: Routledge.

Oldroyd, David. 1996. *Thinking about the Earth: A History of Ideas in Geology*. London: Athlone.

Ornduff, Robert. 1984. "Darwin's Botany." *Taxon* 33: 39–47.

Ospovat, D. 1981. *The Development of Darwin's Theory: Natural History, Natural Theology, and Natural Selection, 1838–1859*. Cambridge: Cambridge University Press, reissue 1995.

Ostrom, J. H. 1969. "Osteology of *Deinonychus antirrhopus*, an Unusual Theropod from the Lower Cretaceous of Montana." *Bulletin of the Yale Peabody Museum of Natural History* 35: 1–165.

1976. "*Archaeopteryx* and the Origin of Birds." *Biological Journal of the Linnean Society of London* 8 (2): 91–182.

Owen, R. 1832. "On the Mammary Glands of the *Ornithorhynchus paradoxus*." *Philosophical Transactions of the Royal Society of London* 122: 517–38.

1834. "On the Ova of *Ornithorhynchus paradoxus*." *Philosophical Transactions of the Royal Society of London* 124: 555–66.

1840. *Odontography; or a treatise on the comparative anatomy of the teeth; their physiological relations, mode of development and microscopic structure in the vertebrate animals*. 3 vols. Vol. 1. London: Hippolyte Bailliere.

1841. "Description of *Lepidosiren annectens*." *Transactions of the Linnean Society of London* 18: 327–71.

1849. *On the Nature of Limbs*. London: Voorst.

1854. "On Some Fossil Reptilian and Mammalian Remains from the Purbecks." *Quarterly Journal of the Geological Society* 10: 420–33.

1860. "Darwin on the Origin of Species." *Edinburgh Review* 111: 487–532.

1863. "On the *Archeopteryx* of Von Meyer, with a Description of the Fossil Remains of a Long-tailed Species, from the Lithographic Stone of Solenhofen." *Philosophical Transactions of the Royal Society of London* 1863: 33–47.

Paley, William. 1802. *Natural Theology; or, Evidences of the Existence and Attributes of the Deity. Collected from the Appearances of Nature*. London: Printed for R. Faulder.

Panchen, A. L. 1994. *Classification, Evolution, and the Nature of Biology,* Cambridge: Cambridge University Press.

Parshall, Karen Hunger. 1982. "Varieties as Incipient Species: Darwin's Numerical Analysis." *Journal of the History of Biology* 15 (2): 191–214.

Paston, George. 1932. *At John Murray's: Records of a Literary Circle 1843–1892.* London: John Murray.

Pearson, K. 1900. *The Grammar of Science,* second edition. London: Black.

Peile, J. 1913. *Biographical Register of Christ's College 1505–1905 and of the Earlier Foundation, God's House 1448–1505.* 2 vols. Cambridge: Cambridge University Press.

Pfeiffer, E. 1888. *Sonnets: Revised and Enlarged Edition.* London: Field and Tuer.

Piveteau, J., editor. 1954. *Buffon: Oeuvres philosophiques.* Paris: Presses Universitaires de France.

Poulton, Edward Bagnall. 1904. "What Is a Species?" *Proceedings of the Entomological Society of London* [1903]: lxxvii–cxvi.

1908. *Essays on Evolution.* Oxford: The Clarendon Press.

Punnett, R. C. 1909. *Mendelism.* New York: Wilshire. (American edition pirated from the second edition with additional papers.)

Rachootin, Stan. 1984. *Darwin's Embryology.* Ph.D. dissertation, Yale University.

Rhodes, F. H. T. 1991. "Darwin's Search for a Theory of the Earth: Symmetry, Simplicity and Speculation." *British Journal for the History of Science* 24: 193–229.

Richards, R. A. 2003. "Character Individuation in Phylogenetic Inference." *Philosophy of Science* 70: 264–79.

2007. "Species and Taxonomy." In *The Oxford Handbook for the Philosophy of Biology,* 161–88. Oxford: Oxford University Press.

Richards, Robert J. 1987. *Darwin and the Emergence of Evolutionary Theories of Mind and Behavior.* Chicago: University of Chicago Press.

1992. *The Meaning of Evolution: The Morphological Construction and Ideological Reconstruction of Darwin's Theory.* Chicago: University of Chicago Press.

2002. *The Romantic Conception of Life: Science and Philosophy in the Age of Goethe.* Chicago: University of Chicago Press.

2004. "Michael Ruse's Design for Living." *Journal of the History of Biology* 37: 25–38.

2008. *The Tragic Sense of Life: Ernest Haeckel and the Struggle over Evolutionary Thought.* Chicago: University of Chicago Press.

Richmond, Marsha. 1985. "Darwin's Study of Cirripedia." In *Correspondence* 4: 388–409.

Ridley, M. 1993. *Evolution.* Cambridge, MA: Blackwell.

Roberts, J. H. 1988. *Darwinism and the Divine in America*. Madison: University of Wisconsin Press.

Royer, C.-L. 1868–69. "Lamarck, sa vie, ses travaux et son système." *La philosophie positive* 3: 173–208, 333–72; 4: 5–30.

Rudwick, M. J. S. 1974. "Darwin and Glen Roy: A 'Great Failure' in Scientific Method?" *Studies in the History and Philosophy of Science* 5: 97–185.

Rupke, N. A. 1994. *Richard Owen: Victorian Naturalist*. New Haven and London: Yale University Press.

Ruse, Michael. 1975a. "Charles Darwin and Artificial Selection." *Journal of the History of Ideas* 36: 339–50.

   1975b. "Charles Darwin's Theory of Evolution: An Analysis." *Journal of the History of Biology* 8: 219–41.

   1975c. "Darwin's Debt to Philosophy: An Examination of the Influence of the Philosophical Ideas of John F. W. Herschel and William Whewell on the Development of Charles Darwin's Theory of Evolution." *Studies in the History of Science* 6: 159–181.

   1979. *The Darwinian Revolution: Science Red in Tooth and Claw*. Chicago: University of Chicago Press.

   1996. *Monad to Man: The Concept of Progress in Evolutionary Biology*. Cambridge, MA: Harvard University Press.

   1999. *The Darwinian Revolution: Science Red in Tooth and Claw*, 2nd ed. Chicago: University of Chicago Press.

   2003. *Darwin and Design: Does Evolution Have a Purpose?* Cambridge, MA: Harvard University Press.

Russell, E. S. 1982. *Form and Function: A Contribution to the History of Animal Morphology*. London: John Murray, 1916; reprint, Chicago: University of Chicago Press.

Schnackenberg, G. 1986. *The Lamplit Answer*. London: Hutchinson.

Schweber, Silvan S. 1980. "Darwin and the Political Economists: Divergence of Character." *Journal of the History of Biology* 13 (2): 195–289.

Sebright, J. 1809. *The Art of Improving the Breeds of Domestic Animals*. London: Howlett and Brimmer.

Secord, J. 1981. "Nature's Fancy: Charles Darwin and the Breeding of Pigeons." *Isis* 72: 162–86.

   1991. "The Discovery of a Vocation: Darwin's Early Geology." *British Journal for the History of Science* 24: 133–57.

   2000. *Victorian Sensation: The Extraordinary Publication, Reception and Secret Authorship of Vestiges of the Natural History of Creation*. Chicago and London: Chicago University Press.

Sedgwick, Adam. 1830. "Presidential Address." *Proceedings of the Geological Society of London (1826–1833)* 1: 204–6.

1831. "Address to the Geological Society." *Proceedings of the Geological Society of London* 1: 281–316.

1845. "Vestiges." *Edinburgh Review* 82: 1–85.

Sheldon, R. W. 2004. *Darwin's Origin of Species: A Condensed Version of the First Edition of 1859.* Victoria: Trafford.

Shillingsburg, Peter L. 2006. "The First Five English Editions of Charles Darwin's *On the Origin of Species.*" *Variants* 5: 221–48.

Sinclair, George. 1826. *Hortus Gramineus Woburnensis; or, An account of the results of experiments on the produce and nutritive qualities of different grasses and other plants used as the food of the more valuable domestic animals; instituted by John, Duke of Bedford.* London: Ridgway.

Sleigh, Charlotte. 2006. *Six Legs Better: A Cultural History of Myrmecology.* Baltimore: Johns Hopkins University Press.

Sloan, Phillip R. 1985. "Darwin's Invertebrate Program, 1826–36: Preconditions for Transformism." In *The Darwinian Heritage*, edited by David Kohn, 71–120. Princeton, NJ: Princeton University Press.

1986. "From Logical Universals to Historical Individuals: Buffon's Idea of Biological Species." In *Histoire du concept d'espèce dans les sciences de la vie*, edited by J. Roger and J. L. Fisher, 101–40. Paris: Fondation Singer-Polignac.

2001. "'The sense of sublimity': Darwin on Nature and Divinity." In *Science in Theistic Contexts*, edited by J. Brooke et al. *Osiris* 16: 251–69.

2002. "Reflections on the Species Problem." In *The Philosophy of Marjorie Grene*, edited by R. Auxier and L. Hahn, 225–55. Chicago: Open Court.

2005. "It Might Be Called Reverence." In *Darwinism and Philosophy*, edited by V. Hösle and C. Illies, 143–65. Notre Dame, IN: Notre Dame University Press.

2006. "Kant on the History of Nature." *Studies in History and Philosophy of Biological and Biomedical Sciences* 37: 627–48.

Sloan, Phillip R., editor. 1992. *The Hunterian Lectures in Comparative Anatomy: May and June 1837.* London: Natural History Museum Publications.

Smocovitis, V. B. 1997. "G. Ledyard Stebbins Jr. and the Evolutionary Synthesis (1924–1950)." *American Journal of Botany* 84: 1625–37.

Snyder, L. J. 2006. *Reforming Philosophy.* Chicago: University of Chicago Press.

Sober, E. 1980. "Evolution, Population Thinking, and Essentialism." *Philosophy of Science* 47: 350–83.

Spencer, H. 1851. *Social Statics.* London: Chapman.

1852. "The Development Hypothesis." In *his Essays: Scientific, Political and Speculative*, 377–83. London: Williams and Norgate.

1865–67. *Principles of Biology.* London: Williams and Norgate.

1884. *The Man versus the State.* London: Williams and Norgate.

1892. *Essays. 3 vols.* New York: Appleton.

1904. *An Autobiography. 2 vols.* New York: Appleton.

Spring, A. F. 1838. *Über die naturhistorischen Begriffe von Gattung, Art und Abart und über die Ursachen der Abartungen in den organischen Reichen.* Leipzig: Fleischer.

Stamos, D. N. 2003. *The Species Problem: Biological Species, Ontology, and the Metaphysics of Biology.* Lexington: Lanham.

2007. *Darwin and the Nature of Species.* Albany: State University of New York Press.

Stebbins, G. Ledyard, Jr. 1950. *Variation and Evolution in Plants.* New York: Columbia University Press.

Stenhouse, J. 1999. "Darwinism in New Zealand, 1859–1900." In *Disseminating Darwinism*, edited by R. L. Numbers and J. Stenhouse, 61–89. New York: Cambridge University Press.

Sterelny, K. 2003. *Thought in a Hostile World: The Evolution of Human Cognition.* Oxford: Blackwell.

Stott, R. 2003. *Darwin and the Barnacle.* New York: Norton.

Strick, J. E. 2000. *Sparks of Life: Darwinism and the Victorian Debates over Spontaneous Generation.* Cambridge, MA: Harvard University Press.

Sulloway, Frank. 1979. "Geographic Isolation in Darwin's Thinking: The Vicissitudes of a Crucial Idea." *Studies in History of Biology* 3: 23–65.

Tennyson, Alfred. 1969. *The Poems of Tennyson.* Edited by C. Ricks. London: Longmans, Green.

Todes, D. 2000. *Darwin without Malthus: The Struggle for Existence in Russian Evolutionary Thought.* New York: Oxford University Press.

Toulmin, S. 2003. *The Uses of Argument.* Updated version. Cambridge: Cambridge University Press.

Tyndall, J. 1879. *Fragments of Science.* London: Longmans.

Uhlmann, E. 1923. Entwicklungsgedanke und Artbegriff in ihrer geschichtlichen Entstehung und sachlichen Beziehung. *Jenaische Zeitschrift für Naturwissenshaft.* 59: 1–107.

van Wyhe, John. 2007. "Mind the Gap: Did Darwin Avoid Publishing His Theory for Many Years?" In *Notes and Records of the Royal Society of London.* Published online at <doi:10.1098/rsnr.2006.0171>.

Vries, H. de. 1910. *Mutation Theory.* Translated by J. B. Farmer and A. D. Darbyshire. 2 vols. Chicago: Open Court.

Vorzimmer, P. 1968. "Darwin, Malthus, and the Theory of Natural Selection." *Journal of the History of Biology* 1: 225–59.

1969. "Darwin's Questions about the Breeding of Animals." *Journal of the History of Biology* 2: 269–81.

1972. *Charles Darwin: The Years of Controversy. The Origin of Species and Its Critics 1859–82.* London: University of London Press.

Voss, Julia. 2007. *Darwins Bilder.* Frankfurt a.M.: S. Fischer Verlag.

Wallace, Alfred Russel. 1876. *The Geographical Distribution of Animals.* 2 vols. London: Macmillan.

1880. *Island Life: Or the Phenomena and Causes of Insular Faunas and Floras: Including a Revision and an Attempted Solution of the Problem of Geological Climates.* London: Macmillan.

1908. "The Present Position of Darwinism." *Contemporary Review* 94: 129–41.

Waters, K. 2003. "The Arguments in the *Origin of Species.*" In *The Cambridge Companion to Darwin,* edited by J. Hodge and G. Radick, 116–39. Cambridge: Cambridge University Press.

Watson, H. C.. 1843. "Remarks on the Distinction of Species in Nature and in Books." *London Journal of Botany* 2: 613–22.

1845. "On the Theory of 'Progressive Development' Applied in Explanation of the Origin and Transmutation of Species." *Phytologist* 2: 108–13, 140–7, 161–9, 225–8.

Weismann, August. 1893. *The Germ-Plasm: A Theory of Heredity.* London: W. Scott.

1999. *August Weismann. Ausgewählte Briefe und Dokumente. Selected Letters and Documents.* Edited by Frederick B. Churchill. 2 vols. Freiburg im Breisgau, Germany: Universitätsbibliothek Freiburg im Breisgau.

Wells, H. G. [1895] 1993. *The Time-Machine.* Edited by M. Moorcock. London: J. M. Dent.

[1896] 1996. *The Island of Dr. Moreau.* London: Orion.

West, David. 2003. *Fritz Müller: A Naturalist in Brazil.* Blacksburg, VA: Pocahontas Press.

Whately, R. 1963. *Elements of Rhetoric.* Carbondale: Southern Illinois University Press. (Reprint of the seventh edition, 1846; first edition, 1828.)

1985. *Historic Doubts Relative to Napoleon Bonaparte.* Berkeley and London: Scholar Press. (Reprint of the first edition, 1819.)

Whewell, W. 1833. *Astronomy and General Physics Considered With Reference to Natural Theology.* London: William Pickering.

1837. *The History of the Inductive Sciences.* 3 vols. London: Parker.

1840. *The Philosophy of the Inductive Sciences.* 2 vols. London: Parker.

1845. *Indications of the Creator.* London: Parker.

[1847] 1967. *The Philosophy of the Inductive Sciences.* Facsimile of the second edition. London: Johnson Reprint Company.

Wilberforce, S. [1860] 1874. "Darwin's Origin of Species." In *Essays Contributed to the Quarterly Review* 1: 52–103. London: Murray.

Wilkinson, J. 1820. *Remarks on the Improvement of Cattle, etc. in a Letter to Sir John Saunders Sebright, Bart. M. P.* Nottingham.

Winsor, M. P. 2006. "The Creation of the Essentialism Story: An Exercise in Metahistory." *History and Philosophy of the Life Sciences* 28: 149–74.

Winther, R. 2000. "Darwin on Variation and Heredity." *Journal of the History of Biology* 33: 425–55.

2001. "August Weismann on Germ-plasm Variation." *Journal of the History of Biology* 34: 517–55.

Wolfe, Gerard R. 1981. *The House of Appleton: The History of a Publishing House and Its Relationship to the Cultural, Social, and Political Events That Helped Shape New York City.* Metuchan and London: The Scarecrow Press.

Wordsworth, W. 1969. *Poetical Works.* Edited by T. Hutchinson and E. de Selincourt. Oxford: Oxford University Press.

Youatt, William. 1834. *Cattle: Their Breeds, Management, and Disease.* London: Library of Useful Knowledge.

Young, R. M. 1969. "Malthus and the Evolutionists: The Common Context of Biological and Social Theory." *Past and Present* 43: 109–45.

1971. "Darwin's Metaphor: Does Nature Select?" *Monist* 55: 442–503.

1985. *Darwin's Metaphor: Nature's Place in Victorian Culture.* Cambridge: Cambridge University Press.

INDEX